S328 Ecology
Science: a third level course

ECOLOGY
Book One
INTERACTIONS

The S328 Course Team

Course Team Chair:	Jonathan Silvertown
Course Manager:	Phil Parker
Authors:	Mary Bell
	Mike Gillman
	Dick Morris
	Phil Parker
	Irene Ridge
	Jonathan Silvertown
	Charles Turner
Course Secretary:	Val Shadbolt
Editors:	Gerry Bearman
	Sheila Dunleavy
	Gilly Riley
	Bina Sharma
	Margaret Swithenby
Graphic Design:	Sarah Crompton
	Keith Howard
	Pam Owen
	Ros Wood
Course Assessor:	Peter Edwards
Book 1 Assessor:	David Streeter
Consultant:	Hilary Denny
Comments:	Eric Bowers
BBC:	Tony Jolly
	Liz Sugden
ACS:	Jon Rosewell

The Open University, Walton Hall, Milton Keynes, MK7 6AA.

First published 1996. Copyright © 1996 The Open University.

Edited, designed and typeset through AppleMac/QuarkXPress by The Open University.

Printed in the United Kingdom by Eyre & Spottiswoode Ltd, London & Margate.

ISBN 0 7492 51816

This text forms part of an Open University Third Level Course. If you would like a copy of *Studying with The Open University*, please write to the Central Enquiry Service, PO Box 200, The Open University, Walton Hall, Milton Keynes, MK7 6YZ. If you have not enrolled on the Course and would like to buy this or other Open University material, please write to Open University Educational Enterprises Ltd, 12 Cofferidge Close, Stony Stratford, Milton Keynes, MK11 1BY, United Kingdom.

S328book1i1.1

BOOK ONE   INTERACTIONS

CONTENTS

ECOLOGY: THE STUDY OF INTERACTIONS   CHAPTER 1

Prepared for the Course Team by Jonathan Silvertown

1.1 Introduction

The science of ecology is the study of interactions between organisms and their environment, where 'environment' is taken to include both the non-living components and the other organisms it contains. We start the Course with a case study of one organism, the rabbit *Oryctolagus cuniculus*, which shows just how far-reaching the influence of one species on its environment may be. The rest of Chapter 1 then looks at a whole range of types of close interactions between organisms.

1.2 Case study: the ecology of the rabbit in Britain

When Beatrix Potter wrote her tales of Peter Rabbit in the early 1900s, the greatest menace that Peter and his real-life relatives faced was the farmer with a gun, the snare and some assorted predators. In the British countryside, rabbits were everywhere and in large numbers. In places, they swarmed so thick that a moving rabbit herd looked like a living carpet. Farmers were well aware of the damage rabbits could do to their crops, and ecologists too had more than an inkling that rabbits played an important role in the ecology of Britain. Just how important they were not to guess. Early in the 20th century, one of the pioneers of ecology in Britain, Arthur Tansley (later, Sir Arthur), set up some experimental exclosures

Figure 1.1 (a) A 25 cm × 25 cm plot of grassland under moderately heavy grazing pressure from rabbits. (b) A similar plot after the exclusion of rabbits and other grazing animals for six years. Each letter represents the shoot of a broad-leaved plant. Different species have different letters. The leaf rosettes marked 'C' belong to stemless thistle *Cirsium acaule* and the one marked 'Cp' belongs to marsh thistle *Cirsium palustre*. The spaces between the shoots shown are filled with sheep's fescue grass *Festuca ovina*.

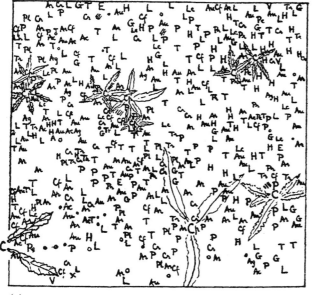

(a)

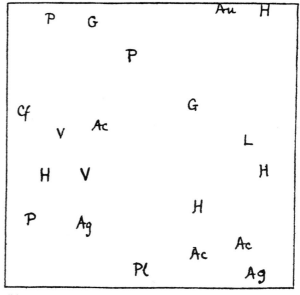

(b)

(areas fenced off to exclude rabbits) around plots of chalk grassland and found dramatic changes to the vegetation after only a few years without rabbit grazing. Sketches he made of unprotected and protected plots are shown in Figure 1.1.

❑ Summarize the differences between plots (a) and (b) that are apparent from Figure 1.1.

■ The grazed plot has:

- many more shoots of broad-leaved plants (you don't need to count them, but there are 441 as compared with only 20 in plot B).
- many more species (21 as compared with 10 in plot B).
- thistles, which don't occur in plot B.
- a smaller area of *Festuca ovina*.
- smaller individuals of the species found in both plots (e.g. 'P', 'H', 'L').

The full impact of rabbits on the ecology of Britain was not to become clear until myxomatosis, a South American viral disease, appeared in south-east England in the autumn of 1953. Within a year, it had spread throughout most of the British Isles, all but exterminating the rabbit. The disease had come from France where, also within a year of its deliberate introduction to an estate near Paris in June 1952, it had spread in a similarly spectacular fashion to the whole country. In France, it was found that the virus was transmitted from rabbit to rabbit on the mouthparts of mosquitoes.

In Britain, research carried out after the outbreak found that the vector transmitting the myxoma virus was the rabbit flea *Spilopsyllus cuniculi*. In the late 1930s, several unsuccessful attempts had been made to introduce myxomatosis to the island of Skokholm, off the coast of Pembrokeshire, in order to control rabbits there. It transpired that these introductions had all failed because the rabbits of Skokholm were free of fleas! It is very probable (though impossible to prove) that the spread of the disease in both France and Britain was encouraged by farmers introducing diseased rabbits to their land as a means of controlling the pest. It was later estimated that the UK agriculture industry benefited by £50 million per annum.

What would be the ecological effects of the extermination of rabbits? In 1954, this was the urgent question that the Nature Conservancy, the government agency chaired by Sir Arthur Tansley and founded just five years previously, had to address. Dr Norman Moore, who was employed as an ecologist by the organization at that time, later recalled:

> 'When it became clear that the initial outbreak of myxomatosis could not be contained it seemed likely that the rabbit, one of Britain's commonest mammals, would become very much rarer if not extinct. Enough was known about its effect as a grazing animal and its importance as prey to many predators for the Nature Conservancy to be deeply concerned about what might happen … .
>
> … My shortlist of species which must be at least partly dependent on the rabbit consisted of Fox, Badger, Stoat, Weazel and Buzzard (all of which preyed on the rabbit) and the Minotaur Beetle (*Ceratophyus typhoeus*) whose larvae feed on

rabbit dung pellets, the Wheatear which nests in rabbit holes and the Stone Curlew which was at least partly dependent on bare ground produced by excessive rabbit grazing. My list of possible competitors consisted of Brown hare, Mountain hare, Red Deer, Roe Deer, and Short tailed Field Vole.'

From N. Moore (1987) *The Bird of Time*, pp. 124–25.

❑ List the kinds of interaction between rabbits and other organisms that are referred to by Moore.

■ **Grazing** on plants, **predation** by other mammals and by buzzards, provision of food to the minotaur beetle (a **detritivore** that eats dead organic matter), provision of **nest sites** for wheatear and stone curlew, **competition** between rabbits and other herbivorous mammals.

Subsequent studies of some of the organisms mentioned by Moore, and of other species, have shown just how complex and far-reaching ecological links may be. We shall now look at some of these studies, at how they were carried out, and at what they can and cannot tell us about the ecological role of the rabbit in Britain.

Table 1.1 Numbers of breeding buzzards (*Buteo buteo*) in the Avon Valley, Devon, and their breeding success before and after myxomatosis.

Pairs	Pre-myxomatosis 1954	Post-myxomatosis	
		1955	1956
resident in April	21	14	12
building a nest	19	13	9
laying eggs	19	1	8
hatching eggs	17	0	5
fledging young	17	0	5
number of young reared	30	0	7

❑ Table 1.1 presents the results of a study of buzzards nesting in the Avon Valley, Devon, where myxomatosis did not arrive until after the breeding season of 1954. What does it suggest about the effect of myxomatosis on buzzard numbers immediately after it arrived, and in the slightly longer term?

■ Myxomatosis appears to have reduced the numbers of buzzards resident in the valley in 1955 by a third and to have had a catastrophic effect on their breeding success. However, in 1956 there was some recovery in breeding success, though not in numbers of resident birds.

❑ Can you think of *two reasons* why the breeding success of the buzzard may have recovered a little in 1956?

■ Either rabbit numbers had begun to recover from the disease, or the buzzards had started to hunt alternative food. In fact, the latter was the case.

Figure 1.2 Changes in the national mean numbers of brown hares shot in Britain (black line) and Denmark (green line).

Neither buzzards nor rabbits recovered their former abundance.

In the years following the crash in rabbit numbers, there were widespread reports of increases in the numbers of brown hares (*Lepus europaeus*). Was this a consequence of reduced competition from rabbits as Moore had forecast? At first, this was thought likely, and this suspicion grew when numbers of brown hares were observed to fall again as rabbits began to make a comeback. However, more detailed study of the situation by Barnes and Tapper (1986) has thrown a good deal of doubt on the idea. First, they found that the numbers of rabbits and hares sighted were positively correlated with each other on English farms that they studied in 1980 and 1981. Next, they looked at bag records for the numbers of rabbits and hares shot in the years 1930/31 (before myxomatosis), 1961/62, 1970/71 and 1980/81. In nearly all cases, there was a positive correlation between rabbits and hares in these records, too. Finally, they compared the average number of brown hares shot per km^2 between 1960 and 1980 in Britain with comparable records from Denmark where rabbits were virtually absent over this period (Figure 1.2).

❑   What do you conclude about the relationship between rabbit and brown hare numbers from Figure 1.2, and from the other evidence gathered by Barnes and Tapper? What must you assume about their sampling methods to reach your conclusion?

■   Figure 1.2 suggests that the decline in brown hare numbers between 1960 and 1980 in Britain was unconnected with any rise in rabbit numbers over that period because it also occurred in Denmark where there were almost no rabbits. Together with the other evidence collected by Barnes and Tapper, this suggests competition with the rabbit was not responsible for the post-1960 decline in the brown hare. The biggest assumption made in interpreting this evidence is that numbers of rabbits and hares recorded shot was proportional to the total numbers of these species.

Barnes and Tapper suggested that the decline in brown hare numbers recorded in Britain and Denmark was due to the intensification of arable farming which occurred in both countries over the period. However, the reported increase in brown hare numbers in the mid-1950s does seem to have been genuine.

❑   If the post-myxomatosis rise in brown hares was not due to a reduction in *competition* from rabbits, what other effect of myxomatosis might have caused it? (Note that myxomatosis *does not* infect hares.)

■   Leverets (young hares) require vegetation as cover to hide them from predators. When rabbits ceased to graze grasslands, there would have been more cover, and this may have increased the survival of leverets.

What were the vegetation changes observed in the years after rabbit numbers crashed? Tansley's results (Figure 1.1), and the sheer numbers of rabbits present before myxomatosis, made it quite clear to the ecologists working for the Nature Conservancy that large changes in the vegetation might be expected to follow the cessation of rabbit grazing. To monitor these changes, A. S. Thomas marked out over 7 km of **transects** (lines used for sampling across an area) in different parts of the country. A sample of the results recorded along one transect at Kingley Vale National Nature Reserve in Sussex is given in Table 1.2.

Table 1.2   Changes in the vegetation along a transect at Kingley Vale National Nature Reserve between 1954 and 1967.

	1954	1957	1961	1967
average height/cm	10.5	16.5	12.5	150
number of species touched	21	20	28	12
cover of individual species:				
sweet vernal grass *Anthoxanthum odoratum*	4	15	35	2
sheep's fescue grass *Festuca ovina*	89	85	76	11
crested hair grass *Cynosurus cristatus*	49	36	40	4
carnation and glaucous sedges *Carex* spp.	22	20	14	3
salad burnet *Sanguisorba minor*	34	36	40	7
bramble *Rubus fruticosus*	0	1	10	61

❏   What were the biggest changes along the transect at Kingley Vale in the first eight years following myxomatosis?

■   The number of species and the amount of sweet vernal grass and bramble *increased*.

❏   What happened in the next six years after that?

■   Bramble increased greatly, all the other species listed in Table 1.2 decreased greatly and the overall number of species dropped dramatically. The height of the vegetation also increased enormously.

With the exception of bramble, all the species listed in Table 1.2 are grassland plants, so the changes shown represent a change in the *type* of vegetation present, as well as in its species composition. Grassland plants were excluded by **competition** from a taller species when rabbit grazing no longer kept the vegetation short. With further time, grasslands such as those once grazed by rabbits at Kingley Vale were colonized by shrubs such as hawthorn *Crataegus monogyna* and dogwood *Cornus sanguinea*. This change is an example of vegetation **succession**, and is a process that in many parts of Britain eventually terminates in the formation of oakwood.

The kinds of changes observed in animal and plant communities following myxomatosis strongly suggest that the rabbit was very important in the ecology of Britain before the epidemic, but what do they *prove*? How do we know that the decrease in buzzards and the change in vegetation at Kingley Vale wouldn't have happened anyway, particularly since the biggest vegetation changes took place some time after rabbits disappeared? The answer is that we can never be sure of the causes for this kind of change unless we can reproduce the effect in an **experiment**. The difficulties in understanding why brown hare numbers changed in the way they did following myxomatosis makes this clear. You might think that myxomatosis was like a huge natural experiment on the ecological effects of the rabbit because it removed almost all rabbits; however, it lacked something that all proper experiments should have: a **control**. A control is a plot or set of plots that are as similar as possible to those receiving a

treatment (such as removal of rabbits), but which are left untreated. The effect of the treatment is then judged by comparison with what happened in control plots.

Rabbits started to make a comeback in the early 1970s, and became locally common again in some places such as the Breckland in Norfolk and Suffolk. Experimental exclosures have been erected there to study the effects of rabbit grazing on the invertebrate fauna of the grassland. The grasshopper *Chorthippus brunneus* feeds on leaves of the grass *Festuca ovina* and the woodlouse *Armadillidium vulgare* feeds on dead plant material of many species. Densities of both were significantly greater inside exclosures than outside in control plots where rabbits grazed (Figure 1.3).

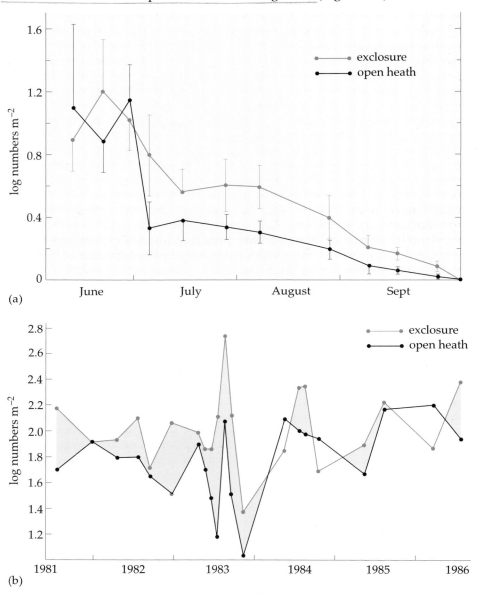

(a)

(b)

Figure 1.3 (a) Population density of the grasshopper *Chorthippus brunneus* inside and outside a rabbit exclosure at Weeting Heath (bars show the standard error). (b) Population density of the woodlouse *Armadillidium vulgare* inside and outside a rabbit exclosure.

Most of the interactions between rabbits and other animals that we have mentioned so far are direct ones that involve a trophic (feeding) relationship. The most catastrophic effect of myxomatosis on another animal that has been recorded actually involved a much more indirect relationship with the rabbit. The large blue butterfly *Maculinea arion* was a rare butterfly of grassland habitats in the south of England that finally declined to extinction in Britain in 1979. The large blue has the most extraordinary life cycle. Its caterpillars feed upon thyme flowers, and when they reach their final instar (stage in life as a caterpillar), they are collected by red ants of the species *Myrmica sabuleti* that carry them to their nest. In the ants' nest, the caterpillar feeds upon ant grubs for many months, repaying the ants with a secretion on which they feed. *Myrmica sabuleti* is confined to close-grazed grassland, because tall vegetation shades the ant hills, making them too cool. Before myxomatosis, conversion of grassland to arable had greatly reduced the habitat of *M. sabuleti* and the large blue. After myxomatosis, many of the remaining sites for the large blue, where the grass had been kept short by rabbits, became unsuitable for the butterfly's host and the butterfly disappeared. Two nature reserves were set up to try to save the species in Britain, but grazing was inadequate at one, and a bad year for the butterfly in 1979 saw its extinction at the other. The large blue still occurs in continental Europe, and English Nature (the modern successor to the Nature Conservancy) is attempting to re-establish colonies at several sites using butterflies obtained from Sweden.

The case of the rabbit and the large blue raises many questions about nature conservation. The grassland habitats in which it occurred depended upon management by grazing. Leaving nature to itself, succession would turn virtually all grassland in lowland Britain to forest. But, you might say, the rabbit would have maintained suitable grasslands naturally – it was only human introduction of myxomatosis that upset the ecological balance. Unfortunately, nothing is so simple! The rabbit is not native to Britain and was itself first introduced in Norman times. Perhaps before settled agriculture in Britain, some grasslands were maintained by other grazing animals that are now extinct or much less common. It is even possible that the introduction of the rabbit played some part in reducing numbers of other herbivores that could not compete with it.

Trying to decide what is 'natural' in a landscape whose ecology has been as much influenced by humans as that of Britain (and indeed most of Europe) is futile. The ecologist's task is to find out why things are the way they are, and how they work. Then we have some hope of making informed choices about how we would like things to be, and some chance of taking appropriate action.

Question 1.1   *(Objectives 2 & 3)*

1   Read back over this case study and make a list in Table 1.3 of all the species mentioned. Divide them up into plants, detritivores, herbivores, carnivores and 'other animals'. For each species, note in the appropriate column of the Table whether:

(a) rabbits, when they were present, normally had a positive or a negative effect on the abundance of the species in the short term (for the plants, the short-term is defined as 10 years).

(b) the observed effect was a direct one (involving no other species) or an indirect one (involving other species).

(c) the evidence for your conclusions is experimental, observational (e.g. based on events following myxomatosis), or based only upon knowledge of the natural history of the species.

Table 1.3 Blank Table for Question 1.1.

| | Effect of the rabbit: | | |
	Positive/negative	Direct/indirect	Evidence
Plants:			
Sweet vernal grass	+	Direct	observation
sheeps Fescuee	(−)		
crested hair grass	(−)		
Carnation/glaucous sedges	(−)		
salad burnet	(+)		
Bramble	+		
Animals:			
detritivores:			
Minotaur beetle	−	Direct	Not lit.
herbivores:			
Hare	+	indirect	
Red deer	+		
Roe deer	+		
Field vole	−		
carnivores:			
FOX	−	Direct	
Badges			
Stoat	=	Direct	obs.
weasel			
Buzzard	−		
other animals:			
wheatear	−	indirect	Not lit
Stone curlew	−		

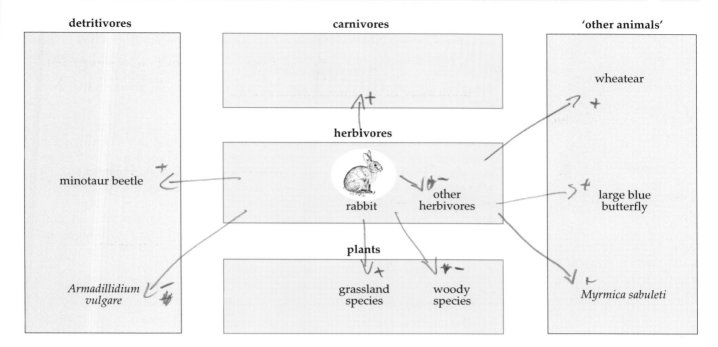

2   When you have completed Table 1.3, look at the diagram in Figure 1.4 which shows the major types of organism referred to in this case study. Draw arrows between the types to show which ones *directly* affected each other (or were expected to do so), and indicate with a plus (+) or a minus (−) by the head of each arrow whether, when rabbits were present, the effects in each direction were positive or negative. Check your answer with Figure A1 in the Answers at the end of this Book.

(a)   In one sentence, summarize the *direct* positive and negative effects on the species involved of the relationships between carnivores and herbivores, and between herbivores and plants that are indicated by the arrows you have drawn on Figure 1.4.

(b)   How do the direct relationships between the rabbit and detritivores and the rabbit and 'other animals' differ from those summarized in (a)?

(c)   How do the *indirect* relationships between the rabbit and detritivores, the rabbit and 'other animals' and the rabbit and plants differ from the direct relationships summarized in (a)?

3   What relationship, if any, would you expect there to be between species *within* the groups shown in Figure 1.4?

Figure 1.4 Hypothetical ecological relationships between plants, detritivores, herbivores and carnivores. (For use with Question 1.1.)

1.3   Some generalizations

A number of generalizations with wide applicability in ecology can be drawn from the case study of the rabbit. First, as we saw in analysing the relationships in Figure 1.4, it is useful to group interacting organisms into **trophic levels** representing carnivores, herbivores and plants because the feeding (i.e. trophic) relationships between these levels are reasonably consistent. Note, however, that detritivores do not fit into this scheme so neatly because some of these organisms feed on animal faeces (e.g. the minotaur beetle) while others feed on dead plant material (e.g. the woodlouse *Armadillidium vulgare*). There are also non-feeding relationships with 'other animals' (e.g. the wheatear) or indirect consequences of feeding relationships (e.g. on the large blue butterfly) that don't allow us to describe all the interactions among organisms purely in terms of trophic levels.

Ecological relationships can get so complicated that ecologists are always on the look out for ways of simplifying the picture in order to improve our understanding without a loss of important detail. When does it suffice to talk about interactions between carnivores, herbivores and plants, and when do we need to be more specific and actually name the species involved: *Buteo buteo*, *Oryctolagus cuniculus* and *Festuca ovina*? These are questions about **levels of organization**, and there are no simple answers. In practice, what ecologists do is to think in terms of a hierarchy of levels of organization:

- the **individual** organism (e.g. the physiology or behaviour of *O. cuniculus*);
- the **population** (e.g. the number of rabbits living in a specified area);
- the **community** (e.g. the group of species with which the rabbit lives);
- the **ecosystem**, which includes all the living components of the system and the physical and chemical components on which life depends (e.g. solar radiation, water, oxygen, carbon dioxide, N, P, K, etc.).

Some ecological questions are more appropriately addressed at the individual level, some at the population level, some at the community level and some at the ecosystem level. For example:

'Why does *Myrmica sabuleti* need nests warmed by the Sun?'

is probably best answered by looking at individuals and specifically at the **physiology** of this species of ant.

'Why do numbers of the woodlouse *Armadillidium vulgare* fluctuate?' (see Figure 1.3b)

needs to be addressed by looking at how and why birth and death rates in the woodlouse **population** are determined.

'Why did the large blue butterfly decline to extinction after myxomatosis?'

is a **population** question which, it turns out, needs a knowledge of interactions in the ecological **community** to answer it.

'How much nitrogen do plants obtain from the faeces of detritivores?'

concerns recycling within the **ecosystem** of a very important mineral element.

This Course is structured around these levels of organization: earlier Books will concentrate on one level while later ones use several levels simultaneously.

Two other terms are very commonly used to describe where a species is typically found. The **habitat** of a species is like its ecological address: the rabbit is found in grassland habitats; the natural habitat of the brown trout is upland streams and lakes; deciduous woodlands are the habitat of the tawny owl. As the impact of human activities on the natural environment intensifies, the distribution of most types of habitat has become increasingly fragmented. Where once in some parts of Britain there were extensive tracts of heathland, chalk grassland and hay meadow for example, there now remain only small remnants. The average size of hay meadows remaining in England is only 4 hectares (ha). As habitats shrink in area, it becomes increasingly important to study the **landscape** that surrounds them because insects, birds and mammals may disperse or forage for food over a wider area than is provided by a single patch of their habitat. For example, a study of the grey squirrel (*Sciurus carolinensis*) in East Anglian woodlands found that in woods without hazel bushes or beech trees, squirrel dreys (which are evidence of squirrels breeding) were more rare per unit area in small woods than large ones (Fitzgibbon, 1993). However, the likelihood of a small wood (about 2.5 ha) containing a squirrel drey increased if there was a hedgerow nearby, and the longer the hedge, the more often the wood contained a drey.

Question 1.2   *(Objectives 1.1 & 1.2)*

Answer the following questions in one sentence each:
(a) Write a definition of ecology, naming an example of an ecological interaction.
(b) What is meant by 'levels of organization'?
(c) What are 'trophic levels'?
(d) What is the relationship between 'habitat' and 'landscape'?

1.4   Types of interaction between species

Interactions between different species take many forms and so some kind of classification of the different types is necessary. The three most important general features of interactions between associated species are:

*   the **benefit** (or disadvantage) derived by each species from the association;
*   the degree of **dependence** of each species on the association;
*   the degree of **specificity** in the association between species.

In an interaction between two species, there are two general methods of assessment:

1   The performance of each species when the other species is present is compared with its performance when the other species is absent, using some kind of short-term indicator such as rate of growth or rate of food intake.

2   The effects of species on each other are measured over a longer period. This involves measuring the survival rate and rate of reproductive success of each species with and without the other species.

❑   Why is the second method the more reliable one?

■   Because short-term benefits and losses may be misleading. The relationship between a pair of species can only persist if both are ultimately able to survive and to reproduce.

Each member of the pair can gain an advantage from the relationship (+), or it can experience a disadvantage (−), or the interaction may have neutral effects (0). There are five ecologically meaningful combinations of these possibilities, namely:

+/+ mutualism	both species gain
+/0 commensalism	one species gains
+/− parasitism, predation or herbivory	one species gains and the other loses
−/− competition	both species lose
−/0 amensalism	one species loses

A **symbiosis** is an association between species that live together and includes most mutualistic and parasitic relationships, but, as we shall see very shortly, it is sometimes difficult to determine the existence of benefit in a relationship.

The degree of dependence of species on those with which they interact varies from **obligate** to **facultative dependence**. For example, common dodder *Cuscuta epithymum* is a heterotrophic plant that entirely lacks chlorophyll and is therefore an obligate parasite of photosynthetic hosts. Yellow rattle *Rhinanthus minor* is obligately parasitic on the roots of other plants, but this species possesses chlorophyll and is only partially parasitic. Truly facultative parasitism is rare, though there are examples among the fungi. The honey fungus *Armillaria mellea* is **saprotrophic**, living on dead wood, but it is also a serious parasite of many woody plants which it kills.

The specificity of ecological interactions varies greatly. The caterpillars of the large blue butterfly are extreme specialists, depending exclusively upon the ant *Myrmica sabuleti*. By contrast, caterpillars of the winter moth *Operophtera brumata* feed on a wide range of trees. The extreme specificity of the large blue's relationship with its ant host explains this butterfly's vulnerability to extinction, while the relative lack of specificity in the winter moth's food requirements must explain why it is so common. Note that just as the benefits to the partners in an interaction need not be symmetrical (i.e. they may be +/−, +/0, −/0), their specificity need not be symmetrical either. For example, good growth in Douglas fir *Pseudotsuga menziesii*, as in most forest trees, is dependent on mycorrhizal fungi associated with its roots (see Section 1.5). It has been estimated that in its native range (western North America) about 250 different species of fungi are specific to *Pseudotsuga*, though of course any individual tree will only harbour a small fraction of these. The fungi in this relationship are much more specific than the tree.

Significant numbers of species of mycorrhizal fungi are officially listed as endangered or vulnerable to extinction in Europe, where it is suspected that air pollution is a threat to them. In Germany, there has been a reduction in nearly all mycorrhizal fungi and a 95% reduction in the chanterelle *Cantharellus cibarius*.

❑   How might the specificity of tree–fungus relationships in European forests affect the impact of the extinction of fungi?

■   If trees are non-specific and can form mycorrhizal associations with many fungal species, extinct fungal species could be replaced by others and forests as a whole might be little affected. On the other hand, if trees depend on specific fungi, the whole forest community could be affected by the extinction of a few fungal species. Of course, if all mycorrhizal fungi are threatened, forests will suffer either way.

Some interspecific associations are easily classified because the benefits and losses to the species involved are obvious from their behaviour. The European cuckoo *Cuculus canorus*, which lays its eggs in the nests of several other species of birds, clearly gains from this relationship because its chicks are raised by the foster parents it chooses for them. This choice has to be a correct one because some species will turn out of the nest eggs that do not resemble their own. The problem for the cuckoo is overcome because individuals specialize in laying their eggs in the nests of particular species and produce eggs with a pattern which mimics that of particular host species. Once successfully hatched by its foster parents, the cuckoo chick is fed by them and shoves chicks or eggs belonging to its host out of the nest.

❑   What type of interaction is this?

■   The relationship between the European cuckoo and its hosts is clearly parasitic +/−.

There are many hundreds of species of birds around the world which, like the cuckoo, practise **brood parasitism.** However, the relationship between brood parasite and host, which so clearly benefits the cuckoo at the expense of its host, is not always so simple or one-sided.

The giant **cowbird** *Scaphidura oryzivora is* a brood parasite living in central America. Unlike the cuckoo, its chicks do not evict host chicks from their nests. The adult female lays her eggs in the nests of several different species of birds, known in Panama by the common name of **oropendolas** (Figure 1.5). The different oropendolas breed in mixed colonies where they construct bag-like nests which hang from the branches of trees in positions where they cannot be reached by vertebrate predators such as opossums. The sites of nesting colonies are fairly permanent and all are frequented by giant cowbirds, but the behaviour of oropendolas towards cowbirds is very different in different colonies. In one type of colony, the host birds try to chase off cowbirds whenever they enter the area and they will destroy eggs found in their nests which do not resemble their own. These hosts are called **discriminators.** The cowbirds that lay their eggs in these nests have eggs which mimic those of their selected host. These chicks are reared by their foster parents when they hatch.

There is a second type of nesting colony in which oropendolas do not scare off cowbirds or discriminate between their own eggs and those of the cowbird. These hosts are called **non-discriminators**. The cowbirds that lay eggs in these nests do not have mimetic eggs. Thus, the same host species behaves very differently towards the same brood parasites in different colonies.

Another difference between the two types of colony was discovered by N. G. Smith (Smith, 1968). The colonies of oropendolas where nests were actively defended against cowbirds were all built in the vicinity of huge nests of wasps and biting bees.

❑   List all the differences which have been mentioned between colonies of discriminators and non-discriminators.

Figure 1.5 Oropendolas nesting (scale: × 0.15).

■   **Discriminators:**   **Non-discriminator:**

scare off cowbirds   ignore cowbirds

remove eggs unlike their own   do not remove eggs unlike
                                            their own

are parasitized by cowbirds   are parasitized by cowbirds
with mimetic eggs   with non-mimetic eggs

have nests near the nests   do not have nests near the nests
of wasps or bees   of wasps or bees

During a four-year study, Smith monitored the survival of oropendola chicks in the nests in the two types of colony. The average fledging success from nests originally containing two eggs is shown in Table 1.4.

Table 1.4   The average fledging success of oropendola chicks from nests in discriminator (+ wasps or bees) and non-discriminator (– wasps or bees) colonies when one cowbird chick was present and when none was present.

	Discriminator (+ wasps or bees)	Non-discriminators (– wasps or bees)
cowbird chick present	28%	53%
cowbird chick not present	53%	19%

❑   Do the data in Table 1.4 shed any light on why discriminators and non-discriminators might behave differently towards cowbirds and their eggs?

■   Yes. Chicks of discriminators do better when there is no cowbird chick present but those of non-discriminators do better when a cowbird chick is present in the nest.

How could the presence of a brood parasite be a severe handicap when there are wasps or bees nesting near the colony but be an actual advantage when there are none? Smith found that the answer lay in the interaction of the cowbirds, oropendolas, wasps and bees with yet another organism – a botfly called *Philornis* sp. which parasitizes oropendola chicks. This fly lays its eggs on young chicks and its larvae burrow into the chick's body, later emerging to pupate at the bottom of the nest. Seven larvae feeding on one chick are enough to kill it. The botfly is common in colonies of non-discriminators and appears to be responsible for most of the chick mortality. But, for some reason which is not clear, botflies are never found in the vicinity of the large nests of wasps and bees where discriminators have their nesting colonies.

Cowbird chicks protect the nestlings of their foster-parents from botflies by picking *Philornis* eggs and larvae from their bodies. Cowbird chicks are much more active than oropendola chicks and they will even catch and eat adult *Philornis* which enter the nest. A non-discriminator's chicks are safe from botflies as long as there is a cowbird chick in the nest.

❑   What happens to a discriminator's chicks if there is a cowbird present?

■   A discriminator's chicks have about half the survival rate (28% *vs*. 53%). This is still better than a non-discriminator *without* a cowbird (19%).

The interactions of all these species appear to revolve around the botfly. If oropendolas can escape the botfly by nesting near wasps or bees and if they do better there, even when they are host to a cowbird, than they do with no bees, wasps or cowbirds, why do not all oropendolas nest near wasps and bees? Perhaps the reason is that there are some extra disadvantages to using these sites which have not been taken into account. A likely one is that discriminators' nesting colonies are densely packed near their protecting insects' nests and the weight of so many birds' nests often causes tree limbs to snap, dashing whole parts of the colony to the ground.

❑   Why is this less likely to happen in colonies of non-discriminators?

■   Their nests do not need to be near wasps' or bees' nests for protection and are consequently less closely packed.

In an experiment, Smith switched young female chicks from nests in discriminator colonies with chicks in non-discriminator colonies to find out whether the difference in behaviour between discriminators and non-discriminators was learned or inherited by oropendolas. When chicks returned as adults to their foster colony to breed, their behaviour towards cowbirds and odd-looking eggs was the same as their foster-parents, not their true parents.

As this example shows, it is often difficult to define who benefits and who loses in an interaction because this may change with circumstances. It is particularly difficult to establish the benefits to partners in the most intimate kind of association between species, called **symbiosis** (from the Greek for 'living together'). This term was first coined in the 19th century by Anton de Bary to refer to organisms that live together for most or all of their lives, though since then the term has been confused by various attempts at redefinition, the most common of which is to treat symbiosis as a synonym for mutualism. In this Course, we will stick to the original definition which covers both mutualism and parasitism; a consideration of a classic case of symbiosis, the lichen, will explain why.

A lichen is a symbiotic association between two species: a fungus and an alga. The 5000 or so fungal species that form lichens never occur without their algal partner though the algae are often found free-living. Some (so-called lichenicolous) fungi acquire algal symbionts by infecting established lichens, killing the fungal partner and taking over its algal cells. The lichen symbiosis is often considered mutualistic because lichens are able to colonize areas such as bare rock where neither partner can grow on its own, and because the partners exchange carbohydrate (from the alga) and mineral nutrients (from the fungus). However, the algal cells in a lichen can lose 60–80% of their photosynthate to the fungal partner and their rate of cell division appears to be severely limited by this, as though they were more the victims of parasitism than the beneficiaries of mutualism. It isn't meaningless to ask whether one or both partners in a symbiosis benefit from the association but it is better not to *define* the term symbiosis in this way because benefit can be changeable and so difficult to determine.

Douglas (1994) has pointed out that a better defining characteristic of symbiotic interactions of the lichen sort is that one (or each) partner confers novel metabolic capabilities on the other.

❑   How does this apply to the lichen symbiosis?

■   The alga provides the fungus with carbohydrate (from photosynthesis) that it isn't able to acquire by itself.

We shall call these **metabolic symbioses** and look at a range of examples in the next Section. Following that, we shall look at some associations that are unequivocally parasitic and then some that are clearly mutualistic. Chapter 2 deals with competition, predation and herbivory.

Summary of Sections 1.1–1.4

The science of **ecology** is defined as the study of interactions between organisms and their environment. The extent and complexity of ecological interactions are exemplified by the case of the **rabbit** whose relationships with other organisms were revealed when myxomatosis all but exterminated the species in Britain. In general, ecological relationships between organisms can be investigated in **experiments**. Organisms may be grouped according to their source of food into **trophic levels** representing **carnivores**, **herbivores** and **plants**. **Detritivores** may feed at several trophic levels.

Ecological questions may be addressed at four distinct **levels of organization**: at the level of the behaviour or physiology of the **individual** organism, at the level of the **population** which is defined as the **number** of individuals inhabiting a specific area, at the level of the **community** which is defined as a group of interacting species, or at the level of the **ecosystem** which includes all the living and non-living components of the system that supports life. Two additional terms describe the physical surroundings of organisms: the **habitat** of a species is like its ecological address (e.g. an oakwood, a pond); the **landscape** refers to the larger geographical context of species and habitats.

Three important properties of interactions between species are: the degree of **benefit** derived by each species, the degree of **dependence** of each species on the other(s), and the degree of **specificity** in associations between species. Benefit is used to classify five kinds of relationship between species: **mutualism** (+/+, both benefit), **commensalism** (+/0, one benefits), **parasitism**, **predation and herbivory** (+/−, one benefits at the expense of the other), **competition** (both lose) and **amensalism** (−/0, one loses).

A **symbiosis** is an association between species that live together and includes most kinds of mutualism and parasitism. The existence of benefit to a symbiont is difficult to determine; it may change and it often depends on circumstances. **Metabolic symbioses** are those symbiotic relationships in which one partner provides the other with a specific substance it cannot manufacture or obtain for itself.

Question 1.3   (Objective 1.3)

Using all the information you have been given about the oropendolas and their symbionts, decide what kinds of association (e.g. +/+, +/0, etc.) exist between (a) oropendolas and botflies, (b) botflies and cowbirds, (c) discriminators and, cowbirds, (d) non-discriminators and cowbirds, (e) oropendolas and wasps/bees, (f) wasps/bees and botflies, (g) cowbirds and wasps/bees, and (h) discriminators and the trees in which they build their colonies.

1.5   Metabolic symbiosis

This Section describes a wide range of symbiotic associations where there is cellular contact between the partners, or where a symbiont is found *inside* the cells of a host. The partners in such symbioses are so intimately associated with one another that ecologically they tend to behave as a single unit. You need not remember all the details of each example but you should be able to remember the metabolic capabilities conferred by different partners on the symbiotic unit.

1.5.1   Lichens (fungus + alga)

Figure 1.6 shows the structure of a typical crustose lichen (a common type where the plant body or thallus is tightly appressed to the substratum, which is usually rock or wood). Other types of lichen are shown in Figure 1.7.

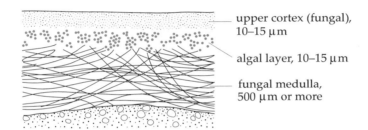

upper cortex (fungal), 10–15 μm

algal layer, 10–15 μm

fungal medulla, 500 μm or more

Figure 1.6   A vertical section through a typical crustose lichen on a rocky substratum, showing the organization of its tissues.

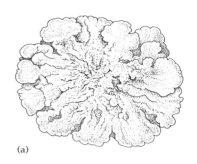
(a)

(b)

Figure 1.7   (a) A foliose (leaf-like) thallus of *Parmelia austrosinensis* (× 2). (b) A fruticose (shrub-like) thallus of *Cladonia rangifera* (× 2).

❑   In terms of volume occupied, which is the dominant partner in a crustose lichen?

◼   The fungus. The alga is usually confined to a single thin layer and constitutes about 5–10% of the total dry weight.

The algae in lichens are simple unicellular or filamentous types. A green alga **Trebouxia** is the commonest in northern latitudes where it occurs in about 70% of lichen species. **Cyanobacteria**, particularly *Nostoc*, occur in about 10% of lichen species and the remainder contain other genera of green algae. There are about 300 genera of lichen fungi but only about 40 genera of lichenized algae. About 500 lichen species possess both green algal and cyanobacterial symbionts. The former provide carbohydrate to the fungus and the latter fix atmospheric nitrogen. **Nitrogen fixation** converts gaseous nitrogen into soluble N compounds. In lichens with cyanobacteria, N is transferred to the fungus in the form of ammonia.

❑   There are about 13 500 lichen species. Which symbionts (algae or fungi) are responsible for this great variety?

◼   The fungi. Algal diversity in lichen symbioses must be low because there are only about 40 genera and 70% of lichens contain just one of these, *Trebouxia*. Precisely how many species of algae are involved is not known.

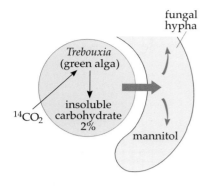

Figure 1.8 The movement of carbohydrate in a typical lichen.

Trebouxia can be cultured separately from its fungal symbionts but is rarely found growing wild without them. *Nostoc* and the green alga *Chlorella* are found in lichen associations but are both also commonly found in the free-living state. Lichen fungi are rarely, if ever, found without their symbionts. When artificially cultured on their own, they look quite different from the usual lichen thallus. Though both partners from lichen symbioses can be cultured separately, a recognizable lichen thallus cannot be synthesized by mixing pure cultures of the two symbionts.

Most lichens grow very slowly and have a remarkable ability to survive in nutrient-poor habitats and extreme climatic conditions where other plants cannot survive; lichens may tolerate extremes of heat, cold and desiccation. So how does this association 'work' and who gets what from whom?

The fungus receives most of its energy, as simple carbon compounds, from the alga. In experiments, between 40 and 70% of the $^{14}CO_2$ fixed by algae was shown to be released as glucose (by cyanobacteria) or as sugar alcohols such as ribitol (by green algae); these substances are taken up by the fungus and immediately converted into the sugar alcohols mannitol or arabitol. The fungus acts as a permanent sink for the products of algal photosynthesis (Figure 1.8).

Where cyanobacterial partners have been shown to fix atmospheric nitrogen, most of the fixed N appears to pass to the fungus. So there is a constant pattern of fixed C or N moving from alga to fungus but little, if anything, passes from fungus to alga. It has been suggested that mineral nutrients are absorbed by the fungus and passed on to the alga, but there is no conclusive evidence to support this.

❑   Suggest another possible benefit to the algae in lichens.

■   Protection from radiation and desiccation by the insulating layer of fungal cortex.

The range of habitats available to algae such as *Trebouxia* is greatly increased by association with fungi in lichens.

1.5.2   Mycorrhizas (fungus + higher plant)

Mycorrhizal symbioses are intimate associations between a fungus and a living plant which involve the transfer of carbon and/or minerals between the symbionts. Fungi involved in mycorrhizas occur in the groups Basidiomycetes, Ascomycetes and in the Phycomycetes. The classification of the mycorrhiza-forming fungi in the last group is difficult because most have not been found in the free-living state; they are sometimes classified separately.

The great variety of fungal symbionts in mycorrhizas is exceeded by the even greater variety of plants found in these symbiotic associations. Mycorrhizas are common in the Angiosperms (flowering plants), Gymnosperms (conifers), Pteridophytes (ferns and club-mosses) and Bryophytes (mosses and liverworts). The type of symbiosis varies from mycorrhizas in which the plant is parasitic upon the fungus or the fungus is parasitic upon the plant to mutualistic relationships between the symbionts. The classification of different types of mycorrhiza is complex, which is not surprising when so many different plants and fungi are involved in such a range of relationships.

For simplicity, two major types of mycorrhiza can be identified on the basis of their structure, though the differences are not absolute. The first type of mycorrhizal symbiosis is the **ectomycorrhiza**, so-called because there is a characteristic fungal sheath around the outside (*ecto* is Greek for 'outside') of the plant root which produces numerous stubby root branches (Figure 1.9). Fungal hyphae form a network called the 'Hartig net' between epidermal and cortical cells, but the hyphae of ectomycorrhizal fungi do not generally penetrate root cells. Ectomycorrhizas are common among forest trees but are rare amongst herbaceous plants. The fungi involved vary in their specificity, but some of the species are confined to associations with trees in particular genera. For example, *Boletus betulicola* occurs only with birch trees (*Betula* spp.).

Ectomycorrhizal fungi are mainly common Basidiomycetes, e.g. *Boletus* spp. and the fly agaric *Amanita muscaria* (Figure 1.10), and it is thought that they are obligately mycorrhizal and cannot grow independently of their 'host'. In the majority, enzymes capable of breaking down complex carbohydrates have only low activity, and carbon compounds are supplied by the tree in the form of simple sugars.

❑   What general factor would you expect to influence the distribution of fungi which form ectomycorrhizas?

■   The distribution of suitable host trees; remember that the fungi cannot survive independently as saprophytes.

The amount of carbohydrate passing from tree to fungus is surprisingly large and may be about $500\,\mathrm{kg\,ha^{-1}\,y^{-1}}$, which is more than one-tenth of that going into wood production. This is a large drain on the tree, so it is reasonable to ask what, if anything, the tree obtains in return. Consider the following observations made by Harley (1971) about trees that normally have ectomycorrhizas:

1   The trees may grow well without mycorrhizas but only on very fertile soil.

2   The intensity of mycorrhizal infection is greater in acid or nutrient-poor soils and especially where phosphate levels are low and light intensity is high.

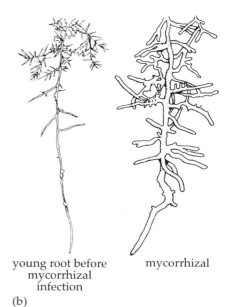

non-mycorrhizal   mycorrhizal
(a)

young root before   mycorrhizal
mycorrhizal
infection

(b)

Figure 1.9 The effects of ectomycorrhizas on root form: (a) in the roots of white pine (*Pinus strobus*) on soils of low nitrogen supply; (b) in roots of beech (*Fagus sylvatica*) (both × 0.5).

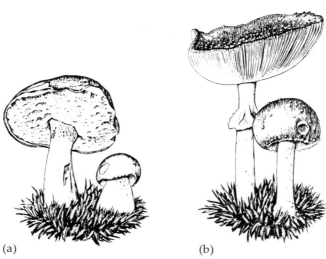

(a)   (b)

Figure 1.10   Fruit bodies of (a) *Boletus edulis* (× 0.25) and (b) *Amanita muscaria* (× 0.5).

3   Mycorrhizal seedlings grow faster and have a greater net production than seedlings without associated fungi (see Table 1.5).

4   Short lateral roots senesce and die less rapidly when associated with a mycorrhizal fungus.

5   Non-mycorrhizal trees appear more susceptible to invasion by dangerous pathogenic fungi, especially in the seedling stage.

Table 1.5   Results of experiments to compare the growth of seedlings or cuttings of trees grown for a season with or without ectomycorrhizal infection. The values indicate growth in one season (g dry wt.).

Plant	− Mycorrhiza	+ Mycorrhiza	Country
Pinus strobus	0.303	0.405	USA
Eucalyptus dives	3.2	5.3	Australia
Eucalyptus pauciflora	3.3	6.2	Australia
Quercus robur	1.14	1.69	Russia

From observations 1–5 and from information given earlier in this Section:

❑   What single piece of evidence indicates that trees benefit from mycorrhizal infection and that photosynthesis does not limit the rate of tree growth?

■   Observation 3 and Table 1.5 show that infected seedlings grow faster than non-infected ones, despite the loss of carbohydrate to the fungus in the former. This must mean growth is not limited by the rate of photosynthesis (carbon fixation) but by some factor which is supplied by the fungus.

❑   Suggest two ways in which ectomycorrhizas appear to promote the growth and survival of trees and cite the observations on which you base your answer.

■   On the basis of observations 1 and 2, it appears that ectomycorrhizas are particularly important on nutrient-poor soils and it is reasonable to suggest that they stimulate growth (observation 3) by increasing the supply of some form of mineral nutrients to the tree. In fact, the fungus absorbs nutrients with great efficiency and, when the external supply is low, trapped nutrients are passed on to the host tree. The longer life of mycorrhizal roots (observation 4) may be a further way of maintaining rates of nutrient uptake. On the basis of observation 5, it appears that mycorrhizal fungi protect the root against invasion by parasitic fungi.

It is probable that the distribution of ectomycorrhizal trees on nutrient-poor soils is dependent upon mycorrhizal infection: without the fungus, the trees may be unable to grow and/or compete with other species for the available nutrients. In fact, afforestation of treeless areas is often dependent upon prior infection of the tree seedlings with a suitable mycorrhizal fungus or inoculation of the soil after planting.

The second type of mycorrhiza is a structurally very diverse group, the **endomycorrhizas**. They involve a wide variety of plants and involve clear cases of both parasitism and mutualism, but have in common that fungal hyphae penetrate root cells. The commonest of these are **vesicular–arbuscular mycorrhizas** (abbreviated to **VAM**) formed by Phycomycete fungi in the family Endogonaceae. VAM fungi are not host-specific and are so widespread that it has often been said that it is easier to name the few plant families and genera that do not form this kind of symbiotic relationship than those that do! VAM fungi penetrate the cells of the plant's roots with hyphae which have swollen storage vesicles and branched bushy growths (or arbuscules) that are visible under the microscope (Figure 1.11). In most plants there is no obvious outward sign of VAM infection in roots, though the fungus forms a fine mycelium in the surrounding soil. There is a net flux of carbohydrate in favour of the fungus and improved uptake of minerals, particularly phosphorus, by the plant. The movement of these substances takes place across the membranes of living hyphae and plant cells. Plant growth may be greatly stimulated by the association, but only on soils that are relatively poor in mineral nutrients. The ability of these plants to survive in the wild and compete with other species on such soils is probably dependent upon mycorrhizal formation.

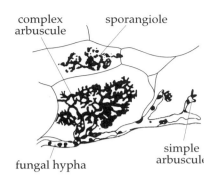

Figure 1.11 Root cells and vesicular–arbuscular mycorrhiza (× 250).

Cassava *Mannihot esculenta*, a tropical root crop, occurs only with its mycorrhizal fungus which probably accounts for its ability to grow in poor soils deficient in phosphate. This makes it a staple food plant in many parts of the wet tropics. Onion plants also benefit from symbiosis with a mycorrhizal fungus; the phosphorus nutrition of mycorrhizal onions is much improved over that of non-mycorrhizal onions and this increases the drought tolerance of the crop.

Another group of endomycorrhizas is found in the heathers and related plants (family: Ericaceae). The fungus found in a large number of these symbioses is a single species of ascomycete, *Hymenoscyphus ericae,* but Basidiomycete fungi also occur in the group. These mycorrhizas bear some resemblance to ectomycorrhizas and have an outer sheath, but hyphae penetrate into root cells of the host. They persist in a cell for a time and are then killed and digested, but exchange of carbohydrate and minerals does occur between root cells and living hyphae. In contrast to VAMs, this type appear to enhance the nitrogen uptake of plants. This may be of particular importance in the acid soils where many ericaceous species grow, e.g. *Rhododendron, Vaccinium myrtillus* (bilberry), *Erica* and *Calluna* (heathers).

Enhanced nitrogen nutrition is particularly important to plants in acid soils because at low pH the slow decomposition in these soils results in low levels of available nitrogen and nitrogen fixation is also often poor. It has also been found that heather (*Calluna vulgaris*) infected with *Hymenoscyphus ericae* mycorrhiza is resistant to concentrations of heavy metals in the soil which are toxic to non-mycorrhizal plants.

In Britain, one family of plants (the Monotropaceae) is related to the heather family in which many species are totally without chlorophyll. *Monotropa hypopitys* (yellow bird's nest), is a waxy, yellow-white plant which grows in woods of beech or pine or on sand dunes among willows. *Monotropa* was once described as a **'saprophyte'** because it was thought to live on decomposing organic matter but it is now known to be part of a peculiar triple relationship: its roots are associated with a *Boletus* (pore) fungus which, in turn, forms ectomycorrhizas with tree species (Figure 1.12).

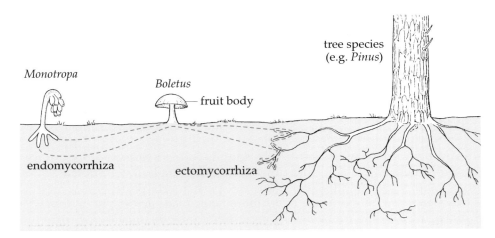

Figure 1.12 The relationship between *Monotropa hypopitys*, a *Boletus* fungus and a tree species (not to scale).

❑ In a situation similar to that shown in Figure 1.12, the tree (a pine) was injected with ^{14}C-labelled glucose and, after five days, ^{14}C was detected in *Monotropa* plants 1–2 m distant. What can you deduce about the dependence of *Monotropa* on (a) the pine tree and (b) *Boletus*?

■ (a) *Monotropa* lacks chlorophyll and must, therefore, receive carbon compounds from another source; this experiment shows that pines may be this source.

(b) As there is no direct contact between *Monotropa* and the pine tree (Figure 1.12), and carbon compounds pass from host trees to the ectomycorrhiza, it appears that the *Boletus* acts as a channel along which carbon compounds from the tree are passed to *Monotropa*. *Monotropa* is often described as an **epiparasite**.

The relationship between members of the orchid family (the Orchidaceae) and mycorrhizal fungi is even more specialized. The many thousands of species in the family Orchidaceae all produce tiny dust-like seeds with minimal food reserves. Successful establishment of seedlings is absolutely dependent upon early infection with a mycorrhizal fungus, but the balance between orchid and fungus (especially a parasitic fungus) is a knife-edged affair. The fungi are mostly Basidiomycetes and they may live as independent saprophytes. Some are virulent pathogens of other higher plants. Many orchid seedlings are overrun by the fungus and die. Successful seedlings restrict the growth of the fungus by means of powerful anti-fungal compounds so that there is limited infection. Once the association is established, the orchid obtains all essential nutrients, including carbon compounds, from the fungus (Figure 1.13) and an orchid tuber gradually develops. Green autotrophic shoots later grow from the tuber, but it is not clear whether the fungus is then digested by the orchid or whether (as in a few orchids which lack chlorophyll) it continues to supply nutrients to the plant.

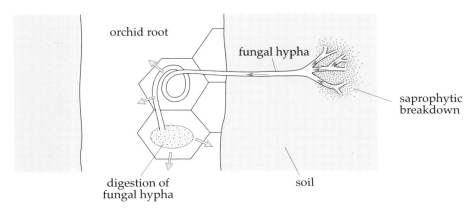

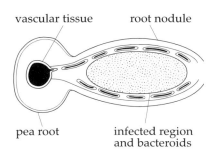

Figure 1.13 An endomycorrhiza in an orchid root involves a saprophytic fungus; arrows show the direction of movement of carbon compounds (not to scale).

Figure 1.14 Cross-section of a legume root nodule to show the arrangement of tissues and the position of bacteroids ($\times$ 60).

1.5.3 Root nodules and plant–bacterial associations

Another type of nutritionally important symbiosis to many plants involves **cyanobacteria** and other **nitrogen-fixing bacteria**. *Rhizobium* bacteria live in soil but fix nitrogen only when they have infected the roots of legumes (members of the pea family, Fabaceae). The bacteria are surrounded by a membranous envelope inside which they form **bacteroids** (Figure 1.14) that occur in nodules on the root. Root nodules contain **leghaemoglobin** which is involved in oxygen transport in the bacteria. The pinkish leghaemoglobin molecule is the product of globin synthesized by the plant and haem synthesized by the bacteria. The bacteroids are incapable of using the nitrogen they fix which is transferred to the plant. The bacteroids obtain carbon compounds from the plant.

❏   What kind of nutritional relationship is involved in the *Rhizobium*/legume symbiosis?

■   The relationship is mutualistic but with the smaller partner (*Rhizobium*) being the more nutritionally dependent.

If a soil is rich in nitrogen, the legumes no longer maintain N-fixing *Rhizobium*; nodules either do not form or are broken down and the bacteria are digested by the roots. Conversely, if the supply of carbohydrate to the bacteria slows down, as it may do in old nodules, in boron-deficient plants or following transfer of plants to the dark, then *Rhizobium* becomes actively parasitic and kills and digests root cells.

❏   How might these N-fixing nodule associations affect the distribution of plants?

■   The survival and/or ability of plants to compete with other species on soils low in nitrogen might depend on the nodules. Nodulated plants can avoid nitrogen deficiency and, even though the microbial organisms require a supply of carbohydrate, the plant's return on this investment is well worthwhile.

Apart from the legumes, eight other families of angiosperms are known to have nitrogen-fixing symbionts, usually belonging to actinomycete bacteria in the genus *Frankia*. Two such species are alder *Alnus glutinosa* (Figure 1.15) and bog myrtle *Myrica gale* which grow in wet, often waterlogged conditions where nitrogen is commonly in short supply. Similarly, sea

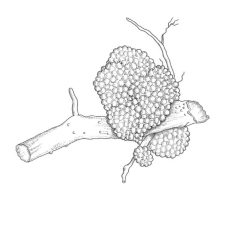

Figure 1.15 Root nodules of *Alnus glutinosa* (× 0.5).

Figure 1.16   *Hippophaë rhamnoides* (× 0.15).

buckthorn *Hippophaë rhamnoides* (Figure 1.16) is a highly successful plant of stabilized but nutrient-poor sand dunes and, since the decline of rabbits following myxomatosis, has been spreading rapidly on European coasts. These nodulated plants are probably even more widespread outside Europe in semi-desert areas and particularly on the white sand of coral islands where *Casuarina,* a nodulated non-legume, is extremely common.

Several unrelated groups of green plants form intimate associations with cyanobacteria (usually species of *Nostoc*). These include liverworts (bryophytes Figure 1.17), the water fern *Azolla,* a free-floating plant which is used as a green manure in tropical rice-growing areas (Figure 1.18), all of the approximately 150 species of cycads (primitive gymnosperms) which harbour their symbionts in coraloid roots above ground (Figure 1.19) and *Gunnera* spp., which are small herbaceous or large, rhubarb-like angiosperms. The cyanobacteria associated with plants fix nitrogen at a higher rate than in their free-living state, releasing ammonia to their host. The hosts supply fixed carbon in return.

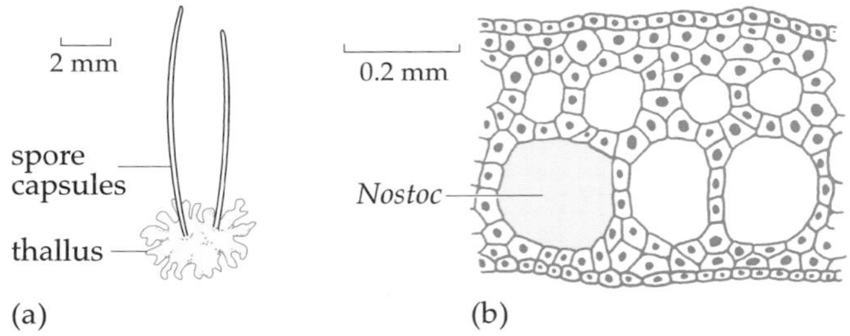

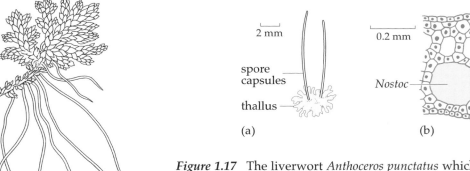

Figure 1.17 The liverwort *Anthoceros punctatus* which grows on wet soil: (a) whole thallus; (b) cross-section of the thallus showing *Nostoc* colonies in cavities.

1.5.4   Algae and invertebrates

Figure 1.18 The water fern *Azolla filiculoides*; blue-green *Anabaena* sp. occur in cavities in the lower sides of leaves (× 2).

Associations with algae occur in over 150 invertebrate genera, including protistans, sponges, platyhelminths, molluscs and cnidarians. The algae, which are mostly green or dinoflagellate species, may live as independent members of the phytoplankton (although they look and behave quite differently when they do); in symbiotic associations, they live and photosynthesize within animal cells. After supplying radioactive $^{14}CO_2$, or bicarbonate, labelled products of algal photosynthesis have been isolated from animal cells and, although algae can utilize and may be supplied with nitrogen compounds excreted by their hosts, it is not at all clear that symbiotic algae have any advantage over the free-living forms.

An example of algal exploitation by an animal occurs in certain sea-slugs (opisthobranch gastropod molluscs) which graze on green seaweeds, sucking out cell contents and incorporating intact chloroplasts into the cells lining the gut (Figure 1.20). The isolated chloroplasts continue to fix CO_2 and supply carbohydrate to the slug for several weeks. An even more extreme example is the marine flatworm (turbellarian platyhelminth) *Convoluta roscoffensis* which obtains all its food from an algal symbiont *Tetraselmis convolutae* and which ceases to feed once it has acquired the algae. *Convoluta* is deep green in colour and through the algal symbiosis virtually turns itself into a plant!

Hydra spp. (cnidarians) containing the green alga *Chlorella* sp. are found in many freshwater habitats; in experiments, it was found that about 50% of the carbon fixed by symbiotic algae passed, as the sugar maltose, to the animals, where it was incorporated into a wide range of compounds (Figure 1.21, overleaf). The fact that algal products pass into animal cells does not, in itself, prove that algae are of great importance for the survival of their hosts. Many of the animals feed actively as carnivores, so the critical point is whether algae provide an essential supplement to this diet or whether they enable animals to survive during periods of temporary food shortage, or have some function unrelated to nutrition.

❑   Look at Figure 1.21a: what do these data show about the importance of green algae in *Hydra*?

■   When food (e.g. brine shrimps) is abundant, the algae appear to make no difference; but when food is short, *Hydra* without algae grow less well than those with symbionts. It appears that algae provide a large part of the animal's food under these conditions.

Another group of cnidarians, the reef-building corals, all have an obligate symbiosis with species of dinoflagellate algae belonging to the genus *Symbiodinium*. Reef-building corals occur only in shallow clear tropical seas: the coral polyps form a massive skeleton of calcium carbonate, and reef building is strictly dependent upon light, but only occurs if corals are not starved of zooplankton. This suggests that coral animals depend on their symbiotic dinoflagellates to perform some essential function – but what? The available evidence suggests that there are three ways in which algae promote coral growth, the relative importance of each varying among different corals:

1   Algae may provide carbon compounds, principally in the form of glycerol, which are used as an energy source by the coral polyps. Muscatine and Porter (1977) calculated that a typical reef-building coral might acquire 40% of the carbon fixed by its symbiotic dinoflagellates and that this could meet 85% of the respiratory requirements of the cnidarian partner in the symbiosis. A few species of soft corals depend entirely on their algae for food and have never been observed to feed carnivorously.

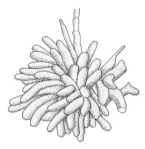

Figure 1.19   A cycad (*Cycas rumphii*) (× 0.01) and its swollen coraloid roots which lie close to the surface of the soil and contain cyanobacteria within the cortical cells (× 0.2).

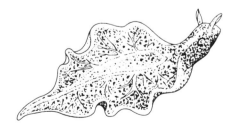

Figure 1.20   *Elysia viridis*; a sea-slug containing chloroplasts, found around British coasts (× 0.5). Its food plant is the seaweed *Codium fragile*.

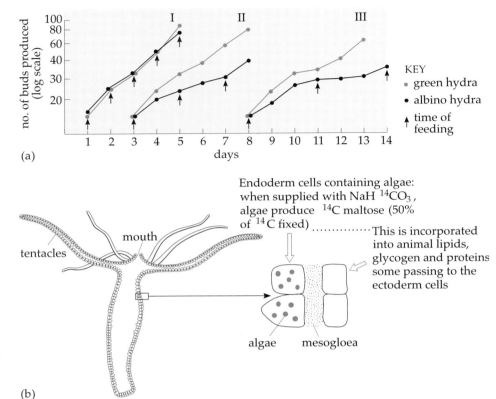

Figure 1.21 (a) Growth of green and albino *Hydra* incubated in the light under different conditions of feeding: (I) fed daily with brine shrimp larvae; (II) fed every second day; (III) fed every third day. (b) Structure of the hydra showing its symbiotic association with an alga (× 4).

2   Algae provide nitrogen compounds for their hosts. Inorganic nutrients from seawater and organic excretory products of the coral can be taken up and converted into substances such as alanine (an amino acid) which is then passed to the coral. Tropical seas are notoriously poor in mineral nutrients and this system effectively provides an internal nutrient cycle which prevents loss of nutrients from the reef; it could be of great importance but is still very poorly understood.

3   Algae may promote calcification of the coral skeleton, either directly or indirectly (the skeleton consists of an organic matrix within which $CaCO_3$ is deposited).

1.5.5 Animals and microbes

The surfaces and guts of animals provide a choice environment for numerous micro-organisms such as bacteria and yeasts (fungi). Most of these do no obvious harm to the animal; some cause disorders of skin and gut but a few perform a useful and sometimes essential function. These functions are of two main types and involve either the digestion of plant food in herbivores or the synthesis of essential dietary supplements in animals with a variety of diets.

The digestion of cellulose

Vertebrates lack the ability to synthesize their own cellulose-digesting enzymes (cellulases). Vertebrate herbivores therefore depend upon micro-organisms in their guts, including bacteria and flagellated protistans, to digest cellulose in the cell walls of their plant food. Microbial fermentation is anaerobic and consequently produces short-chain fatty acids such as acetate that can be utilized by the host. In aerobic conditions, the products of microbial fermentation are only carbon dioxide and water which would be of no benefit to the host. The process of microbial fermentation of cellulose is a slow one, requiring guts that are especially long or capacious. In **ruminants** such as cows, sheep and deer, food begins fermentation in a chamber called the rumen before it even reaches the stomach.

Smaller herbivores have a higher metabolic rate than large ruminants and have an energy demand that could not be met by the slow digestion of cellulose-rich food in large guts. Smaller herbivores may choose diets containing less cellulose, but several of these species also recycle their food by eating their own faeces (**coprophagy**, from Greek *kopros*, dung and *phagein*, to eat) after allowing some time for microbial decomposition. Many rodents and, particularly, lagomorphs (rabbits and hares) are coprophagous; rabbits recycle from 54% to 80% of their faeces and may suffer from protein deficiency if this is prevented. The rabbits produce two kinds of faeces sequentially: (a) soft pellets which are produced exclusively in the caecum and reconsumed in the burrow, and (b) hard pellets that are deposited outside the burrow. There is good evidence that urea is secreted into the caecum where it is utilized as a source of nitrogen for protein synthesis by bacteria that use undigested carbohydrate as a carbon source. Soft faeces contain about 25 to 28% protein, mostly bacterial protein, so the caecal symbionts effectively convert undigested roughage and nitrogenous excretory products into valuable protein for the rabbit! Woodlice and *Dipodomys* (kangaroo rat) probably obtain copper by coprophagy when it is released from indigestible organic complexes by microbial action.

Some insects are able to synthesize cellulases, but a group of termites (Figure 1.22) known as the 'lower termites' have flagellates or a mixture of bacteria, amoebae and ciliates in the hind part of the intestine thare essential to their ability to digest the wood on which they feed. Another group of termites go a step further: faecal material is assembled inside the nest to provide **'fungal gardens'** where fungi in the genus *Termitomyces* break down highly indigestible lignin from the wood. The fungi and the products of lignin degradation form an important food supplement for these insects. This feeding at 'second hand' with the aid of fungi is found also in leaf-cutter ants which cultivate fungal gardens inside their nests, and among wood wasps and bark-beetles, where females deposit fungal spores inside the burrows in which eggs are laid; the larvae feed on the wood-digesting fungi in the burrows.

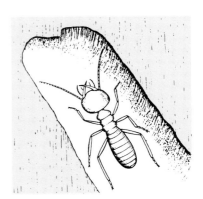

Figure 1.22   A soldier termite (× 2).

Dietary supplements

In addition to carbohydrates, fats, proteins, salts and water, animals need a variety of substances usually lumped together and called 'vitamins'. These act in various ways, often as enzyme cofactors; most animals cannot synthesize all their required vitamins but must obtain some from their food. In many organisms, however, certain vitamins are supplied in part by gut micro-organisms: humans obtain certain B vitamins in this way. Some insects obtain vitamins from yeasts and other micro-organisms which live in the digestive tract. In addition, many insects such as aphids, ants, flour-beetles and some blood-sucking insects possess special groups of cells called **mycetocytes** which contain a variety of micro-organisms including bacteria and yeasts.

The insects listed below all possess mycetocytes:

Pediculus (blood-sucking lice);

Cimex (bed-bugs);

Glossina (blood-sucking tsetse fly);

Lasioderma and *Sitodrepa* (flour-beetles);

aphids such as the black bean aphid *Aphis fabae*.

❑ What feature of their diet is common to all the insects listed above? Why should mycetocytes be of special importance to these insects?

■ All have a specialized and restricted diet. Unlike insects with a more varied diet, the listed insects have no opportunity to obtain extra vitamins from other sources except via their mycetocyte symbionts.

Vertebrate blood is, in fact, deficient in B vitamins and when *Pediculus,* the human body louse, was reared after surgical removal of mycetocytes, no growth occurred unless vitamins were added to the food. Flour-beetles raised without their symbiotic yeasts require vitamins and essential amino acids but when infected with yeasts can live on a completely vitamin-free diet. The plant sap on which bugs such as aphids feed is relatively low in nitrogen and this is mainly in the form of non-essential amino acids. The bacterium *Buchnera aphidicola* found in the mycetocytes of the black bean aphid *Aphis fabae* provides the essential amino acids missing from the aphid's diet.

Summary of Section 1.5

The **lichen** symbiosis is an association between a fungus and an alga. The green alga ***Trebouxia*** occurs in the majority of lichens in northern latitudes, but there are about 40 other genera of lichenized algae or bacteria including nitrogen-fixing **Cyanobacteria**. **Nitrogen fixation** converts gaseous nitrogen into soluble N compounds. Many of the algae and cyanobacteria found in lichens also occur in the free-living state. The fungi do not occur in the free-living state and are more specific as well as more dependent on their symbionts which pass 60–80% of their photosynthate to the fungus.

Mycorrhizal symbioses involve associations between fungi and most kinds of plants. The two major kinds of mycorrhiza are the **ectomycorrhiza,** common in temperate trees and usually involving toadstools such as the fly agaric, and the **endomycorrhizas** of which the commonest kind is the **vesicular–arbuscular mycorrhiza** (**VAM**). Mycorrhizal fungi acquire large quantities of carbohydrate from their plant symbiont. Infected plants may obtain nutritional benefit, particularly in the form of P, and also some protection against pathogenic fungi and heavy-metal toxicity. A few plants such as *Monotropa* are parasitic on their mycorrhizal fungus and many, particularly the orchids, cannot grow without them.

Nitrogen-fixing bacteria in the genus *Rhizobium* occur in nodules on the roots of most species in the pea family. *Rhizobium* or other nitrogen-fixing bacteria are found in a several other plants including alder.

 Algal–invertebrate symbioses occur in a wide variety of animals including protistans, sponges, platyhelminths, molluscs and cnidarians. Reef-building corals (cnidarians) are nutritionally dependent upon algal symbionts in the genus *Symbiodinium*. A wide range of micro-organisms are involved in **animal–microbial** symbioses which are particularly important in the guts of herbivores and some parasites.

Question 1.4   (*Objectives 1.2 & 1.4*)

What organisms are involved in the following symbioses, and what metabolic capabilities are acquired by the larger member in each symbiosis from the smaller one?

(a) Lichens, (b) ectomycorrhizas, (c) vesicular–arbuscular mycorrhizas (VAM), (d) root nodules, (e) green *Hydra* spp., (f) reef-building corals, (g) ruminants, (h) wood-eating lower termites, (i) blood-feeding insects.

Question 1.5   (*Objective 1.3*)

West, Fitter and Watkinson (1993), studying the rare annual grass *Vulpia ciliata* in East Anglia, found that the plant's roots were infected by VAM fungi at three sites, but that it grew well without these symbionts at a fourth site. They tested two hypotheses about the importance of VAM for *V. ciliata*: (1) that VAM infection improved the phosphorus nutrition of plants; and (2) that VAM protected plants' roots from infection by pathogenic fungi, particularly *Fusarium oxysporum*.

Figure 1.23 shows the relationship found between the inflow of phosphorus to plants, measured by changes in the leaf concentration of P between plants harvested from experimental plots in April and May, and the percentage length of roots in the plots infected by VAM fungus.

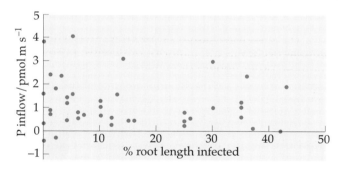

Figure 1.23   The relationship between the inflow of phosphorus to plants of *Vulpia ciliata* and the % root length infected by VAM fungus.

The fungicide benomyl was applied to experimental field plots and the dry weight of plants, the degree of VAM infection and the relative abundance of *F. oxysporum* were measured and compared with control plots (Table 1.6, overleaf). Using the data presented, answer the following questions, citing the evidence for your conclusions in your answers.

(a) Is hypothesis (1) supported?

(b) Is hypothesis (2) supported?

(c) What additional experiments or measurements would you do to test your conclusions?

Table 1.6   Effect of treatment with the fungicide benomyl on the growth of the grass *Vulpia ciliata*, its infection by VAM fungus and the abundance of the pathogenic fungus *Fusarium oxysporum* in the soil clinging to roots. Differences between the treatment and control are: n.s. = not significantly different, *$P < 0.05$.

	Control	Treatment	Difference
dry weight of shoots (mg)	16	14.5	n.s.
% root length infected by VAM fungi	35	5	*
relative abundance of *Fusarium oxysporum*	10	0.02	*

1.6 Parasitism

Mutualism and parasitism include some of the most intimate symbiotic associations between organisms. Though in theory it ought to be simple to distinguish between mutualistic (+/+) and parasitic (+/−) interactions, the very intimacy of these relationships often makes it difficult to draw a dividing line between them in practice.

Broadly speaking, the term **parasitism** is applied when one organism not only derives structural support or transport from another but also derives some nourishment from the other over a significant part of its lifetime. Metabolic dependence of a parasite on its host may not be confined to nutritional requirements; it may also include reliance to a varying extent on the host's digestive enzymes, or reliance on certain host reactions or internal environmental conditions to trigger off developmental stages, or even, in some cases, complete synchronization of the parasite's maturation with that of the host. **Viruses** are the ultimate parasites, integrating themselves into the DNA of their host's cells and taking over the host's replicating machinery to multiply themselves.

Mutualism also often involves a nutritional relationship between the symbionts.

❑   In what way does this nutritional relationship differ from that of parasite and host?

■   Mutualists in close cellular contact are nutritionally *inter*dependent whereas only the parasite is nutritionally dependent on its host.

The degree of dependence on the host is also related to the exact site of the parasite on or in its host. For instance, **ectoparasites** clinging to the exterior surface of their hosts may extract some nutriment from the external medium as well as directly from the host, whereas internal parasites (**endoparasites**) such as *Plasmodium,* the protistan blood parasite which causes malaria, are totally dependent for all their requirements, including oxygen, on their host. Dependence involving enzymes is perhaps best illustrated by tapeworms which are able to take in food material only as simple molecules which can be absorbed through their outer surface. Thus, they depend totally on the host's system to break down carbohydrates, fats and proteins into small molecules. In some cases, animal endoparasites themselves may pass out enzymes which help to break down the tissue near to their bodies.

An important feature of symbioses is their specificity. Parasites, and particularly **pathogens** (bacteria, fungi and viruses that cause disease), are often highly specific and may attack only one species of host, or even just one genotype. It is common for different cultivars of commercially grown crops (e.g. wheat or potatoes) to differ in their susceptibility to particular diseases. Faced with several diseases which may potentially attack a crop, it may be wise for farmers to grow several different cultivars or to choose strains that are not genetically uniform.

1.6.1 Endoparasites of animals

Many parasites live in the vertebrate digestive tract or in other cavities that open to the exterior of the host, feeding on materials captured, eaten and digested by the host. Flukes and tapeworms (Platyhelminthes), many roundworms (Nematoda), and numerous protistans (single-celled organisms) have this habit. The vertebrate gut is perhaps one of the most hazardous of habitats for the existence of parasites because it undergoes regular physiological and physicochemical changes, all of which present some problems to invading organisms. For mechanical reasons, the mouth, oesophagus and stomach only rarely act as habitats for parasites. One of the most attractive sites along the alimentary canal is the duodenum: the much-folded inner lining provides ideal shelter for parasites. Conditions are mildly acidic and the region is rich in highly nutritious material. It is not surprising that in many animals this area often supports a varied collection of parasitic species. In fishes, gills are another much-favoured site.

Some parasites in the bodies of vertebrates live in internal cavities, such as the blood vessels, or in the liver. As a source of nutrients, blood varies in relation to the morphology and physiology of the parasite using it. Among the most successful are single-celled protistans such as *Plasmodium*, the malarial parasite (Figure 1.24) which invades red blood cells, and *Trypanosoma* spp. (Figure 1.25, overleaf), the causative organism of sleeping sickness in humans and of related animal diseases, which are found in the blood plasma. Both are transmitted by insect vectors which are also essential **secondary hosts** in the parasites' life cycles.

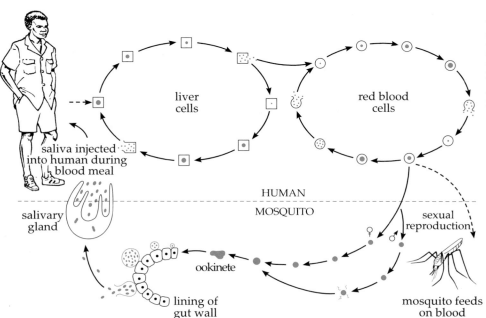

Figure 1.24 Life cycle of *Plasmodium vivax*, the malarial parasite. The definitive host is humans, the secondary host mosquitoes (*Anopheles* spp.). *P. vivax* reproduces sexually in its mosquito host (not to scale).

❑ Suggest a method of controlling diseases caused by parasites with secondary hosts.

■ A method which is often tried in areas where malaria and sleeping sickness occur is to exterminate the secondary hosts which act as vectors of the disease.

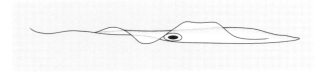

Figure 1.25 *Trypanosoma* sp. (× 4000).

Yet other important blood parasites are *Schistosoma* spp., the blood flukes (Figure 1.26), which are comparatively large and live in the mesenteric blood vessels of the rectum or bladder. They have a gut and a well-developed enzyme system; for them, the protein in the blood plasma is a particularly nutritious diet.

Although there are parasitic species belonging to nearly all the major animal phyla, the most important groups are the Protista, the Platyhelminthes, Nematoda and the Arthropoda (insects, mites, crustaceans, etc.). All these groups include both ectoparasites and endoparasites. Nearly all endoparasites and some ectoparasites spend their entire 'active' lives upon or within a host. Some parasites make use of a host for only a part of their life cycle; some are parasitic only as larvae; and others, e.g. fleas, become parasitic as adults.

Among the most-specialized endoparasites are the tapeworms (phylum Platyhelminthes, class Cestoda) which live in the intestines of vertebrates as adults; a representative species is shown in Figure 1.27.

❑ What kinds of organs are (a) notably absent and (b) much in evidence in *Echinococcus*?

■ (a) There are no sensory organs (eyes, ears, etc.) and no gut or obvious excretory organs (there are excretory canals running longitudinally). Food is absorbed as small molecules in solution over the whole surface of the body and excretory products pass into the excretory canals and out via the excretory pores.

(b) Reproductive organs are much in evidence and each proglottid (segment) is hermaphrodite (i.e. it has both male and female reproductive organs). Self-fertilization is the rule (although cross-fertilization can occur between adjacent tapeworms in the gut) so the worm can reproduce perfectly well without a mate and, indeed, is little more than a reproductive factory; vast numbers of eggs are produced and shed, inside gravid proglottids (body segments), with the host faeces.

The existence of a secondary host in the life cycle is typical of platyhelminth parasites. The life cycle of *Echinococcus* is shown in Figure 1.28. This species is small; some tapeworms which infect large mammals (including humans) are several metres long! However, unless the host is in poor physical condition or is seriously undernourished, the tapeworm does little harm but may cause discomfort. An exception is a disease of fish in the carp family called **ligulosis** caused by the tapeworm *Ligula* which has a

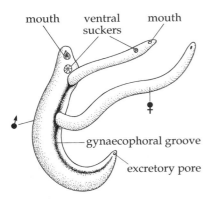

Figure 1.26 *Schistosoma* (× 8).

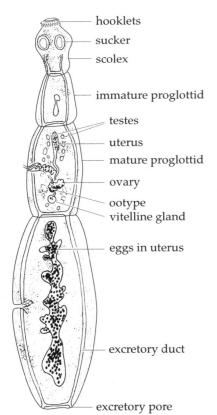

Figure 1.27 *Echinococcus granulosus*, the dog tapeworm (~ ×10).

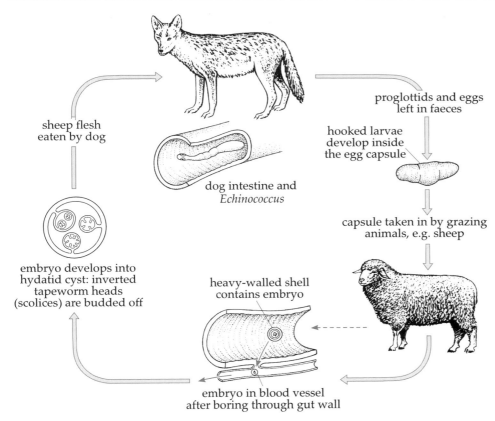

sheep flesh
eaten by dog

dog intestine and
Echinococcus

proglottids and eggs
left in faeces

hooked larvae
develop inside
the egg capsule

capsule taken in by grazing
animals, e.g. sheep

embryo develops into
hydatid cyst: inverted
tapeworm heads
(scolices) are budded off

heavy-walled shell
contains embryo

embryo in blood vessel
after boring through gut wall

Figure 1.28 The life cycle of the dog tapeworm, *Echinococcus granulosus*. Dogs are the definitive host and the secondary host may be a species of grazing mammal, or even humans. The parasite causes considerable damage to the secondary host both when hooked larvae penetrate the gut wall and when hydatid cysts form in the tissues; the cysts, within which the heads, or scolices, of future adult tapeworms develop, may reach the size of an orange.

brief adult life in fish-eating birds. While living in the body cavity of fishes, the tapeworms grow so large that other organs are displaced and may atrophy; parasites reach more than 10% of the total weight of the host and have been recorded up to 30%. The gonads atrophy completely so that the fish become sterile; when the fish die, they may then be eaten by the final hosts (birds).

The precise location of endoparasites within a host depends not only on their structural adaptations and specific behaviour but also on interactions with other parasites. Figure 1.29 shows the distribution of protistan parasites of the genus *Eimeria* in the gut of a fowl; they are causative organisms of the disease coccidiosis.

❑ Given that conditions over the whole length of the duodenum and intestines are tolerable for the three *Eimeria* species which live there, suggest an explanation for their restricted distribution as shown in the fowl.

■ A reasonable hypothesis is to assume that there are minor differences along the intestine and, if other species are present, each occupies the part where conditions are optimal for it. Where some factor is in limiting supply, e.g. space or nutrients, and two species require this limiting factor, then there is competition and only one species is likely to survive. As *Eimeria necatrix* overlaps throughout its range in the large intestine with *E. maxima*, it could be assumed that there are ample space and nutrients for both; but if the fowl were in a starving condition so that nutrients were limiting, it is probable that only one species could occupy the large intestine.

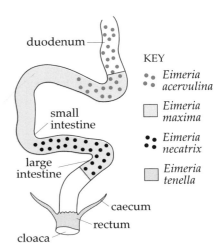

duodenum

small
intestine

large
intestine

cloaca

caecum

rectum

KEY

*Eimeria
acervulina*

*Eimeria
maxima*

*Eimeria
necatrix*

*Eimeria
tenella*

Figure 1.29 Distribution of *Eimeria* spp. in the gut of a fowl (not to scale).

❑ Suggest an explanation for the restricted distribution of *Eimeria tenella*.

■ Conditions in the rectum and caeca are likely to be very different from those in the duodenum and intestines (e.g. more solid material and different levels of O_2). Restriction of *E. tenella* to this region is probably due to particular physiological requirements and not to competition from other species.

Parasites may affect the distribution of their hosts, too. In Canada, moose and white-tailed deer have different distributions in summer but where they come into contact in winter, moose succumb to a parasitic infection which is commonly carried by the deer but which is lethal to moose.

❑ Moose and white-tailed deer are potential competitors for food in winter. How would you classify the symbiosis between deer and their parasite in such a situation?

■ It has been suggested that the white-tailed deer's parasites confer an advantage on it because they reduce competition from moose when food is short in winter. This would suggest that the relationship could be mutualistic. In summer, the moose migrate northwards and do not come into contact with white-tailed deer. Thus, in summer the deer–parasite relationship may be quite different.

One or two studies suggest that the presence of a parasite is reflected in a differential distribution of a host species. For example, around the coasts of the USA, oysters are infected by a fungal tissue parasite *Dermocystidium*, which cannot flourish in waters of low salinity. This is believed to be the reason why oyster populations in the less-saline waters of the southern United States achieve a higher density than they do further north. However, on the Atlantic coast of North America, larger oyster populations are discovered in more northerly waters of high salinity because the lower temperature affects the activity of the parasite more adversely than it does that of the host.

Occasionally, parasites can affect the distribution of whole populations through mass mortality. *Nitzschia sturionis* is a gill fluke (a monogenean trematode) of sturgeons; it damages the gill tissues and consumes blood. Although the parasite is not uncommon on sturgeons in the Atlantic, Mediterranean, Black and Caspian Seas, no **epidemics** have been noticed in these areas. Some *Acipenser stellatus*, one of its sturgeon hosts, were transferred to the Aral Sea and this led to a dangerous epidemic due to *Nitzschia* among *Acipenser nudiventris*, the native species there; they became infested with hundreds of gill flukes each and suffered from asphyxia and anaemia. The stock of this commercially valuable sturgeon was so much affected that fishing for it was totally prohibited. Evidently *A. nudiventris*, never in contact with *Nitzschia* before, had no immunity to the parasite.

1.6.2 Pathogens of plants and animals

Pathogens are parasitic disease-causing viruses, bacteria and fungi and are found in virtually every population of animals and plants. Because they are so common (and so small), their presence is often overlooked except when an outbreak causes widespread sickness or death in the host population. At other times, the death of young or weak individuals may be caused by disease but this goes unrecorded because such individuals appear to be the casualties of overcrowding or insufficient food which makes them more susceptible.

Like other endoparasites, most pathogens can multiply only inside a host. Transmission from one host to another is a crucial step in the life cycle. Figure 1.30 illustrates the life cycle of a **rust fungus**. These fungi typically have sexual and asexual generations on different host species.

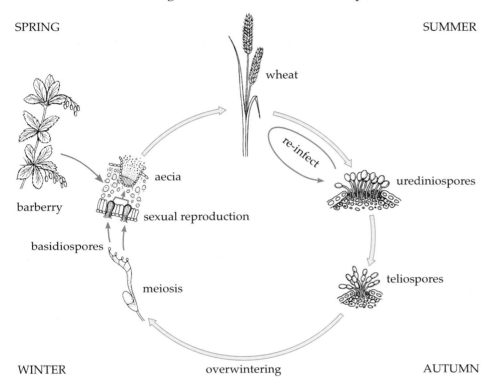

SPRING   SUMMER

wheat

re-infect

barberry   aecia   urediniospores

sexual reproduction

basidiospores   teliospores

meiosis

WINTER   overwintering   AUTUMN

Figure 1.30 The life cycle of the wheat rust fungus *Puccinia graminis*. The primary host is wheat *Triticum vulgare* and the secondary host is barberry *Berberis vulgaris*, a hedgerow or garden shrub. Rust spreads rapidly among wheat plants in summer by means of the wind-blown red urediniospores. It overwinters on wheat straw as thick-walled black teliospores which germinate in spring after fusion of nuclei followed by meiosis and the production of basidiospores. These are dispersed by wind and germinate only on barberry leaves, producing first a mycelium and then structures called spermogonia. Sexual reproduction occurs here by the fusion of 'male' spores with the tips of special 'female' hyphae. Binucleate cells are produced which divide to form cluster cups or aecia on the undersides of leaves and, from these, chains of aeciospores are released which infect wheat plants.

Outbreaks of disease, like epidemics of other parasites, may occur when the rate of transmission of a pathogen from infected to non-infected hosts increases, allowing the pathogen to multiply rapidly. The density of hosts is an important factor in the transmission of the parasite and outbreaks of disease are common in wild populations of animals and plants when they reach high densities.

❑   Of what practical importance might this be to farmers and foresters?

■   High densities of livestock, crops or trees are more likely to suffer an epidemic disease than are low densities of the same organisms.

Pathogens, particularly viruses, are often transmitted between different hosts by herbivorous or blood-sucking insects which may cause more damage to their food plants or vertebrate hosts by introducing infection than they cause by their direct consumption of its tissue. An example is **Dutch elm disease** which is caused by a fungus *Ceratostomella* that is transmitted by two species of elm bark-beetle. Both the beetles and the disease were present in Britain before the outbreak in the 1970s which seems to have been caused by a new strain of the pathogen introduced from America. Recall that myxomatosis, which had such a devastating effect on the rabbit in Western Europe, was also an introduced disease.

In America, chestnuts *Castanea* spp. were virtually wiped out by a new virulent strain of the **chestnut blight** fungus *Endothia parasitica* which appeared in about 1960. All three of these epidemics – Dutch elm disease, myxomatosis and chestnut blight – were unusually severe because they were caused by pathogens or strains of a pathogen which had not occurred in the area before. In the more usual situation when a parasite does not kill but merely weakens its host, the effect on the host population must depend on environmental conditions, physical, chemical and biotic. A weakened host is more likely to die in severe climatic conditions; it is more likely to be killed by predators, or fail to rear young, or starve when food is short. So, in 'difficult' conditions, the presence of a parasite may just be sufficient to tip the scales against a host and, conversely, a mutualistic association such as an ectomycorrhiza may tip the scales in favour of host survival.

1.6.3 Ectoparasites of animals and plants

By definition, these parasites live outside the body of their host from which they derive nourishment. The organs of attachment in these parasites are obvious features of their morphology and often secure the parasite to a very localized area of the host's body. Two species of louse (insects, order Anoplura) affect humans: one usually lives among the pubic hairs (*Phthirus pubis*, the crab louse) and the other occurs in two forms, one of which infests the head and the other the body and clothing (*Pediculus humanus* – Figure 1.31).

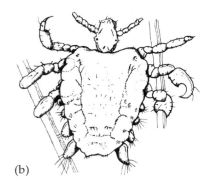

(a)   (b)

Figure 1.31 Lice occurring on humans: (a) *Pediculus humanus*, the head or body louse; (b) *Phthirus pubis*, the crab louse (both × 15).

These lice, and many of the ectoparasitic ticks and mites (arachnids, order Acarina), suck blood by means of piercing and sucking mouthparts. Figure 1.32 shows the sheep tick: note the sac-like body which becomes enormously distended after feeding. Some mites have biting mouthparts and feed on scurf and oily secretions of the host; **feather mites** (Figure 1.33), which live among bird feathers, are of this type. They show little host specificity but are very particular about the part of the host they live on – the type of feather and the exact part of a feather (cf. human lice).

❑ What environmental factors, which vary in different parts of the bodies of birds or mammals, might determine the distribution of lice or feather mites on the body?

■ The most likely factors are temperature and humidity. Temperature is usually highest close to the skin but the temperature of extremities may be lower than the rest of the body. The relative distribution of insulating layers (hair, clothes or feathers) also affects local temperature and humidity. Distribution also depends on the spacing of hair and the ability of the host to groom and remove parasites from that part of the body.

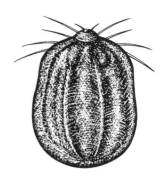

Figure 1.32 The sheep tick *Ixodes ricinus*, a common and widespread ectoparasite which feeds on blood (× 4).

The commonest ectoparasites in aquatic environments are crustaceans, notably cirripedes (barnacles), isopods and copepods (on gills). Figure 1.34 shows two examples of cirripede, *Rhizolepus* being totally dependent on the host for food while *Anelasma* is only partially so and has rudimentary external feeding organs. The ectoparasitic cirripedes described have a high degree of host specificity.

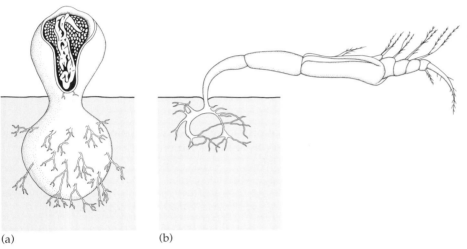

(a)   (b)

Figure 1.34 Ectoparasitic cirripedes: (a) *Anelasma squalica* (× 1.75), which projects from the skin of sharks just in front of the dorsal fin; (b) *Rhizolepas annelidicola* which attaches to the scale worm *Laetmatonice* and has neither mouth nor anus.

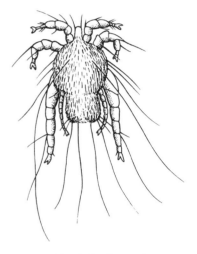

Figure 1.33 A feather mite *Megninia* sp., which feeds on scurf and exudations from the skin of birds (× 10).

❑ From Figure 1.34, suggest an explanation for this specificity in view of the fact that terrestrial mites and ticks usually have low host specificity.

■ The cirripedes feed through special root-like extensions which actually penetrate into host tissues; presumably, special enzymes are necessary for this and are effective only on one or a few host species. This kind of tissue invasion may also provoke defensive reactions in potential host species and so prevent the establishment of parasites. Mites and ticks have less-specialized feeding organs which are suitable for many species of host and would be unaffected by defensive reactions of host tissues.

There are several species of angiosperms which parasitize other higher plants and operate in a way similar to the ectoparasitic cirripedes (Figure 1.35). These parasitic plants have an external shoot but roots are either rudimentary or absent and instead there are absorbing organs which penetrate host tissues.

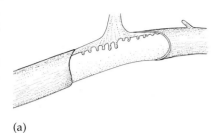

(a)

(b)

❑   From general knowledge, name one such parasite that occurs on aerial shoots and one that occurs attached to roots.

■   The best-known aerial parasite is mistletoe *Viscum album*, and another is dodder *Cuscuta* spp. (see Figure 1.35a, b).

Root parasites are more common in the UK and include the broomrapes *Orobanche* spp. (Figure 1.36) which lack chlorophyll and several species of the family Scrophulariaceae which do have chlorophyll, e.g. yellow rattles *Rhinanthus* spp. (Figure 1.37) and eyebrights *Euphrasia* spp.; *Striga* spp. are a genus of root parasites which are serious arable weeds in Southern Europe. Some mistletoes depend on the host chiefly for water and mineral nutrients but other mistletoes and scrophulariacean parasites also obtain carbohydrates from them. These green parasitic plants are sometimes called **hemiparasites** because they conduct their own photosynthesis: all are relatively non-specific and occur on a wide variety of host species. Broomrapes show complete nutritional dependence on their host. These, and other totally parasitic higher plants, are usually restricted to one or two related host species.

Figure 1.35 Aerial ectoparasitic plants: (a) section of part of an apple branch (× 0.4) parasitized by mistletoe and showing the absorbing organs; (b) common dodder *Cuscuta epithymum* (× 1), which usually grows on *Calluna* and *Ulex* and has no leaves or roots (water and nutrients are absorbed via 'suckers' which penetrate host tissues).

Figure 1.36 *Orobanche hederae*, ivy broomrape (× 0.15).

Figure 1.37 *Rhinanthus minor*, yellow rattle, a hemiparasite of grass roots (× 0.15).

Summary of Section 1.6

Parasites include **pathogens** such as bacteria and **viruses**, **ectoparasites** that cling to the exterior of their hosts and **endoparasites** that live in the gut, liver, blood or other tissue. Endoparasites frequently have a complex life cycle involving a **secondary host** which aids in the multiplication or transmission (or both) of the parasite. Pathogens are usually quite host-specific. Many **rust fungi** have a sexual generation that infects one plant species and an asexual generation that infects a different one. Some green plants are **hemiparasites**, acquiring water and nutrients from the roots of another plant.

Question 1.6   *(Objectives 1.2 & 1.5)*

(a) What role do secondary hosts play in the life cycle of endoparasites? Name two examples of endoparasites and their secondary hosts.

(b) Name two diseases in which the pathogen is transmitted by an insect vector.

(c) Name two examples of ectoparasitic animals and two examples of ectoparasitic plants.

1.7 Mutualism

The metabolic symbioses and parasitic associations discussed in Sections 1.5 and 1.6 involved intimate physical contact between organisms and, more precisely, an exchange of materials between cells; all were concerned in some way with nutrition – the provision of carbon compounds, mineral nutrients or vitamins, for example. In many cases, there was also physiological dependence of one organism on another where some essential process was carried out by a symbiotic partner (e.g. *Monotropa* or orchid seedlings and endomycorrhizas). In this Section, we consider a range of associations where organisms live on or close to other organisms and obtain advantages such as additional defence, feeding, transport or some combination of these. However, there is usually no *physiological* interdependence. Each organism can perform all essential bodily functions but may still depend on its partner for survival.

1.7.1 Pollination of flowers by animals

The relationship between flowers and their animal visitors is probably the most conspicuous kind of symbiosis to be found between plants and animals, with clear benefits to both. Insects are the most common animals to visit flowers and will be dealt with in most detail here. However, particularly in the tropics, birds and bats also pollinate a number of species. Bat-pollinated flowers such as *Parkia* spp. (Figure 1.38) often have pendulous inflorescences with anthers and stamens exposed where they will come into contact with a visiting bat's head or body. Release of pollen tends to occur at night when bats are attracted to the flowers by strong odours and copious nectar. By contrast, bird-pollinated flowers such as species of *Aquilegia* (Figure 1.39) are visited by day, tend to lack scent but are brightly coloured (often scarlet) and are tubular or trumpet-shaped. Hummingbirds in the Americas and sunbirds in Africa and Asia are two groups of flower-visitors that are highly dependent on nectar. Hummingbirds migrate the length of the USA and are therefore strictly seasonal visitors. In the mountains of Colorado, migrating hummingbirds visit flowers of scarlet gilia (*Ipomopsis aggregata*). At about the time hummingbirds disappear from the plant's habitat, *Ipomopsis aggregata* changes its flower colour to a lighter shade that is more attractive to hawkmoths which are an alternative pollinator.

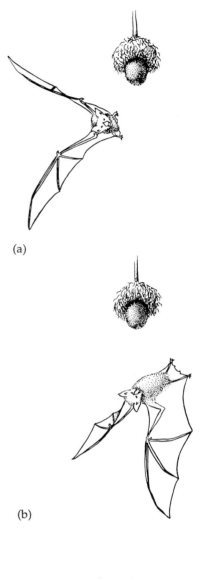

(a)

(b)

(c)

Figure 1.38 *Parkia* flowers and bat pollinators (× 0.1).

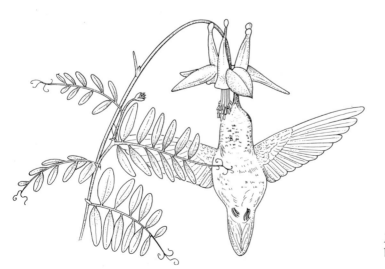

Figure 1.39 *Aquiligea* sp., a bird-pollinated flower (× 0.5).

Insects such as honey bees and bumble bees and their larvae, and adult (but not larval) butterflies and moths, may depend entirely upon flowers for their food and many flowers depend upon them for pollination and to carry their pollen to other plants. Despite these strong dependencies, in most cases the relationship between plant and insect is not highly specific. Meadow buttercup *Ranunculus acris*, for example, is visited by species of thrips, bugs, butterflies, bees, wasps and beetles.

These kinds of pollinating insects, though not permanently specializing upon one species of flower, will tend to forage upon one kind at a time, seeking out more flowers of a kind from which they have already been rewarded with nectar or pollen. Nectar generally contains sugars but often has several other constituents too, including amino acids and lipids. Nectars that contain lipids often also contain an antioxidant compound such as ascorbic acid (vitamin C) which acts as a preservative. Wild thyme *Thymus polytichus* and red campion *Silene dioica* are two flowers that have these two constituents in their nectar.

Nectar functions as a **reward** for visiting insects but many also consume pollen which is a rich source of protein and an essential ingredient of 'royal jelly', without which honey bee nests would not produce any queens. Some flowers produce no nectar and are visited only for their pollen. Production of both nectar and extra pollen is the metabolic cost to a plant of having some of its pollen transported to another flower and of receiving pollination itself. Pyke (1991) measured the cost of nectar production in an Australian perennial herb *Blandfordia nobilis* by comparing the number of seeds set by plants that he pollinated but protected from birds, honey bees and ants that normally remove nectar, with hand-pollinated plants from which nectar was removed daily. When nectar was removed, flowers secreted more nectar, so that the total nectar production by these flowers was three times that of control flowers. Plants from which nectar was removed produced significantly fewer seeds than controls, suggesting that nectar production did have a cost as well as a benefit in terms of reproduction.

Insects are initially attracted to flowers by colour, including wavelengths in the ultraviolet, and/or by scent. Different colours tend to attract different groups of insects and the shapes of flowers tend to make nectar accessible to some and not to others. For example, butterflies are particularly attracted to red, blue or yellow flowers, and moths to white ones. Bee vision has been studied by testing the ability of bees that have been trained to visit one colour to distinguish it from another. The spectrum visible to honey bees extends from the ultraviolet through violet, blue, green and yellow to orange; but they cannot see red. Flowers that appear a particular colour to us may therefore appear to be another colour or inconspicuous to bees. For example, the flowers of creeping cinquefoil (*Potentilla reptans*) are yellow to the human eye, but they reflect ultraviolet too and appear to bees as a mixture of colours from the opposite ends of their visible spectrum. This colour is called 'bee-purple' by analogy with the colour purple which we see when colours from the opposite end of the human visible spectrum (blue and red) are mixed. Flowers that attract bees appear to us as blue, yellow or white.

Flower shape also influences insect visitors. Lepidoptera often visit flowers that are tubular in shape. The length of the insect's proboscis may be too short to reach the nectaries at the bottom of the corolla tube of some species, so restricting the flowers on which they can feed. Bumble bees which also feed with a proboscis may overcome this problem by biting a hole through the bottom of a flower to get at nectar. This 'robs' the flower of its nectar without providing pollination.

Beetles, flies and wasps tend to visit drab-coloured or white flowers that are cup-shaped and easily entered or that have their flowers arranged in flat-topped inflorescences like those of the carrot *Daucus carota* (Figure 1.40). Such simple flowers are visited by a very large variety of insects. One study in Utah recorded over a period of four years that 334 species belonging to 37 families visited the flowers of a commercial carrot crop grown for seed. In NW Europe, the wild carrot is particularly attractive to adult parasitic wasps, 40 species of which have been observed visiting this flower. Larvae of these wasps are parasitic in the bodies of other insects (see Chapter 2).

Wild carrots whose flowers (but not roots!) resemble the cultivated kind were the subject of an experiment to test the attractiveness of these flowers. The tiny flowers of carrots (and other plants in the family Umbelliferae) are massed in a flat-topped inflorescence called an umbel with their anthers and stigmas protruding slightly from the surface, allowing pollen to be easily brushed off or deposited by alighting insects. The central flower in the umbels of many (but not all) carrots is a deep red, forming a conspicuous spot in the centre of the white inflorescence. The presence of the red flower has puzzled investigators for over 100 years, though Darwin suggested that it enhanced the attractiveness of the umbel to flies. Eiskovitch (1980) placed umbels with and without red flowers in an observation box containing house flies and observed which kind was first visited. In 19 out of 21 trials, flies landed first on an umbel with a red centre. When the experiments were repeated, this time using two discs of filter paper (one with an anaesthetized fly in the centre and the other without), the disc with the fly was significantly more attractive than the disc without.

❑ What aspect of fly behaviour do carrot umbels with a red centre appear to benefit from?

■ Flies land where they see other flies. It is a plausible hypothesis that the red spot is a fly mimic and hence encourages landing.

Mimicry of insects reaches a high point among the orchids in the genus *Ophrys* which resemble a variety of pollinators. The scent of the fly orchid *Ophrys insectifera* (Figure 1.41) attracts male solitary wasps of the species *Gorytes mystaceus* and *G. campestris* which try to copulate with the flower and in the attempt remove sticky masses of pollen grains known as **pollinia**. Pollen is transferred from these to the stigma of other orchids when next the male wasp is deceived.

❑ What kind of symbiotic association exists between *Ophrys insectifera* and *Gorytes* spp?

■ The relationship is +/0. The orchid gains but the wasp evidently does not. This one-sidedness is typical of relationships involving mimicry.

Figure 1.40   *Daucus carota* flower (× 0.5).

Figure 1.41   Fly orchid (× 0.5).

Though fly orchids occur in Britain and *Gorytes* is thought to pollinate them here, the related bee orchid *Ophrys apifera* appears to be self-pollinated in Britain.

Many tropical orchids in Central America depend upon male Euglossine bees for pollination. Females are not known to visit the flowers. These orchids have no nectar but produce fragrance and oils which attract male Euglossines and which are collected by them and stored in chambers in their hind legs. The fragrance and oils are thought to be precursors of pheromones which attract females into the male's territory. None of these quite specific relationships between orchids and bees provide the insect with nutrition, but adult *Heliconius* butterflies, which as larvae are specialist feeders on passion flower vines, do have a specialized nutritional relationship with a quite different group of climbing plants.

Adult *Heliconius* collect nectar and pollen, both of which they use as food, from cucurbit vines in the genus *Psiguria*. These plants are widely scattered about the forest where they live and their long-lasting inflorescences produce a single new flower every day or so for many months on end. One *Psiguria* plant kept in a greenhouse produced 10 000 flowers, containing the equivalent of 145 g of dry sucrose and 20 g of pollen in one year. Individual *Heliconius* butterflies appear to know where *Psiguria* plants are to be found and visit them daily on a regular collecting tour. The constant supply of nutritious food which *Psiguria* provides is thought to be responsible for the longevity of female *Heliconius* adults which survive and reproduce for a period of 4–6 months (Gilbert, 1980). The plants are dioecious (with separate males and females), so pollen transport is essential.

Some of the closest associations between pollinators and plants are to be found in the symbiosis of **yuccas** and the **yucca moths** *Tegiticula* spp. (Figure 1.42). *Tegiticula maculata is* the only insect known to pollinate *Yucca whipplei* which is a large rosette plant of arid areas in California and NW Mexico. Rosettes are semelparous (die after flowering) and take six to seven years to reach flowering size. When flowering takes place, a large stalk is produced bearing flowers that quickly wither and drop if they are not pollinated. A female *T. maculata* enters a flower where she collects a number of pollinia (pollen masses) into a ball and then flies off to another plant (Proctor and Yeo, 1973; Aker and Udovic, 1981). Here she enters a flower, inserts her ovipositor into an undeveloped ovary and deposits an egg. She then uncoils her mouthparts in which the pollinia are held and drags these back and forth over the top of the stigma. Following this, the whole process is repeated in the same flower, or in a different flower on the same plant. Pollination initiates the development of a seed pod in which there are several ovaries containing developing seeds. Some of these are consumed by the larva of the yucca moth but others survive. When the larva has completed its growth in the seed pod, it emerges from the pod to pupate in the soil, leaving behind its feeding galleries which provide evidence that it was there. Mature seed pods without galleries are very rare and yucca moth larvae only occur in the seed pods of yuccas.

The specificity of the *Yucca–Tegiticula* relationship is quite untypical of plant–pollinator relationships in general, but it is not totally unique. So far as is known, all 800 or so species of figs (*Ficus* spp.) depend for cross-pollination upon a similar relationship with tiny fig-wasps that lay their eggs in fig syconia which are round, hollow structures with an apical pore and containing large numbers of tiny flowers. Each fig species is pollinated

(a)

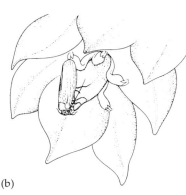

(b)

Figure 1.42 (a) Yucca plant (× 0.15) and (b) yucca moth (× 1).

by a different wasp species belonging to the family Agaonidae, which breed nowhere else. Pollinating fig-wasps are often greatly outnumbered by other species of wasps (in related families) that oviposit in figs, but provide no pollination and are therefore purely parasitic on the fig and its pollinators (Bronstein, 1992). Figs are tropical or mediterranean in distribution (except for those domesticated individuals of the popular houseplants, the weeping fig *Ficus benjamina* and the rubber plant *Ficus elastica*!), but one other highly specific plant–pollinator mutualism is known and actually occurs in Britain. The globe flower *Trollius europaeus* is a plant of the buttercup family found in wet places in mountain areas of northern Britain (Figure 1.43). Its large, tightly closed, globe-shaped flowers are pollinated by flies of the genus *Chiastocheta* that lay their eggs in the flower-head and whose larvae mine various parts of this as fruit develop. In Finland, where this symbiosis was studied by Pellmyr (1989), three species of *Chiastocheta* laid their eggs during flowering and carried pollen between plants. A fourth species laid its eggs after flowering and parasitized the plant without providing any pollination. All four species of *Chiastocheta* have been recorded in Britain (Pellmyr, 1992), but the symbiosis has yet to be studied here.

1.7.2   Cleaning associations

Grooming is a widespread activity among animals in the sense of removing debris, food particles and, in particular, parasites from the body. Some animals groom themselves or other members of the same species but there are animals that 'make a living' as specialized cleaners of different species.

In Africa, large vertebrates such as the wildebeest that cannot groom themselves are often accompanied by birds that remove ticks and insects from the body. These 'tick-pickers' appear to perform a useful function for their hosts and may, in addition, serve as lookouts, uttering characteristic warning cries when predators approach. The hosts make no attempt to dislodge or attack the birds and the Nile crocodile even opens its jaws for tooth cleaning by its cleaner bird. The birds usually clean only one or a group of related species so there is clearly 'recognition' between cleaner and host and special patterns of behaviour are involved.

About 40 species of fish, six species of shrimps (Figure 1.44) and one crab are recognized as cleaners and the most highly adapted of these species, which depend solely on cleaning for food, occur in warm tropical seas. Below is an account of the activities of a cleaner wrasse *Labroides dimidiatus* (Figure 1.45), a reef fish from the Indian Ocean.

A pair of wrasse occupied a permanent 'cleaning site' on the reef. These small cleaners (about 10 cm long) are conspicuously striped, have tweezer-like teeth, and when approached by a large fish go into a complex see-sawing dance routine. With large predatory fish such as groupers or moray eels, the dance appeared to be more vigorous than for smaller or herbivorous fish. After this preliminary, the large fish became perfectly still with head lowered and gill covers raised (see Figure 1.45). The wrasse immediately set to work cleaning meticulously over the body, including the delicate gills, and removing external parasites and dead skin around infected areas. When nudged by the wrasse, the large fish would relax that part of the body allowing the wrasse access to the mouth, gills or under fins as appropriate. Finally, the fish closed its mouth, opened it, shook its

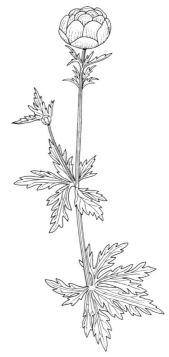

Figure 1.43 *Trollius europaeus.*

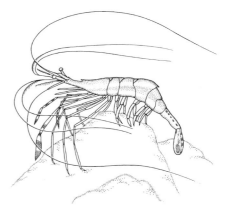

Figure 1.44 The Pederson shrimp *Periclimenes pedersoni*, a specialized cleaner shrimp (~ × 1).

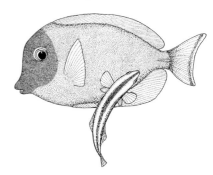

Figure 1.45 The cleaner fish *Labroides dimidiatus* working over a surgeon fish (× 0.2).

body a few times and immediately the wrasse moved away. Large queues of fish waited at the cleaning station and as many as 50 clients could be dealt with in an hour.

❑ Suggest an explanation for the conspicuous patterning and the see-saw dance of the wrasse.

■ The patterning and dancing of the wrasse ensure, first, that it is noticed. In addition, they signal to the large fish that the wrasse is a cleaner and not a dangerous fish or a potential prey. Recall that the dance is most vigorous with predatory fish which would normally eat small fish. Similar adaptations occur in cleaner shrimps which have vivid colours and often wave antennae and sway the body when attracting attention.

❑ Is the behaviour of large fish during cleaning of any significance in this association? Give reasons for your answer.

■ The behaviour of large fish is of considerable significance. First, it facilitates cleaning by the wrasse; secondly, it is quite different from 'normal' behaviour; all tendencies to move about, attack or retreat are suppressed. This symbiosis works only because of complex behavioural adjustments between the partners.

❑ Suggest an experiment to determine whether cleaning is a non-essential 'luxury' for large fish or is important for survival and could affect their distribution.

■ The most obvious experiment would be to remove all cleaner fish from a large area and then observe the effects upon larger 'client' fish. In fact, this was done by Limbaugh (1961) on a reef off the coast of California; he observed that, after a fortnight, many large fish had left the area and those which remained were in very poor condition with numerous parasites and festering sores. This and other evidence suggest that the location of cleaners is of major importance in determining the distribution of many species of fish.

1.7.3　Defence and shelter associations

Some form of protection in an area where there is also a supply of food is a necessary requisite for all defenceless animals and is often a major factor determining their small-scale distribution. Living in a **shelter association** with well-protected animals is one way of achieving this. The British woodlouse *Platyarthrus hoffmannseggii* (Figure 1.46), a blind white species, lives almost exclusively in the protected environment of ants' nests, chiefly those of the common yellow ant *Lasius flavus*; the woodlouse eats ant faeces. Exactly why this woodlouse is not attacked by the ants or whether removal of faeces is at all useful for nest hygiene is not known.

Defence associations between ants and plants are common, particularly in the tropics, although there are also temperate examples. Two common British species, bracken fern *Pteridium aquilinum* and common vetch *Vicia sativa*, possess **extra-floral nectaries**, located at the base of their leaf stalks, which attract ants to the plant. Whether this association with ants protects the plants or not cannot be taken for granted, as another case illustrates. Wood ants (*Formica rufa*) are voracious carnivores that prey upon

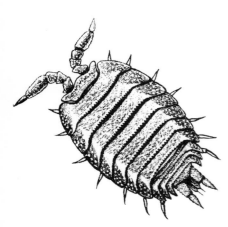

Figure 1.46 The woodlouse *Platyarthrus hoffmannseggii* (× 10) which lives in ants' nests.

herbivorous insects for which they forage in the canopies of trees near their nests. Whittaker and Warrington (1985) found that sycamores (*Acer pseudoplatanus*) foraged by wood ants grew significantly better than sycamores with no wood ants, but a similar study of birch (*Betula pendula*) (Mahdi and Whittaker, 1993) found little effect of wood ants on the growth of this tree species, even though the ants removed herbivores and this reduced leaf damage. The reason why sycamore benefited from the activities of wood ants while birch did not may be because the herbivore communities on the two trees were very different. Sycamore is not a native tree in Britain and has relatively few species of insect herbivores associated with it compared to birch which is native and has many. Both tree species harboured aphids, and in both cases one or more aphid species increased in abundance when tended by ants, while other species of aphids decreased in abundance.

❑ What effect would you expect a mutualism between wood ants and aphids to have upon tree growth where both insects were present, as compared to where only the aphid was present, all other things being equal?

◼ By increasing aphid numbers, ant attendance would decrease the growth of the tree.

Mahdi and Whittaker (1993) suggested that in sycamore, which had only one species of ant–mutualistic aphid, the negative effect of wood ants on the herbivore community as a whole was larger than their positive effect, so tree growth benefited. In the more species-rich and complex community of herbivores on birch, positive and negative effects of the presence of wood ants were more evenly balanced and so little benefit was measurable in the tree.

Ant–plant interactions in the tropics are numerous and varied, involving shelter, food and defence in all combinations and to all degrees (Huxley and Cutler, 1991). *Myrmecodia* is a genus of **epiphytic** plants (growing on branches or trunks of other species) found in the region of Borneo and New Guinea, that develops a swollen, tuberous stem filled with cavities which become occupied by ants (Figure 1.47). In this association and the many like it found in tropical epiphytes, it is thought that the ants sheltered by the plant provide it with a valuable source of nutrients from the waste that collects in the nest.

Stinging ants in the genus *Azteca* inhabiting trees of the tropical American tree genus *Cecropia* provide a different kind of benefit to their host. Queen ants make their nests in the hollow stems of young saplings and a colony develops inside the plant. Ant workers patrol the branches of the plant, collecting glycogen-rich globules called **Müllerian bodies** that are produced by the plant at the base of petioles. The benefit to *Cecropia* of providing 'bed and board' to ant colonies appears to be that the ants defend the plant from insect herbivores. More especially, by pruning any vines touching their tree, *Azteca* prevents the tree from becoming overgrown by climbing plants which are very numerous in the well-illuminated forest gaps where *Cecropia* species typically grow (Janzen, 1969).

Vine-pruning behaviour may indirectly benefit the ant symbionts that do it because it benefits the tree, but Davidson, Longino and Snelling (1988)

Figure 1.47 An epiphyte *Myrmecodia* sp. with a cross-section through its tuber showing the cavities that shelter ants (× 0.2).

suspected that there might also be a more direct benefit to the ants; they suggested that ant colonies compete for access to tree hosts, and that vines and the leaves of neighbouring plants provide bridges that can be used by invading ants. To test the hypothesis that vine-pruning reduces this threat to resident ant colonies, Davidson *et al.* (1988) conducted a field experiment with *Triplasis americana*, a small tree growing in the Peruvian Amazon rainforest. This tree is inhabited by the ant *Pseudomyrmex dendroicus* which often prunes a circular area clear of all other vegetation 1–2 m in diameter around its host.

Figure 1.48 Damselfish *Amphiprion percula* and its anemone partner (× 0.3).

In the experiment, wires were used to bring 12 *T. americana* trees into contact with the foliage of neighbours. The presence of alien ant species was monitored on these and on 12 isolated trees that were used as experimental controls. In the first six weeks of the experiment, trees in the treatment group were invaded by an average of 14 times as many worker ants of alien species as trees in the control group. One year after the start of the experiment, *P. dendroicus* had been replaced by other ants on all or a part of seven of the twelve treated trees, but none of the control trees had been taken over. Once again, this example demonstrates that symbiotic interactions are often more complex than they at first appear, and that experiments are vital to understanding them.

There are numerous examples of defence and shelter associations in the sea: small fish often live among the stinging tentacles of jellyfish (cnidarians). A coating of mucus helps them avoid being stung. There appear to be no other special adaptations of host or residents in these associations but it has been suggested that, for the young of some species (e.g. haddock, whiting and cod), this kind of protection is essential for survival. A more specialized interaction occurs between damsel (or clown) fishes (e.g. *Amphiprion* spp.) and large sea-anemones of tropical reefs. The damselfishes spend their whole life among the tentacles of a particular anemone (Figure 1.48) and produce a mucous secretion which inhibits discharge of the anemone's stinging cells (nematocysts). As damsels are poor swimmers, they would probably not survive long without their well-armed partner, but it has been suggested that the fish are also useful to the anemone as scavengers, removing digestive wastes. One species, *Amphiprion polymnus*, has been observed in aquaria to bring food to the anemone but whether this is really a 'food offering' or food storing (which the anemone exploits by eating) is unclear.

Figure 1.49 Three shrimp-fish *Aeoliscus strigatus* and two ling-fish *Diademichthys deversor* shelter head-down among the spines of the sea-urchin *Diadema* (× 0.25).

Figure 1.50 (right) The burrowing echiuroid worm *Urechis caupo* and some of its lodgers – an arrow goby *Clevelandia ios*, a pea-crab *Scleroplax granulata*, a scale worm *Hesperonoe adventor* and a clam *Cryptomya californica* (× 0.5).

Figure 1.51 (below) A pistol shrimp *Alpheus djiboutensis* and a gobiid fish *Cryptocentrotus lutheri* (× 1): the shrimp digs and continually maintains (i.e. shores up and so prevents the collapse of) a short burrow and feeds on detritus in the sand. The goby obtains protection in the burrow and appears to act as a lookout, twitching its tail when a predator approaches, at which signal both shrimp and goby retreat into the burrow.

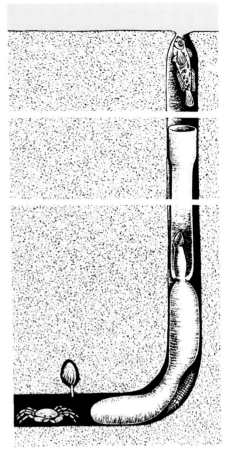

Dozens of other examples could be cited; in fact, nearly all well-protected nooks created by marine animals; the spines of sea-urchins (see Figure 1.49), branches of corals, the burrows of bottom-living invertebrates (see Figures 1.50 and 1.51), are exploited by other animals, which may be highly adapted in shape, physiology or behaviour to one particular habitat.

❑ How would you classify the benefits to symbionts in the marine shelter associations described so far in this Section?

■ There is certainly exploitation of the larger host animals by the lodgers but, in most cases, it is not certain whether this has any effect on the host. If there is no effect, this is a +/0 or commensal association.

For other associations, there is clearer evidence of mutualism. Figure 1.52 shows a partnership common in sandy parts of the Red Sea and Indian Ocean. Hermit crabs have a soft vulnerable abdomen and always seek protection in abandoned mollusc shells. A common British species, *Eupagurus bernhardus*, often carries a sea-anemone *Calliactis parasitica* on its shell and the anemone's stinging tentacles give added protection to the crab. As the crab moves about, the anemone obtains food from the small organisms and detritus which are stirred up and usually droops over so that its tentacles sweep the sea-floor (Figure 1.52).

When *E. bernhardus* outgrows its shell and must move to a larger one, it taps the anemone which then relaxes the basal disc, whereupon the crab inserts a claw under the base and lifts the anemone onto the new shell. Some anemone species are found only on hermit shells (e.g. *Adamsia* sp.) and die if their host dies, so this kind of association clearly affects anemone distribution and probably affects the survival and distribution of hermit crabs.

Figure 1.52 The sea-anemone *Calliactis parasitica* with its hermit crab host (× 0.4).

1.7.4 Epibionts

Epibionts are animals or plants that use other species for living space, without directly parasitizing or necessarily harming them. Plants that do this are called **epiphytes** and **epizoites** are sedentary animals that live in the same way. It was originally thought that almost the only thing that epibionts obtained from their host was space to live, but there is now evidence that some obtain a substantial proportion of their organic nutrient or mineral requirements from waste or excretory products of the host, or from nutrients which simply diffuse from or are leached out of host tissues. Epibionts may benefit or sometimes harm their hosts, but they are not parasites because they do not obtain their nutriment from living host tissue. Cases of mutualism between epibionts and their hosts occur (e.g. the relationship between the hermit crab and sea-anemone mentioned above), but not all such relationships are mutualistic.

The majority of species in the largest plant family, the orchids, are epiphytes living on trees in the tropics (Figure 1.53). Outside the tropics, the commonest epiphytes are lichens, mosses and ferns (Figure 1.54). Drought is a potential problem for all epiphytes, which are consequently most common in areas of high rainfall. In the warm, wet conditions of tropical rainforests, the leaves of many plants are colonized by a film of algae and lichens which reduce the amount of light reaching the leaf surface. It has been suggested that so many tropical trees have independently evolved leaves with a drip-tip at their end (Figure 1.55) because this helps drain the leaf surface after rain, making the surface less hospitable for **epiphylls** (epiphytes that live on leaves).

The bromeliads, the family to which the pineapple belongs, contains many tropical and sub-tropical epiphytes displaying a variety of structures that aid in the capture of water. The tank bromeliads have a rosette of leaves whose overlapping leaf bases form a container that traps water. Water is absorbed through roots that grow up between the leaf bases or by specialized leaf hairs. In some species, a tank may hold as much as five litres of water and is home to an aquatic community of tree frogs, mosquito larvae and other animals and plants including bladderworts (*Utricularia* sp.) found in no other habitat. Another group of epiphytic bromeliads, in the genus *Tillandsia*, have thin leaves densely covered in a felt of fine, water-absorbing hairs that acquire both water and dissolved nutrients directly from atmospheric moisture. Although mature plants have no roots, this mechanism is so effective that some *Tillandsia* can be found growing on cacti in deserts. One of these, *Tillandsia recurvata*, contains the nitrogen-fixing bacterium *Pseudomonas stutzeri* which is likely to be a useful source of nitrogen in an otherwise very nutrient-poor habitat. *Tillandsia caput-medusae* is one of several species in the genus that harbour ants in the bulbous cavities formed by their swollen leaf bases (Figure 1.56). These inhabitants may also aid the plant's nutrition.

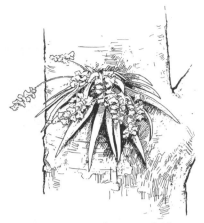

Figure 1.53 Epiphytic orchid. (× 0.2)

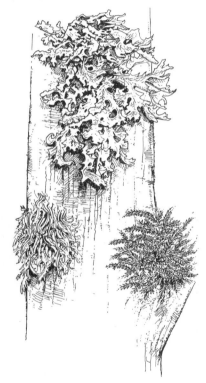

Figure 1.54 Epiphytic lichen and mosses (× 0.2).

Figure 1.55 Leaf of a tropical tree (*Ficus benjamina*) with drip-tip (× 1).

Figure 1.56 *Tillandsia caput-medusae.* This species is a popular house plant, one of several *Tillandsia* spp. sold as 'air plants' because they require no soil to grow (× 1).

Epibionts may benefit the host. The jewelbox clam *Chama pellucida* lives attached to rocks below the tidal zone off the coast of southern California. Its shell is normally covered by a dense growth of algae and sessile animals belonging to several different phyla. In a field experiment, Vance (1978) removed the epibionts from the shells of clams in experimental plots and compared their survival with that of clams covered in epibionts in adjacent control plots. The chief predator on *Chama* was the starfish *Pisaster giganteus*. Over a period of about 50 days, *Pisaster* took an average of 21 clams from experimental plots compared to only one clam per plot from the controls.

Figure 1.57 The scorpaenid fish *Minous inermis* largely covered by the hydroid *Podocorella minoi*.

Some epibionts are highly host-specific and may occur only on one host species, and there may also be detrimental or beneficial effects on the host. The numerous epiphytic algae which grow on other seaweeds must, inevitably, reduce light penetration and reduce rates of production by the host. On the other hand, the fish *Minous inermis* from the Indo-Pacific Ocean is always found with a dense cover of epizoite hydroids (Figure 1.57) which occur nowhere else; it is thought that the hydroids provide camouflage for the fish and that, for some reason, the fish provides an exceptionally favourable habitat for the hydroids.

Summary of Section 1.7

Plant–pollinator relationships are usually mutualistic because a **reward** (usually nectar or pollen) simultaneously attracts animals to flowers and ensures that they deliver pollen and/or carry it away to another flower in search of another reward. In a few plants, **mimicry** attracts insects which receive no reward for their visit, while some insects rob flowers of their nectar without pollinating them. Most plant-pollinator relationships are quite non-specific, though there are a few known cases of extreme specificity and dependence between plant and pollinator: **figs and fig-wasps**; **yuccas and yucca moths**; and *Trollius europeaus* and *Chiastocheta* spp.

In **cleaning associations**, one animal removes and consumes the parasites or waste food of another. This is particularly common in coral reef communities where fish may depend upon this service. **Defence and shelter associations** are also very common in the sea. There are many terrestrial examples of **ants protecting plants** from competition or predation, though such associations do not always prove beneficial to the plant when they are investigated experimentally. **Epibionts** are animals (**epizoites**) or plants (**epiphytes**) that use other species for living space. Epibionts do not parasitize their supporting host, though sometimes they may cause harm (e.g. **epiphylls** on leaves) or protection from predators.

Question 1.7   (*Objectives 1.2, 1.3, 1.5 & 1.6*)

(a) What characteristics are typical of flowers visited by (i) bats, (ii) birds, (iii) butterflies, (iv) bees, (v) beetles, flies and wasps?

(b) How would you classify the relationship between *Tegiticula* and *Yucca whipplei*?

(c) Three species of moth in the genus *Prodoxus*, which is related to *Tegiticula*, also occur on *Yucca whipplei* but none pollinates the plant and they are known as bogus yucca moths. The larvae of one species of *Prodoxus* feed in the pith but not the seeds inside developing yucca pods, and those of the other two feed in different parts of the tall stem of the inflorescence. How would you classify the relationships between the different species of *Prodoxus* and *Y. whipplei*?

(d) How would you classify the relationship between the different *Prodoxus* spp. and *Tegiticula maculata*?

(e) What is an epibiont, and what distinguishes these organisms from parasites?

(f) Give two examples of epiphytes and two examples of epizoites.

(g) Give two examples of defence or shelter associations.

(h) Give two examples of cleaning associations.

1.8   The ecological theatre and the evolutionary play

One of the founders of modern ecology, G. Evelyn Hutchinson, gave this highly graphic title to a famous essay in which he discussed the relationship between ecological and evolutionary processes (Hutchinson, 1965). Organisms are like actors brought together in an ecological arena where the consequences of their interactions with one another are played out over evolutionary time. Darwin himself made the point that natural selection, the force that drives evolutionary change, arises from the interactions between organisms, and between organisms and their environment (see Box 1). Of course, we now define ecology in precisely this way; as the study of those interactions. The evolutionary history of symbiotic relationships throws light upon their three important general features: benefit, dependency and specificity.

Box 1   Evolution by natural selection

Evolution is simply the change, with time, in the characteristics of a population. When populations diverge sufficiently from one another to become reproductively isolated, new species are formed. Although the idea that organisms evolved was unorthodox at the time Darwin published *The origin of species by means of natural selection* in 1859, the notion was by no means new. Darwin's grandfather, Erasmus, was one of the many evolutionists whom Darwin acknowledged as his precursors in '*An historical sketch*' of the idea with which he prefaced his book. The huge advance that Darwin made over his predecessors in *The origin* was to propose a convincing *mechanism* by which evolution could occur. That mechanism, natural selection, is still the only force we know of that can bring about adaptive evolutionary change.

Natural selection requires three conditions in order to change a characteristic with time:

• Individuals in a population must vary in the character.

• The variation must be inherited.

• Variation in the character must affect the **fitness** of individuals.

The fitness of an individual is just a measure of its success in leaving offspring and thereby transmitting its genes to future generations. If a population contains heritable variation for a character that affects fitness, variants with a higher fitness will leave more offspring than those with lower fitness and favourable variants of the character will spread, generation by generation. An excellent example is the evolution of **virulence** in the myxoma virus and of resistance to the virus in the European rabbit after the disease and host first encountered each other.

When the first outbreaks of myxomatosis occurred in Britain, the fatality rate was at least 99%, but within ten years the majority of strains of myxoma virus showed a significantly lower mortality per infected rabbit (i.e. lower virulence).

❏   Why should the myxoma virus evolve lower virulence?

■   If a rabbit dies too soon after it is infected, the virus is unlikely to be transmitted to another host before death. A less virulent virus strain will kill its host more slowly, giving it more time to be transmitted. Less virulent strains will therefore replace more virulent ones.

The rabbit also evolved in response to the appearance of myxomatosis. Virus strains which caused 90% mortality in rabbits from Norfolk populations in 1966 caused only 21% mortality in Norfolk rabbits ten years later. In Australian rabbit populations that were repeatedly affected by myxomatosis, it was found that survivors increased their resistance after each successive exposure to the disease (Fenner and Ross, 1994).

In both cases – the evolution of reduced virulence in the myxoma virus and the evolution of increased resistance in the rabbit – natural selection brought about greater **adaptation** between the organisms. Adaptation is a rather loose but very important concept describing the 'fit' between an organism and its environment (which of course includes other organisms). You might think that this process of adaptation would continue until the virus caused no mortality in its host. However, it appears that this probably won't happen because virus strains of intermediate virulence have a higher fitness than those with very low or very high virulence. Because of this, the rabbit and the myxoma virus are now engaged in an evolutionary contest against each other. (Further information about evolution and natural selection can be found in the Open University Course, S365 *Evolution*.)

The ecological interactions discussed in this Chapter provide many illustrations of how ecological relationships may influence the evolution of the species involved in them. The simplest cases are where we can compare a species or a population that has a symbiotic partner with species or populations that have evolved without the partner. For example, there are island populations of *Cecropia peltata* in the West Indies that are devoid of ants and which also lack Müllerian bodies. Mainland populations of *C. peltata* possess both. The question arises here as to whether the West Indian populations of *C. peltata* ever had Müllerian bodies, or whether they have lost them. In this particular instance, the question is easily answered; since island populations must be derived from mainland populations with Müllerian bodies and most *Cecropia* species also have them, the island populations must have lost them.

❑   Why should *Cecropia* plants without Müllerian bodies have an evolutionary advantage over those with them when ants are absent?

■   Müllerian bodies are energy-rich and must therefore exact some physiological cost upon a plant that produces them. If they do not help plants defend themselves (because there are no ants to attract), then a plant that produces fewer or smaller Müllerian bodies will probably grow better than a normal individual because it has more energy available. If this better growth is translated into better survival and more offspring than normal, the genotype with reduced Müllerian bodies will eventually replace the normal one. Over time, generation-by-generation, individuals with reduced Müllerian bodies would always be favoured over those with more until no plants produce them.

In another defence–shelter relationship between the ant *Pseudomyrmex* and *Acacia* trees, it has been found that *Acacia* spp. which have ants associated with them lack toxic chemicals that deter herbivores but that species without such an association do possess these compounds. Since toxic compounds are a much more common defence system in plants than ant protection, it is likely that the ancestors of the ant-defended *Acacias* did once possess such compounds, but then lost them.

In order to trace the evolutionary history of the appearance or loss of any trait, we need to construct a **phylogenetic tree** that shows how living species with or without the trait are related to each other by descent (a more detailed discussion of phylogenetics is provided in S365 *Evolution*). Many of the ants involved in defence relationships with plants belong to the group Pseudomyrmecinae. A phylogenetic tree for this group (Figure 1.58) shows that the obligate plant–ants belonging to the Pseudomyrmecinae often have non-symbiotic close relatives, suggesting the relatively recent evolution of the symbiosis. The phylogeny suggests that the obligate plant-defending symbiosis evolved independently at least 12 times in this group of ants. The next question is whether the evolution of the ant partner in this symbiosis was closely coupled with the evolution of the plant partner. No phylogenetic tree is available for the plants, but if the evolution of the two symbionts was tightly coupled, then you would expect considerable specificity in which ants inhabited which plants.

❑   Is this the pattern observed in Figure 1.58, and what do you conclude from the observed pattern?

■   The pattern does not suggest high specificity. Many of the ants inhabit more than one plant genus and some plant genera (particularly New World acacias) are inhabited by two or more quite distantly related ants. This suggests that ants have switched between plant hosts that had already evolved the symbiotic habit.

Phylogenetic analysis has been applied to the leaf-cutter ants and simultaneously to the symbiotic fungi which they raise in fungus gardens in their nests (see Section 1.5.5). The phylogenetic tree for the five ant species shown in Figure 1.59 is quite remarkably congruent with the tree for the fungal symbionts, which all belong to the basidiomycetes, or gill-fungi (like the common mushroom).

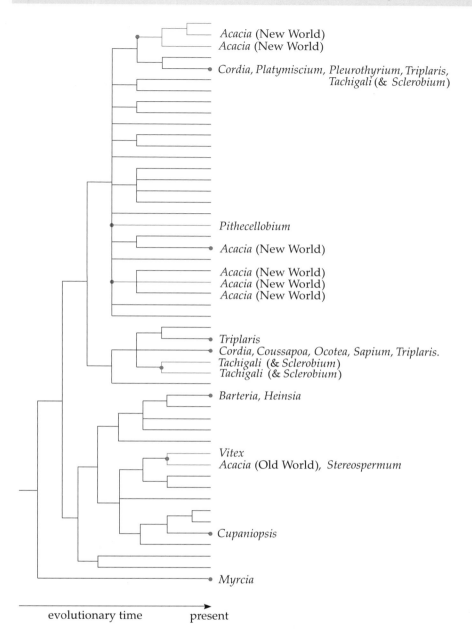

Figure 1.58 Estimated phylogenetic tree for the Pseudomyrmecine ants. Each branch tip in the tree represents a species. These species are obligate symbionts of plants, and twelve possible independent origins of this trait are shown by dots.

evolutionary time → present

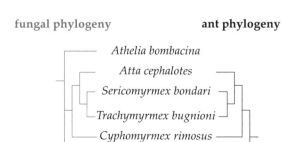

Figure 1.59 Estimated phylogenetic trees for five species of leaf-cutter ants and for their fungal symbionts.

❏   What does the congruence between the phylogenetic trees for the leaf-cutter ants and their fungi tell us about these symbiotic associations?

■   It must mean that the associations are extremely old and stable ones and that each ant species has **coevolved** with its fungus, rather than evolving separately and acquiring its partner later.

The stability of these symbioses probably depends upon the habit of queen leaf-cutter ants starting the garden of a new nest with a small piece of fungus taken from the nest where they were born. From fossil evidence, it appears that these symbioses are between 25 and 40 million years old. The coevolution of leaf-cutter ants with their symbionts is to be contrasted with the much looser evolutionary relationship between the Pseudomyrmecinae and their hosts (Figure 1.58).

At the time of writing (1995), the phylogenetic analysis of symbioses is a new but burgeoning area of research. In the legumes, for example, it now appears that the formation of nitrogen-fixing root nodules may have arisen separately three times in the evolutionary history of the family (Doyle, 1994). The phylogeny of the aphids has been found to be congruent with that of the symbiont *Buchnera aphidicola* which occurs in their mycetocytes, implying a single origin of the aphid symbiosis which must be between 160 and 280 million years old. Whiteflies and scale insects harbour two other groups of bacterial endosymbionts that have a separate evolutionary origin to *Buchnera* in aphids. The symbiosis so essential to the sap-feeding habit in all these insects has therefore evolved at least three times (Moran and Baumann, 1994). Because each fig species is pollinated by a different species of fig-wasp, it comes as no surprise that the phylogenies of these plants and their pollinators have been found to be completely congruent with each other.

We already know that benefit is a changeable property of symbiotic relationships and that even obligate symbionts need not be particularly specific. The phylogenies of symbioses that are available now also show that while some symbioses may be ancient and stable, in other cases there may be multiple origins, new ones can form and partnerships can change fairly frequently in evolutionary time. The evolutionary history of symbiotic relationships helps explain their specificity. And conversely, the dependence and benefit of relationships helps explain their evolution.

Summary of Section 1.8

Evolution is the change with time in the characteristics of a population. Evolution by **natural selection** occurs when a character varies between individuals, this variation is inherited, and the variation affects an individual's success in leaving offspring (its **fitness**). Natural selection leads to greater **adaptation** between an organism and its environment, which can include other organisms. The evolution of reduced virulence in the myxoma virus is an example. Other symbiotic relationships may also result in natural selection for greater adaptation between partners, or **coevolution**. The best evidence that this has occurred comes from **phylogenetic trees** which reconstruct the evolutionary relationships among species. In some cases, such as that of the leaf-cutter ants and their garden fungi, the phylogenies of the symbiotic partners are congruent and demonstrate that coevolution has occurred. In other cases, there is no congruence, proving that an association is recent in the evolutionary history of the species involved.

Objectives for Chapter 1

This Chapter should have given you an overview of the richness of ecological interactions between species, but we have still scarcely touched upon some of the most important and widespread ecological interactions of all; we shall deal with competition, predation and herbivory in the next Chapter.

After completing Chapter 1, you should be able to:

1.1 Recall and use in their correct context the terms shown in **bold** in the text. (*Questions 1.1, 1.2, 1.4, 1.6 & 1.7*)

1.2 Give a definition of 'ecology', illustrated by named examples of ecological interactions between species. (*Question 1.2*)

1.3 Interpret data on interactions between species, and name and classify the interactions in terms of the $+/-/0$ system. (*Questions 1.1, 1.3–1.5 & 1.7*)

1.4 Name the types of organisms involved in the different symbioses discussed and recall their degrees of benefit, dependence and specificity. (*Question 1.4*)

1.5 Name and describe examples of parasites, pathogens, epibionts, cleaning associations and defence–shelter associations. (*Questions 1.6 & 1.7*)

1.6 List the characteristics of flowers that are important in their relationships with pollinating animals. (*Question 1.7*)

References for Chapter 1

Aker, C. L. and Udovic, D. (1981) Oviposition and pollination behaviour of the yucca moth *Tegeticula maculata* (Lepidoptera: Prodoxidae), and its relation to the reproductive biology of *Yucca whipplei* (Agavaceae), *Oecologia*, **49**, 96–110.

Barnes, R. F. W. and Tapper, S. C. (1986) Consequences of the myxomatosis epidemic in Britain's rabbit (*Oryctolagus cuniculus* L.) population on the numbers of brown hares (*Lepus europaeus* Pallas), *Mammal Review*, **16**, 111–16.

Bradley, R., Burt, A. J. and Read, D. J. (1981) Mycorrhizal infection and resistance to heavy metal toxicity in *Calluna vulgaris*, *Nature*, **292**, 335–37.

Bronstein, J. L. (1992) Seed predators as mutualists – ecology and evolution of the fig pollinator interaction, *Insect–Plant Interactions*, Vol. 4 (ed. E. Bernays), pp. 1–44, CRC Press Inc, 2000 Corporate Blvd NW, Boca Raton, FL 33431.

Davidson, D. W., Longino, J. T. & Snelling, R. R. (1988) Pruning of host plant neighbours by ants: an experimental approach, *Ecology*, **69**, 801–8.

Douglas, A. (1994) *Symbiotic Interactions*, Oxford University Press.

Doyle, J. J. (1994) Phylogeny of the legume family: An approach to understanding the origins of nodulation, *Annual Review of Ecology and Systematics*, **25**, 325–49.

Fenner, F. and Ross, J. (1994) *Myxomatosis. The European Rabbit.* (ed. H. V. Thompson and C. M. King), pp. 205–39, Oxford University Press.

Fitzgibbon, C. D. (1993) The distribution of grey squirrel dreys in farm woodland: the influence of wood area, isolation and management, *Journal of Applied Ecology*, **30**, 736–42.

Gilbert, L. E. (1980) Ecological consequences of a coevolved mutualism between butterflies and plants, in L. E. Gilbert and P. H. Raven (eds), *Coevolution of Animals and Plants* (2nd edn), Texas University Press.

Gotto, R. V. (1969) *Marine Animals: Partnerships and Other Associations*, English Universities Press.

Granall, U. and Hofsten, A. V. (1976) Nitrogenase activity in relation to intracellular organisms in *Sphagnum* mosses, *Physiologia Plantarum*, **36**, 88–94.

Harley, J. L. (1971) *Mycorrhiza* (Oxford Biology Reader 12), Oxford University Press.

Hassall, M. and Dangerfield, J. M. (1990) Density-dependent processes in the population dynamics of *Armadillidium vulgare* (Isopoda: Oniscidae), *Journal of Animal Ecology*, **59**, 941–58.

Hinkle, G., Wetterer, J. K., Schultz, T. R. and Sogin, M. L. (1994) Phylogeny of the attine ant fungi based on analysis of small subunit ribosomal RNA gene sequences, *Science*, **266**, 1695–97.

Hutchinson, G. E. (1965) *The Ecological Theater and the Evolutionary Play*, Yale University Press.

Janzen, D. H. (1969) Allelopathy by myrmecophytes: the ant *Azteca* as an allelopathic agent of *Cecropia*, *Ecology*, **50**, 147–53.

Limbaugh, C. (1961) Cleaning symbiosis, *Scienific American*, **205**, 42.

Mahdi, T. and Whittaker, J. B. (1993) Do birch trees (*Betula pendula*) grow better if foraged by wood ants?, *Journal of Animal Ecology*, **62**, 101–6.

McBee, R. H. (1971) Significance of intestinal microflora in herbivory, *Annual Review of Ecology and Systematics*, **2**, 165–76.

Molina, R., Massicotte, H. and Trappe, J. M. (1992) Specificity phenomena in mycorrhizal symbioses: community–ecological consequences and practical implications, *Mycorrhizal Functioning* (ed. M. F. Allen), pp. 357–423, Chapman & Hall.

Moore, N. W. (1987) *The Bird of Time. The Science and Politics of Nature Conservation*, Cambridge University Press.

Moran, N. and Baumann, P. (1994) Phylogenetics of cytoplasmically inherited microorganisms of arthropods, *Trends in Ecology and Evolution*, **9**, 15–20.

Muscatine, K. and Porter, J. W. (1977) Reef corals: mutualistic symbioses adapted to nutrient-poor environments, *Bioscience*, **27**, 454–60.

Pellmyr, O. (1989) The cost of mutualism: interactions between *Trollius europaeus* and its pollinating parasites, *Oecologia*, **78**, 53–9.

Pellmyr, O. (1992) The phylogeny of a mutualism – evolution and coadaptation between *Trollius* and its seed-parasitic pollinators, *Biological Journal of the Linnean Society*, **47**, 337–65.

Proctor, M. and Yeo, P. (1973) *The Pollination of Flowers*, Collins.

Pyke, G. H. (1991) What does it cost a plant to produce floral nectar?, *Nature*, **350**, 58–9.

Smith, D. C. (1973) *Symbiosis of Algae with Invertebrates* (Oxford Biology Reader 43), Oxford University Press.

Smith, N. G. (1968) The advantage of being parasitized, *Nature*, **319**, 690–94.

Vance, T. R. (1978) A mutualistic interaction between a sessile marine clam and its epibionts, *Ecology*, **59**, 679–85.

West, H. M., Fitter, A. H. and Watkinson, A. R. (1993) Response of *Vulpia ciliata* ssp. *ambigua* to removal of mycorrhizal infection and to phosphate application under natural conditions, *Journal of Ecology*, **81**, 351–58.

Whittaker, J. B. and Warrington, S. (1985) An experimental field study of different levels of insect herbivory induced by *Formica rufa* predation on sycamore (*Acer pseudoplatanus*) III. Effects on tree growth, *Journal of Applied Ecology*, **22**, 797–811.

COMPETITION, PREDATION AND HERBIVORY   CHAPTER 2

Prepared for the Course Team by Jonathan Silvertown

2.1   Interspecific competition

Competition is the most widely studied and most argued-about interaction between species. It occurs where the individual performance (e.g. growth) or population size of two or more species is limited by the same resource. **Limiting resources** may be food, nest sites, refuges from predators, soil nutrients, light and water for plants or anything else whose limited supply can lead one species to affect another negatively when the resource is consumed.

In the past, some ecologists have regarded the presence of competition as axiomatic wherever species with similar requirements for food or other resources occur together. After all, they argue, if the two oaks *Quercus petraea* and *Q. robur* both require the same mineral nutrients, water and light to grow, the species must negatively affect each other when they grow in the same place, mustn't they? Others have been more sceptical and have insisted upon experimental evidence that resources are limiting and that competition for a limiting resource exists. Many field experiments to test for the existence of competition have now been carried out, in a great variety of communities and among a wide range of types of organism. Do these experiments suggest that most co-occurring species compete, or should we still be sceptical and insist on experimental evidence of competition in every instance where it is suspected? Read on.

2.1.1   Mechanisms

The definition of competition adopted in Chapter 1 requires two species to have negative effects upon each other. This definition has the advantage of simplicity but, on its own, the knowledge that an interaction between species is −/− is not enough to understand the ecology of the interaction. We also need to know something about the *mechanism* underlying the competitive interaction. Two types of competitive mechanism are commonly distinguished: (1) **interference competition** and (2) **exploitation competition**. Investigations of competitive mechanisms sometimes show that −/− interactions actually involve a third species, and that competition is more **apparent** than real. Each of these situations is discussed below.

Interference competition

In interference competition, one or both of the species physically interferes with the other's access to the resource. For example, Williams (1981) studied interactions among animals inhabiting a coral lagoon on the coast of Jamaica. She found that the three-spot damselfish (*Eupomacentrus planifrons*) and two species of sea-urchin (*Echinometra viridis* and *Diadema antillarum*) all grazed an algal 'lawn' that grew on branches of the staghorn coral. Williams observed three-spot damselfish pick up and physically remove sea-urchins from coral clumps where they were feeding.

Figure 2.1   *Necrophorus humator* (× 1.5), a burying beetle and the bluebottle *Calliphora vomitora* (× 1.6).

Figure 2.2   A fruiting body of the fungus *Coprinus heptemerus* on a rabbit pellet (× 2.5). This fungus has a strong inhibitory effect on the growth of other species in rabbit dung.

Interference competition appears to be the norm among the organisms that encrust hard surfaces such as rocks and the undersides of corals in shallow coastal waters. These organisms include sponges, ectoprocts, ascidians, serpulids, bryozoans and cnidarians. There is very little uncolonized space in such communities and these clonal animals compete for what space there is by growing over the top of each other, and in the case of some cnidarians by using stinging cells called nematocysts to attack neighbours.

An interesting kind of interference competition has been observed between flies and carrion beetles (*Necrophorus* spp.) (Figure 2.1) competing for food on the corpses of dead woodmice (*Apodemus sylvaticus*). Flies of the genus *Calliphora* generally lay their eggs on a corpse before carrion beetles arrive. In experiments, Springett (1968) found that beetles placed on a corpse already occupied by fly larvae were unable to reproduce there. However, in the wild where carrion beetles do reproduce successfully, they carry up to 40 mites (*Poecilochirus necrophori*) clinging to their bodies. This relationship between the beetles and mites, in which one organism depends upon another for transport, is an example of a type of symbiosis called **phoresy** (from the Greek 'to carry'). When a carrion beetle arrives at a corpse, the phoretic mites alight from the beetle, seek out and consume the fly eggs and small larvae that are the carrion beetle's potential competitors. In Springett's experiments, carrion beetles were able to reproduce successfully in competition with flies only when phoretic mites were present on the corpse.

Many of the soil-inhabiting and wood-rotting fungi involved in decomposition produce **antibiotics** and other substances that interfere with the growth of competing fungi and bacteria. For example, Harper and Webster (1964) studied the interaction among three fungi *Ascobolus viridulus*, *Coprinus heptemerus* (Figure 2.2) and *Pilobius crystallinus* in rabbit pellets. *Coprinus* inhibited the fruiting of both *Pilobius* and *Ascobolus*, and *Pilobius* was also inhibited by *Ascobolus*. When grown on agar plates separated by cellophane, *Ascobolus* was still inhibited by *Coprinus*, demonstrating that some diffusable substance of low molecular weight was responsible for the interaction (Ikediugwu and Webster, 1970).

There are many studies claiming that some higher plants produce **allelopathic chemicals** that interfere with competitors. While it is possible to show in the laboratory that a substance extracted from one plant inhibits the growth of another, demonstrating that this occurs naturally in the field is much more difficult. The main problem is that every habitat contains other organisms, particularly decomposer micro-organisms, that affect the chemical environment and ecology of the competitors. A cautionary example is the case of *Salvia leucophylla*, a shrub that occurs in arid grasslands in California. This shrub is usually surrounded by a zone, which may extend 60–90 cm beyond the canopy of the plant, that is bare of other vegetation. Muller *et al.* (1964) carried out extensive studies of this plant and found that *Salvia* produces terpenes, which are volatile organic chemicals that in laboratory experiments were highly toxic to seedlings of many of the grasses one would expect to find growing around *Salvia*. Terpenes were found in the air and on the soil surface around *Salvia* bushes, and the growth of grass seedlings planted in soil sampled from around bushes was inhibited. All this seemed to add up to pretty conclusive evidence that the bare zone around *Salvia* bushes was caused by allelopathic interference with the plant's competitors. Until, that is, Bartholomew (1970) carried out a field experiment which excluded rabbits from small, fenced plots placed inside bare zones. Growth of vegetation in these plots was compared with growth in control plots that were fenced on only three sides (giving access to rabbits). Controls were placed within bare zones and outside them. The controls within the bare zones produced only 5% of the biomass found in controls outside. By contrast, as much vegetation grew in the exclosures as in the control plots outside the bare zones. Despite all the circumstantial evidence that bare zones around *Salvia* were caused by allelopathy, it appears that the real cause was selective grazing by rabbits.

Exploitation competition

In exploitation competition, the effects of competing species on each other are exerted entirely through the depletion of the resource. In winter, tits (*Parus* spp.) form mixed-species flocks which forage for insects, spiders and seeds in the branches of trees. The goldcrest (*Regulus regulus*) may also forage with them. Do these species compete for food and, if so, by what mechanism? In coniferous woods in Sweden, Alatalo *et al.* (1985, 1987) recorded how goldcrests and three tit species divided the time they spent foraging between tree trunks and branches in the inner canopy and foraging on twigs and needles in the outer canopy. These observations were used as controls for experiments in another set of plots where some competing species were removed.

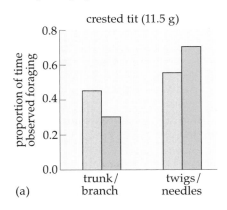

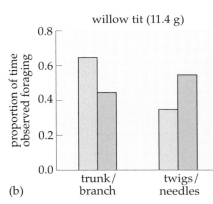

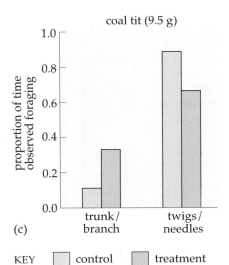

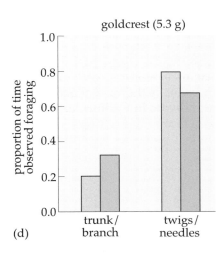

KEY   control   treatment

Figure 2.3 The foraging locations of (a) crested tits, (b) willow tits, (c) coal tits and (d) goldcrests on pine trees in woodland in central Sweden. Grey columns show the proportion of time birds were observed foraging in different locations when feeding in mixed flocks (experimental control). Green columns show the proportion of time crested and willow tits were observed foraging in different locations when goldcrests and coal tits were experimentally removed, and the time spent in different locations by goldcrests and coal tits when crested and willow tits were removed. Differences between experimental and control results for each species are all statistically significant. The average weight (g) of adult birds of each species is shown. (Experimental data from Alatalo *et al.* 1987 and 1985.)

❏ Figure 2.3 shows how the four species divided their foraging time between the inner and outer canopies of pine trees in control plots. Which two species were most often seen in the inner canopy, and which two in the outer canopy? How does this pattern relate to the weight of the different species?

■ Of the four species, crested tits and willow tits spent the most time in the inner canopy (although crested tits actually spent over half their time in the outer canopy), while goldcrests and coal tits spent the most time in the outer canopy. These patterns coincide with the birds' weights. The heavier birds foraged on the trunk and branches, while the lighter species foraged on more slender twigs and needles.

Two removal experiments were carried out to determine whether competition between species feeding in different parts of the canopy affected foraging patterns. In the first experiment, goldcrests and coal tits were removed from experimental plots, and the foraging of crested and willow tits was observed some months later. In the second experiment, willow and crested tits were removed and the foraging patterns of goldcrests and coal tits were observed later. The results of these experimental treatments are shown by the green columns in Figure 2.3.

❏ (a) How did removal of goldcrests and coal tits affect the foraging patterns of crested tits and willow tits?

(b) How did removal of crested tits and willow tits affect the foraging patterns of goldcrests and coal tits?

■ (a) After the removal of goldcrests and coal tits, crested and willow tits spent more time foraging in the outer canopy compared to controls.

(b) After the removal of crested and willow tits, goldcrests and coal tits spent more time foraging in the inner canopy compared to controls.

These experiments showed quite clearly that the bird species feeding in the two parts of the canopy were in competition with one another, but was the mechanism interference or exploitation competition? In social interactions that have been observed between these species, the larger species are able to supplant the smaller ones by attacking them, suggesting that interference competition may play a role in keeping goldcrests and coal tits away from the inner canopy. However, Alatalo et al. (1987) argue that only exploitation competition can be involved in keeping crested tits and willow tits away from the outer canopy when goldcrests and coal tits are present, because the latter two species are too small to interfere with them. This argument is plausible, but the real way to resolve the question is to measure how resources are depleted by each competitor with and without the other species present.

Field competition experiments seem to suggest that interspecific competition is rarely symmetrical in its effects: competing species are seldom evenly matched. A possible explanation for this is that pure exploitation competition is unusual, and interference may often give one species the edge. This is especially true for plants.

❏ Imagine two species of plant in competition with each other for light. One is twice as tall as the other and shades it with its canopy. Is this exploitation or interference competition, or some combination of the two?

■ The taller plant is certainly capturing light that is then unavailable to the shorter species, and this is exploitation competition, albeit very one-sided because the short plant cannot deprive the tall one of light. However, this interaction isn't a simple case of exploitation competition because plants utilize only a small fraction of incident light. The rest is transmitted, reflected or absorbed and re-irradiated as heat. In other words, the shorter plant is deprived of much more of the resource (light) than the amount its competitor actually uses. There is therefore interference competition, too.

Competition between plants for light is really a combination of exploitation and interference competition. This conclusion has some important implications: the question might be put 'what are leaves really for?'. Have plants evolved the area and number of leaves they happen to have because this is what is needed to capture resources for themselves or because this is what is needed to deny them to their competitors? The question is not as fanciful as it might sound because it is susceptible to experimental investigation. Plant breeders have produced a leafless variety of pea that can be compared with an ordinary leafy variety with which it is otherwise identical. Leafless peas are quite viable because they still have photosynthetic tissue in their stems and stipules (small 'wings' on the stem). When grown with their own type at a range of densities of 16, 25, 44, 100 and 400 plants m^{-2}, Snoad (1981) found that leafless peas yielded better than leafy varieties at densities of 44 m^{-2} and above. This suggests that leaves are a disadvantage rather than an asset to seed production! Imagine, though, leafless peas growing in competition with leafy ones. Unfortunately, this experiment was not done, but it is easy to see that in competition with other plants that have leaves, leafless plants would be at a major disadvantage. Leaves function as organs of interference competition at least as much as they function as organs of resource capture. Within a plant, leaves are often arranged in a pattern that minimizes overlap and self-inflicted interference.

Apparent competition

A minus–minus interaction between species can reflect competition between them, or may sometimes hide a more complex relationship that involves a third species that is a parasite or predator of two hosts that only appear to be competitors.

❏ How would you characterize the relationship:

1st host → parasite → 2nd host in terms of the +/− classification?

■ It is simply two host–parasite relationships −/+, +/− with the parasitic species in the middle being common to both, i.e. −/+/−.

If we ignore, or do not know about, the role of the parasite in such a relationship between hosts, it will look like a straightforward −/− interaction between competitors. This is called **apparent competition**. Well-established cases of apparent competition are very few. One likely example occurred when the variegated leafhopper *Erythroneura variabilis* invaded the San Joaquin valley in California in 1980. *E. variabilis* increased rapidly to pest proportions in vineyards, while at the same time the abundance of the grape leafhopper, *Erythroneura elegantula*, declined. Competition experiments were conducted with the two leafhopper species, caging them in mixed and single-species populations at a range of densities in bags placed around vine leaves (Settle and Wilson, 1990). The effect of competition between leafhoppers was measured by counting the number of eggs produced per female per day in each cage. Females insert their eggs inside the vine leaf. The experiments showed that egg production per female of both species fell as the number of leafhoppers per cage was increased. This fall in egg production occurred just as much when all leafhoppers in a cage belonged to the same species as when the species were mixed.

❑   What do the results of this caging experiment suggest about competition between the two leafhopper species?

■   First, the fact that egg production fell as density rose demonstrated that competition was taking place. Secondly, the fact that egg production by either species was affected as much by a rise in the density of its own species (in single-species cages) as by a rise in density of the other species (in mixed-species cages) suggested that the two species were equally effective competitors against each other.

If the two leafhopper species competed equally with each other, why did one species, *Erythroneura variabilis,* appear to have so dramatically reduced the abundance of the other, *E. elegantula*? Settle and Wilson (1990) suggested the explanation might lie in the effect of *E. variabilis* on the amount of parasitism of *E. elegantula* by a parasitoid that attacks them both; a wasp called *Anagrus epos*. This parasitoid lays its eggs in the eggs of both leafhoppers, but attacks *E. elegantula* more successfully because *E. elegantula* lays its eggs nearer the surface of the vine leaf than does *E. variabilis*. Figure 2.4 shows how the proportion of *E. elegantula* parasitized by *A. epos* rose as the relative density of the other leafhopper species, *E. variabilis,* increased.

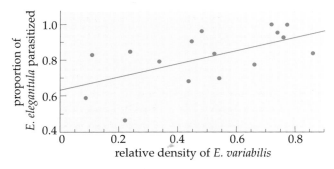

Figure 2.4   The relationship between the proportion of *Erythroneura elegantula* parasitized by *Anagrus epos* and the relative density of the leafhopper *E. variabilis*.

A likely example of apparent competition among plants involves the heteroecious rusts, which are a group of parasitic fungi that infect two different, invariably quite unrelated host plant species during their life cycle (Figure 1.30). For example, in the eastern USA, the white pine blister rust *Cronartium ribicola* infects white pine *Pinus strobus* and the currant *Ribes ribicola*. *Ribes* survives infection, though nearby *P. strobus* that become infected by spores released from currant bushes may be killed by the disease. Rice and Westoby (1982) argue that many heteroecious rusts that infect herbs and shrubs also infect trees that may potentially shade them, in which case the rust may simply be an agent of interference competition that benefits the smaller plant in competition for light with the larger one.

This seems implausible because the communities in which the two hosts of the rust grow will invariably contain many other plant species, immune to the rust but able to compete with both hosts. *Ribes* isn't likely to gain a long-term advantage from killing white pine because some other species of tree will just take its place. Where the direct competitive relationships between the alternate hosts of a parasite are rather unspecific compared to the highly specific effects of the parasite on the two hosts, the interaction between hosts may best be characterized as apparent rather than actual competition.

2.1.2 Evidence

How important is interspecific competition in nature, and what is the evidence? A number of attempts have been made to answer this question by surveying the results of field experiments. Two of the first such attempts, carried out separately but published in the same scientific journal in the same year, were by Schoener (1983) and Connell (1983). Because each study involved more than one species, there are two ways of looking at the overall findings: (1) the percentage of *studies* that found interspecific competition; (2) the percentage of *species* that were affected by interspecific competition. Schoener looked at 164 studies of animals and plants and Connell at 72, but both found that nearly all studies reported significant evidence of interspecific competition. Of the species in Schoener's survey, 76% were affected by interspecific competition, while more than half of those in Connell's survey were affected. Goldberg and Barton (1992) carried out a survey of 89 field studies of competition between plants, finding that significant interspecific competition was reported in 79% of cases. Denno *et al.* (1995) surveyed studies of 193 pairwise interactions between herbivorous insects, most of which were experimental and carried out in the field. Competition was found in 147 of the interactions (76% of the sample), with mechanisms dividing approximately equally between exploitation and interference.

These four surveys looked at different sets of experiments performed on rather different sets of species (aquatic/terrestrial, plant/animal, etc.), but all came to remarkably similar conclusions: the overwhelming majority of published studies that have looked for competition in the field have found it. However, we should not take this conclusion completely at face value. Underwood (1986) scrutinized 95 field studies of interspecific competition with a view to their statistical rigour and experimental design and found that 16% of them were either not true experiments because they lacked controls or were carried out in unnatural conditions from which little could be deduced about the true field situation. A further 29% of the studies were insufficiently replicated.

The four surveys by Schoener, Connell, Goldberg and Barton, and Denno *et al.* broke down their results into much more detail than has been described here, comparing studies of different trophic levels, or terrestrial species with studies of aquatic ones for example. This kind of survey is called 'vote-counting' because it counts up how many instances there are of the phenomenon of interest (interspecific competition in our case) among different groups (animals, plants, aquatic organisms, etc.). Such an approach can be biased by the fact that, for example, there have been more terrestrial than freshwater field studies of competition. A more sensitive method for comparing 'how much' competition there is in different groups is to look at the magnitude of competitive effects, not just whether they are significant or not. This approach is also more likely to weed out badly designed studies of the kind identified by Underwood. The approach is called '**meta-analysis**' and has been carried out on a sample of 46 field studies by Gurevitch *et al.* (1992). The sample was a relatively small one because only studies that measured the biomass (weight of living tissue) of competitors were included. This restriction was necessary to make it possible to compare studies on quite different organisms with each other on some common basis.

In each study, the effect of competition on a species was measured as the difference between its mean biomass when competitors were removed (the treatment) and the mean biomass when competitors were present (the control). This was standardized to give a measure of competitive effect called *d*, by dividing the difference of the means by the standard deviation of all the measurements (Figure 2.5).

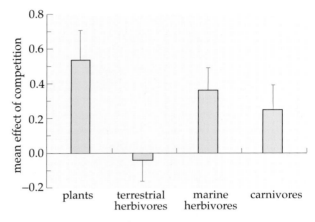

Figure 2.5   The mean effect of interspecific competition (*d*) in field studies of plants, herbivores and carnivores. Error bars indicate 95% confidence limits (see *Project Guide*).

❑   What is the relative strength of interspecific competition (*d*) in the three trophic levels shown in Figure 2.5?

■   Interspecific competition was strongest among plants, absent in terrestrial herbivores (arthropods) though strong in marine herbivores (molluscs), and intermediate among carnivores.

We shall explore a possible explanation of these differences among trophic levels in Book 2.

2.1.3   The niche

If interspecific competition is as prevalent as field experiments suggest, we may ask how so many species are able to coexist with one another. Why don't superior competitors simply exclude inferior ones, reducing ecological communities to combinations of just a few mutualistic species or to mixtures of species that do not interact? How species coexist is actually one of the biggest questions in ecology and we will examine it more fully in Book 2, Chapter 3. Here, we shall look at just one possible explanation; that the influence of competitors on each other is reduced by a division of the limiting resource into separate **niches**. A niche is that part of a limiting resource which one particular species is able to exploit better than its competitors. If each species in a community has its own niche, the intensity of competition may be reduced to a level at which they can coexist with one another.

The tits illustrate how this might occur. In the winter months when food is scarce, five species of tits (*Parus* spp.) can be seen foraging for food in broadleaved woodlands in Britain; the blue tit (*P. caeruleus*), the great tit (*P. major*), the marsh tit (*P. palustris*) and two of the species studied by Alatalo *et al.* in Sweden; the willow tit (*P. montanus*) and the coal tit (*P. ater*). (In Britain, the crested tit, *P. cristatus*, is confined to coniferous woods in northern Scotland.) The five species are similar in many respects. All nest in holes, all eat insects in winter and seeds throughout the year and all feed their young on caterpillars in spring (Perrins, 1979). Is it possible that the five tits have sufficiently distinct niches that they can all coexist on limited food?

❏   How might tit species divide food resources between them?

■   The experiments and observations by Alatalo *et al.* suggest birds may reduce competition for food resources by foraging in different parts of the tree canopy.

Lack (1971) studied the foraging behaviour during winter of the five tit species in Marley Wood, Oxford. His results are shown in Figure 2.6.

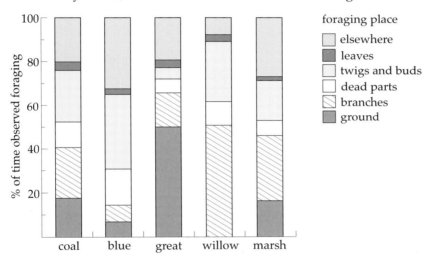

coal tit

blue tit

great tit

willow tit

marsh tit

Figure 2.6   Lack's data on foraging tits during winter in Marley Wood, Oxfordshire. (All marginal drawings × 0.3.)

❏   To what extent, if at all, do Lack's results support the idea that tits occupy different feeding niches during winter?

■   There were some clear differences between species in their feeding sites. The great tit spent half its time on the ground, where little time was spent by any other species. The willow tit spent half its time on branches, where the marsh tit was the only other bird to spend much time. The blue tit spent more time than other species on dead parts, twigs and buds.

In a different study in the Forest of Dean, Betts (1955) measured the size of insect prey taken by four of the tit species (Figure 2.7).

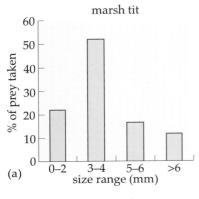

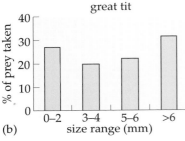

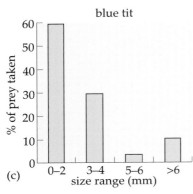

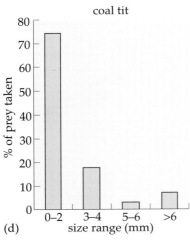

Figure 2.7   The size distribution of prey taken by four species of tit.

❑ To what extent, if at all, do Betts' results support the idea that tits occupy different feeding niches?

■ There were some clear differences between species in the size of insect prey they took. Coal tits and blue tits took mainly small prey (especially 0–2 mm), marsh tits took a wider range of sizes but concentrated on prey in the 3–4 mm range, while great tits took prey of all sizes including the largest (>6 mm) that other species mostly neglected.

These differences in feeding habits by no means exhaust the ecologically important differences between tit species. They also differ in the mix of seed/insect food they take, the types of seed they eat, the nesting sites they favour, and their relative abundance in different habitats (e.g. conifer *vs.* broadleaf woodland). Each of these characteristics defines a range of resources that is described as a **niche axis** or a **niche dimension**. Niche axes are used to define the niche of a species quantitatively so, for example, the proportion of the diet made up of insects can be measured on a scale of zero to one, and the value for each tit species can be plotted on the horizontal axis of a graph. On the vertical axis of the same graph, we could plot the average height (or range of heights) in the tree canopy at which birds forage. The resulting graph shows in two dimensions how the niches of different tits compare. There is no reason to stop at a two-dimensional graph, and we could add another dimension (i.e. graph axis) for type of nesting site, another for prey size and so on. The result would be difficult to visualize if there were more than three dimensions, but any niche described in this way may be thought of as an ***n*-dimensional volume**. In other words, the niche is some amount of resources (the volume) defined by some number (*n*) of niche axes.

This quantitative way of describing niches is useful for comparing species with each other, but it raises some important questions:

- What should the niche axes be (e.g. food size, food type)?
- How many niche axes should be measured (i.e. what should the value of *n* be)?
- How much overlap between niches (i.e. competition between species) can occur without a species excluding its competitors?
- Are all niche axes equally important?

There are no simple answers to these questions and ecologists tend to answer them rather pragmatically. Niche axes tend to be chosen on the basis of a knowledge of the species' ecology and of which resource limits its population size (or which is thought to do so!). The number of niche axes that should be measured is an even knottier problem. One solution is to measure as many as you think might be important, and then to select the combination of axes that produces the smallest overlap between species by using a statistical procedure called principal components analysis (Austin, 1985). In a study of eight species of plants in limestone grassland in Yorkshire, Mahdi *et al.* (1989) measured six niche dimensions: the **phenology** of each species (i.e. seasonal timing of growth), and the soil depth where each species grew, the pH, and the level of available N, P and K. Principal components analysis showed that 80% of the differences between species could be measured on just two composite axes; the first axis was influenced mainly by phenology and the second by pH (Figure 2.8).

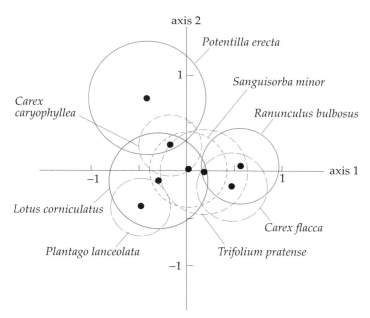

Figure 2.8 Niche overlap of eight perennial plant species on calcareous grassland, determined by principal components analysis. Circles show 95% confidence limits for the position of each species.

❑ Does the niche diagram in Figure 2.8 suggest that the eight plant species have different niches?

■ No, there is a great deal of overlap between virtually all species except one (*Potentilla erecta*).

If the overlap between species in Figure 2.8 had been less, we would have needed to answer the question 'how much overlap is allowed?' before we could have reached a conclusion about whether the niche differences were sufficient to permit coexistence. (Note that the 95% confidence limits shown in the Figure do not answer this question, they just allow us to judge whether niche differences between species are *statistically* significant or not. They tell us nothing about how big those differences should be to hold *ecological* significance.) As the results stand, they bear at least two possible interpretations: (1) there is no niche separation between the species; (2) the wrong niche axes were measured, and there may be niche separation on some other set of axes. With negative results like this, we can never confidently reject the second interpretation because there is no way of deciding in advance which niche axes are important, even when we know what resource limits the species. There are too many ways in which a limiting resource can be divided up. Very few studies like that of Mahdi *et al.* (1989) have been undertaken, but perhaps when they have been we shall be able to decide such issues as whether there is niche separation in limestone grassland plant communities on the balance of a large volume of evidence, as has been done with the issue of how often interspecific competition occurs in the field.

Despite its limitations, the niche concept is useful, particularly because we can use it to describe how competitors influence each others' use of resources. A species' **fundamental niche** is the set of resources that it could potentially utilize in the absence of competing species. The **realized niche** is the set of resources actually used when competing species are present.

❏ The difference between a species' fundamental and realized niches is best determined by experiment. In what experiment already described was such a difference demonstrated?

■ The field competition experiments carried out by Alatalo *et al.* demonstrated that the feeding niches of tits and goldcrests were limited by the presence of competitors, and that therefore their fundamental niches were larger than their realized niches.

Summary of Section 2.1

Two types of competitive mechanism are **interference competition**, in which one competitor physically interferes with another's access to resources, and **exploitation competition**, in which competitors affect each other only through the depletion of a limiting resource such as food. **Apparent competition** occurs when two species appear to interact negatively with one another (−/−) only because they share a natural enemy (such as a parasitoid) that attacks both of them. **Antibiotics** produced by fungi and **allelopathic** chemicals produced by some plants both play a role in interference competition.

Surveys of interspecific **competition experiments** are remarkably consistent in showing that competition is important in the majority of cases studied. Competition between species may be reduced if competitors specialize on different types or different parts of a limiting resource. A species' **niche** is defined as that part of a limiting resource that it is able to exploit better than its competitors. The coexistence of competing species is possible if their niches are sufficiently different. A species' **fundamental niche** is the set of resources that a species could potentially utilize in the absence of competing species. The **realized niche** is the set of resources actually used when competing species are present.

Question 2.1   (*Objective 2.1*)

(a)   Briefly define and give one example of each of the two types of mechanism of competitive interaction.

(b)   What is meant by the term 'apparent competition'? Give an example.

(c)   What is meant by the term 'limiting resource'?

(d)   What is the relationship between a species' niche and its limiting resources?

(e)   What is the distinction between a species' fundamental niche and its realized niche?

Question 2.2   (*Objective 2.2*)

Ecologists studying the marine invertebrate fauna at Santa Catalina Island in California found that two groups of molluscs were negatively associated with each other in the sub-tidal zone. Marine snails (gastropods) were most abundant on cobble substrata while small clams (bivalves) were most abundant among submerged boulders. Marine snails are mobile and feed by grazing algae from the surface of rocks. Clams are sessile filter-feeders, often found attached inside rock crevices.

Three species of invertebrate predator were also present; a lobster *Panulirus interruptus*, a small octopus *Octopus bimaculatus* and a large whelk *Kelletia kellitii*. The following observations and experiments were made:

Observations:

(i)   All three predator species preferred clams to snails as prey.

(ii)   The three predator species ranged over both types of substratum, but all were significantly more common among boulders than among cobbles.

(iii)   Some clams were always able to escape predation because predators could not enter small rock crevices. Exposed clams were covered in epibionts that reduced their vulnerability to predators (Chapter 1). Snails had neither kind of protection from predation.

(iv)   The density of snails and predators on boulder substrata differed between sites where clams were present and sites where they were absent, as shown in Table 2.1.

Table 2.1   For use with Question 2.2.

Density of:	Bivalves present	Bivalves absent	$P<$
all snails m^{-2}	20.7	42.9	0.05
all predators m^{-2}	0.112	0.032	0.05

Experiment:

(v)   In an experiment carried out in areas of cobble substrata, bivalves were added to some plots and the effect of this on the densities of predators and dead snails 65 days later was compared with densities in unmanipulated control plots (Table 2.2).

Table 2.2   For use with Question 2.2.

Density of:	Bivalves added	Control	$P<$
dead snails m^{-2}	448.0	123.0	0.05
all predators m^{-2}	0.160	0.034	0.001

Three hypotheses may account for the negative association between snails and clams: competition; habitat selection; and apparent competition.

(a)   What is the evidence for and against the competition hypothesis?

(b)   What is the evidence for and against the habitat selection hypothesis?

(c)   What is the evidence for and against the apparent competition hypothesis?

(d)   Overall, which of the three hypotheses is the best supported, and what further experiments should be performed to test it?

Apparel

Remove pred dog
From the experi

2.2 Predation

Predators are animals that eat and kill other animals (although sometimes ecologists talk of herbivores preying upon plants). We shall look at two important groups of predators: **parasitoids** which are insects whose larvae develop in the bodies of other insects and **carnivores** which comprise animals belonging to many groups including birds of prey, cats (big and small), carnivorous fish and spiders. Depending on whether carnivores eat herbivores or other carnivores, they feed at different trophic levels. In Section 2.2, we shall concentrate on those called **top carnivores** which feed in the highest trophic position.

2.2.1 Parasitoids

There are probably few insect species that are not attacked by parasitoids (Figure 2.9). Even parasitoids themselves have hyperparasitoids that attack them. Most species of parasitoid belong to the Diptera (true flies) or to the Hymenoptera (ants, bees, wasps and sawflies). The larvae of parasitoids live inside the bodies of other insects and in this respect they are like parasites; but because, unlike parasites, parasitoids invariably kill their hosts, their effect on the host is that of a predator. Parasitoid behaviour is also predatory. Adult females have to search for their prey, often using the smell of their host or the smell of their host's food to locate them. The parasitoids that attack fruit flies (*Drosophila* spp.) are attracted by smell to rotting fruit which is where *Drosophila* females lay their eggs and where their larvae feed. It has been shown experimentally that several parasitoid species are able to learn new chemical or visual cues associated with the presence of a host.

Many parasitoid species attack only one host species or a restricted number of them. Animals that feed on a single species of organism are described as **monophagous**, those that feed on a few species are described as **oligophagous** and those that feed on many are **polyphagous**. (Some authors use these terms to refer to the number of families rather than the number of species on which an organism feeds.)

❑   What difference might it make to the relationship between two herbivorous insect species occurring in the same habitat whether their parasitoids were monophagous or not?

■   If their parasitoids were *not* monophagous, the same parasitoid species might attack both host species. This could have a variety of effects on the relationship between hosts, including the possibility that it might create apparent competition. A possible example of this was described in Question 2.1.

Some parasitoid species attack their victim's eggs, others their larvae, pupae or adults. The genus *Trichogramma*, belonging to the chalcid wasps all of which are parasites, is an important group of parasitoids that attack the eggs of many kinds of insects, particularly those of moths (Figure 2.10). One species alone, *Trichogramma evanescens*, attacks the eggs of more than 150 species belonging to seven different orders of insects. When a female *Trichogramma* discovers an insect egg, she walks over its surface before inserting her ovipositor (an egg-laying tube at the tip of her abdomen) into it to deposit an egg of her own. Female *Trichogramma* that have oviposited in an egg leave behind an odour trail on its surface that deters other

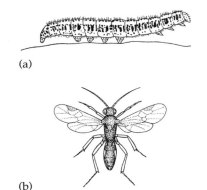

(a)

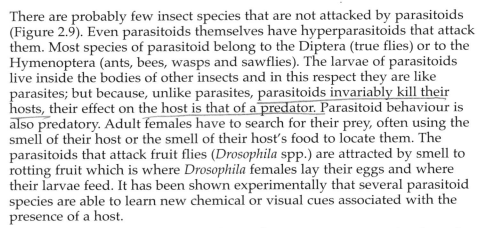

(b)

Figure 2.9   Caterpillar of (a) the large cabbage white butterfly *Pieris brassicae* (× 1.25), and its parasitoid *Apanteles glomeratus* (× 7).

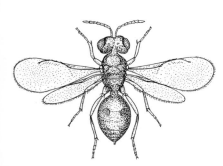

Figure 2.10   A chalcid wasp (× 10).

females from ovipositing in the same host. This deterrence is quite common in parasitoids. It has been demonstrated experimentally by washing the odour off an egg, whereupon other *Trichogramma* females will oviposit.

❑ Why should a female *Trichogramma* avoid ovipositing in an egg that has already been parasitized?

■ If two *Trichogramma* eggs are laid in the same host, there is likely to be competition between them for food. A later-arriving competitor will usually lose against an earlier-arriving parasitoid larva because the earlier one has a head-start in development and growth.

Various other aspects of parasitoid behaviour also increase the likelihood that a host will provide the parasitoid's larva with enough food for development. Depending upon whether the host is big enough when it is attacked to support the development of the parasitoid larva, a parasitoid egg may lie dormant until the host has grown larger or, if the host is large enough when attacked, it may hatch and begin consuming the host immediately. Parasitoids that lay their eggs in mobile hosts such as caterpillars usually paralyse them with a sting prior to oviposition, although the paralysis may be only temporary, later permitting the host to continue feeding. Parasitoid eggs have to overcome the host's immune system in order to develop and in a number of species the ovipositing female is known to inject the host with a symbiotic virus that prevents host tissue from encapsulating her egg.

Larvae of *Apanteles* parasitizing caterpillars of the large cabbage white (Figure 2.9) feed only upon their host's fat reserves, avoiding its vital organs and thus allowing the caterpillar to continue feeding as long as possible. Another parasitoid attacking the large cabbage white, *Pteromalus puparum*, lays its eggs only in pupae and has been observed waiting for several hours next to caterpillars that are about to pupate. Aphids have no larval or pupal stage but juveniles and adults are heavily attacked by parasitoids, particularly towards the end of the summer by which time numbers of both parasitoid and host have been able to multiply greatly. Braconid wasps in the genus *Aphidius* lay their eggs singly in aphids. Parasitized aphids become 'mummified', fastened to a leaf by silk that is spun by the *Aphidius* larva when it pupates inside the corpse (Figure 2.11). When it hatches from its pupa, the adult *Aphidius* escapes through a circular hole that it cuts in the aphid mummy.

❑ Of what economic importance are parasitoids and a knowledge of their ecology?

■ Parasitoids attack insect pests of economic importance (e.g. aphids) and a knowledge of the ecology of parasitoids may be used to design methods of **biological control** for these pests.

The biological control of pests uses the manipulation of a pest's natural enemies to reduce the pest population and thereby to reduce the damage the pest causes. As discussed in Chapter 1, myxomatosis was introduced into France and Australia, and was possibly spread unofficially in Britain as a biological control agent against the rabbit.

❑ What ecological characteristics would be desirable in a parasitoid that was to be used as a biological control agent?

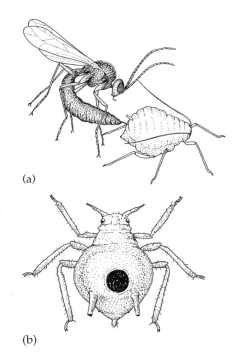

(a)

(b)

Figure 2.11 (a) Aphid *Aphidius* and (b) an aphid mummy (× 10).

■   It should be monophagous, attacking only the target pest, and it should be efficient at finding and attacking the pest.

To understand the relationship between parasitoids and their hosts properly, we need to know how each affects the numbers of the other. We shall therefore return to the topic in Book 2.

Question 2.3   *(Objective 2.2)*

Many species of ant attend aphid colonies and collect the 'honeydew' that aphids secrete. In order to test the hypothesis that ant attendance benefited the aphids, attendance by the ant *Lasius niger* on the bean aphid *Aphis fabae* living on creeping thistle was studied by Vökl (1992). Two parasitoids of *A. fabae, Trioxys angelicae* and *Lysiphlebus cardui* were found and a number of hyperparasitoids that parasitize *T. angelicae* and *L. cardui* were also present (Figure 2.12). Rates of parasitization of the aphid and hyperparasitization of its parasitoids in ant-attended and ant-free aphid colonies are given in Table 2.3. In addition, experiments showed that:

(i)   Aphids parasitized by *L. cardui* produced a significantly greater volume of honeydew than unparasitized aphids.

(ii)   Significantly more ants attended parasitized aphids than unparasitized controls.

(iii)   Ants attacked searching females of the parasitoid *T. angelicae*, but not females of *L. cardui*.

Figure 2.12 Relationships between the ant *Lasius niger*, the bean aphid *Aphis fabae* living on creeping thistle, *Cirsium arvense*, two parasitoids of *A. fabae, Trioxys angelicae* and *Lysiphlebus cardui*, and hyperparasitoids.

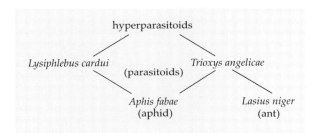

Table 2.3   Effect of ants attending aphid colonies living on creeping thistle on parasitoids and hyperparasitoids. Differences between ants present/absent are all statistically significant ($P < 0.01$). – means no data.

	Observation	Ants in aphid colony: present	absent
Ti	% of aphids parasitized by *L. cardui*	85	35
Tii	% of aphids parasitized by *T. angelicae*	1	8
Tiii	no. of *L. cardui* females captured foraging in aphid colonies	82	23
Tiv	% of *L. cardui* hyperparasitized	17	75
Tv	% of *T. angelicae* hyperparasitized	–	90

(a)   Do these results support or refute the hypothesis that *Lasius niger* benefited the colonies of *A. fabae* that it attended? Give evidence to support your answer.

(b)   What was the nature of the relationship between *Lasius niger* and the parasitoid *Lysiphlebus cardui*? Give evidence to support your answer.

(c)   What was the nature of the relationship between *Lasius niger* and the parasitoid *T. angelicae*? Give evidence to support your answer.

(d)   What was the nature of the relationship between *L. niger* and the hyperparasitoids? Give evidence to support your answer.

2.2.2 Carnivores

Carnivores are important in most communities, though larger species are highly vulnerable to human disturbance and have been hunted to extinction or near so in many countries. In Roman times, the bear was exported live from Britain for Roman circuses but this species was probably extinct by the 10th century. The last wolf recorded in Britain was killed in Scotland about 1740 and it is now nearly extinct in the rest of northern Europe, though there are reports of an increase in the former Soviet Union. By the beginning of the 20th century, the pine marten, polecat and wildcat were almost gone from Britain and the otter was considered rare, although it is now recovering in parts of Britain. Exceptions to this tale of woe afflicting mammalian carnivores are shrews, the mole, the stoat, the weasel, the fox and the badger although all but shrews have at some time been persecuted by gamekeepers or hunting.

❑   Why might larger carnivores such as the wolf and bear be more susceptible to extinction than much smaller ones such as the fox, the stoat and the weasel?

■   For at least two reasons to do with their population ecology: (1) large carnivores have much lower population densities than small ones and are therefore more vulnerable to hunting in the first place; (2) large animals have lower birth rates (young per female per year) than small ones so their numbers are slower to recover from hunting.

The fashion for field sports in Britain became widespread in the 19th century and lasted until the end of the First World War. Birds of prey (also called **raptors**) bore heavy casualties from the guns of gamekeepers and sportsmen so that by 1920 there were no populations of marsh harrier, honey buzzard, goshawk, osprey or white-tailed eagle left breeding in Britain. Breeding populations of buzzards, hen harriers and red kites were confined to small areas of the country where there were no shooting estates and golden eagles were extinct everywhere except in the Highlands of Scotland where numbers were greatly reduced. Populations of some of these species began to recover as hunting pressure was reduced but a new hazard began to afflict many birds of prey in the 1950s.

Organochlorine pesticides such as DDT used as a general insecticide and dieldrin used as a seed dressing were introduced on a widespread scale in the late 1940s and 1950s. At the same time widespread declines were observed in the breeding success of the merlin, peregrine, sparrowhawk and other birds of prey. Research revealed that breeding success was declining simultaneously elsewhere in the world where organochlorine compounds were in use and that residues from these compounds were accumulating to high concentrations in the body fat of birds. Birds with high levels of organochlorine residues were found to lay eggs with thin shells that often failed to hatch.

marsh harrier

honey buzzard

goshawk

white-tailed eagle

Where measurements were made of organochlorine residues in plants, herbivores and carnivores from the same place, concentrations were found to increase at each trophic level.

❏ Suggest an explanation for the increase with trophic level in the concentration of organochlorine pesticide residues.

◼ Herbivores concentrate residues in their bodies and when they are eaten by carnivores the residues are concentrated still further.

Among raptors, concentrations were highest in those species such as the merlin and peregrine that eat insectivorous birds and lowest in species such as hen harrier and kestrel that eat herbivorous mammals.

❏ Suggest an explanation for the difference in concentration of organochlorine residues between bird-eating and mammal-eating raptors.

◼ Raptors that eat insectivorous birds feed at the fourth trophic level (plants → insects → insectivorous birds → raptor) while mammal-eating raptors feed at the third trophic level (plants → mammalian herbivore → raptor). The extra trophic level causes an additional degree of residue concentration.

The ecological research which uncovered the connection between organochlorine pesticides and the breeding failure of raptors helped to bring about restrictions on the use of these pesticides in Britain and elsewhere, although there are still countries where they are widely used.

What are the consequences for the rest of the ecological community when the carnivores are removed? This question has caused some controversy because carnivores sometimes seem to take the very young, the old or sick individuals from a prey population in preference to healthy, breeding adults.

❏ Why should the effects on prey of removing carnivores depend upon what kind of individuals carnivores kill?

◼ If carnivores only kill individuals that are likely to die from other causes anyway (e.g. the old or sick), then predation may have little effect on the size of the prey population.

The question of whether predators have any effect on prey numbers is particularly relevant to attempts to control the rabbit in Britain because so many rabbits die from myxomatosis. Does the presence of predators, or their removal by gamekeepers, make any difference to rabbit numbers? In order to answer this question, Trout and Tittensor (1989) carried out a survey of 14 sites in southern England that differed in the degree of predator removal carried out by local gamekeepers and farmers. Each month of the year, rabbit numbers at each site were counted along transects that were walked at twilight when rabbits emerge to feed. The number of predators removed (mainly foxes, stoats and cats) was used to place each site into a category of removal effort. Average rabbit numbers throughout the year at sites where predator removal effort was high or low are shown in Figure 2.13.

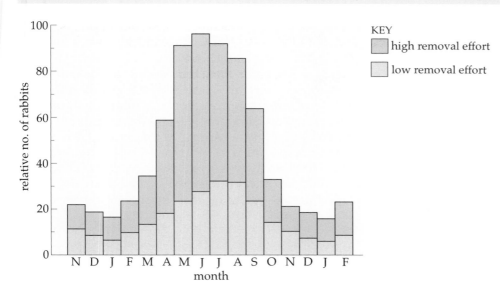

Figure 2.13 Average rabbit numbers throughout the year at sites in southern England where the effort put into predator removal was high or low. Differences between treatments were statistically significant between May and August inclusive.

❑　What can you deduce from the data in Figure 2.13 about the effect of predators on rabbit numbers?

■　Predators had a large effect on the size of the summer peak in rabbit numbers but had no statistically significant effect on the overwintering population.

A wider survey of spring rabbit numbers and predator-removal effort at 450 sites all over England and Wales confirmed the results in Figure 2.13; the presence of predators does significantly affect rabbit numbers and removing predators does cause a local increase. Before these studies were carried out, it had been suggested that the huge populations of rabbits which occurred in Britain prior to myxomatosis were chiefly the result of widespread predator removal by gamekeepers. This now seems quite likely, especially if Australian experience with the rabbit is also true in Britain. Studies in Australia suggest predators are able to keep rabbit numbers in check if those numbers are already low, but that once rabbit numbers become high predators make little impact. In other words, once rabbit numbers escape from control by predators they may later remain high. This is why, contrary to initial expectation, myxomatosis and predation both play a role. Myxomatosis outbreaks tend to occur at high density and predators affect rabbit numbers at low densities, and perhaps *only* at low densities.

Is the relationship between the rabbit and its predators typical of other relationships between carnivores and their prey, or is this an unrepresentative, artificial situation because it is so shaped by human intervention which first introduced the rabbit, then killed the rabbit and its predators and finally introduced myxomatosis? There are few places where carnivore ecology can now be studied in anything like natural conditions. In tropical rainforest, the large mammals are the first animals to disappear when there is human disturbance, well before anything like full deforestation takes place. Though carnivores and their prey have both been

hunted with the gun for a century-and-a-half, the two ecosystems where large carnivores and their prey still survive are the Arctic and the savannah in parts of Africa.

In parts of northern Canada, it has been found that wolves and bears account for more than 70% of the mortality of caribou, elk, moose and white-tailed deer calves. Moose numbers in particular seem to be limited by wolf predation and reach high densities only where there are other species of prey for bears and wolves, or where wolves are absent. In northern latitudes, a number of mammalian herbivores cycle in abundance. The classic example is the snowshoe hare (*Lepus americanus*) which shows approximately ten-year cycles that are also reflected in the lynx (*Lynx canadensis*) which is its chief predator (Figure 2.14). In northern Europe, voles (*Microtus* spp.) show 3–4-year cycles, although in southern Scandinavia and Britain cycles are less pronounced or absent. Precisely why some herbivore populations cycle and others (often of the same species) do not is not fully understood. One idea is that where predators depend on a restricted range of prey species, fluctuations in prey abundance produce fluctuations in predator numbers, too. A fall in prey is followed by a fall in predators but, because predator numbers lag behind prey numbers, when prey numbers begin to recover predation is low and prey are able to build up to high density.

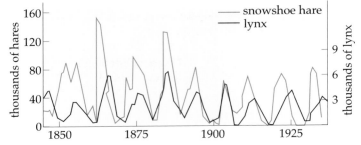

Figure 2.14 The snowshoe hare–lynx cycle.

❑ Why might you expect prey numbers *not* to cycle when predators have several alternative species of prey?

■ Predators that have several alternative species of prey can sustain themselves by **switching** from one species to another, depending upon which is most available. When numbers of a particular prey species begin to increase, predators switch back to eating them again.

A study of predation on a non-cycling population of field voles (*Microtus agrestis*) in southern Sweden found that foxes, domestic cats and common buzzards switched between voles and rabbits in this way. In cycling populations, it is not clear whether the reason that prey numbers eventually fall again is due to a shortage of food for the herbivore at high density or whether it is due to predator numbers catching up, causing an increase in predation that forces prey numbers down again. It could, of course, be some combination of the two.

In relatively intact African savannah ecosystems, there are several carnivorous species (e.g. lions, leopards, cheetahs and wild dogs) and many herbivorous prey and it is possible to ask whether predation has any significant impact on herbivores compared to the impact of competition among herbivore species. The many species of large ungulates (mammals with a cloven hoof) that graze the Serengeti in East Africa have long been regarded as one of the best examples of a community in which species occupy clearly separate feeding niches, differentiated by habitat, by species' food plants and by the plant parts each species prefers to graze. The community is divided into migratory species such as wildebeest (*Connochaetus taurinus*), zebra (*Equus burchelli*) and Thompson's gazelle (*Gazella thompsonii*) and non-migratory ones such as impala (*Aepyceros melampus*), topi (*Damaliscus korrigum*) and buffalo (*Syncerus caffer*). During the wet season (November–June), wildebeest, which are by far the most numerous of the ungulate species in the Serengeti, graze the short-grass plains in the south. At the beginning of the dry season in July, the wildebeest move north in huge numbers into the Mara region on the border of Tanzania and Kenya. In the space of just a few days, the arrival of the wildebeest in the Mara adds perhaps a million new mouths to the resident community of grazers.

Sinclair (1985) set out to test three hypotheses about the effect of the arrival of wildebeest on other ungulates in the Mara by comparing the kinds of habitat they used when wildebeest were present and when they were absent (> 600 m away). The preferred habitat of the wildebeest is areas of short green grass. The three hypotheses and their predictions about the effect of wildebeest on the distribution of other animals are summarized in Table 2.4 and are further explained below.

Table 2.4   Three hypotheses and their predictions about the effect of the presence of wildebeest on the distribution of other animals.

Hypothesis	Use of wildebeest habitat should:	Distance from wildebeest herds should be:
1 wildebeest and other ungulates compete for food	decrease	greater than expected by chance
2 other ungulates herd near wildebeest as shelter from predation	increase	closer than expected by chance
3 wildebeest facilitate grazing by other ungulates	increase	greater than expected by chance

Hypothesis 1: wildebeest and other ungulates compete for food

Ungulate species in the Serengeti do not interfere with each other, but exploitation competition could cause other animals to avoid wildebeest herds and habitat.

Hypothesis 2: other ungulates herd near wildebeest as shelter from predation

Wildebeest are the preferred prey of all the large carnivores. In the Mara, wildebeest were eaten by lions at twice the frequency that was to be expected from their relative numbers. Other ungulates might benefit from a lower predation risk by grazing near wildebeest because members of a herd are generally less vulnerable to predation than loners, and this is all the more so when the other animals in the herd belong to a species that predators prefer.

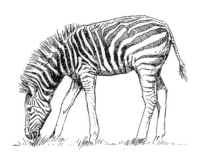

zebra

topi

Thompson's gazelle

impala

Hypothesis 3: wildebeest facilitate grazing by other ungulates

Because most of the Serengeti grazers are selective in their choice of food, some preferring short grass to long for example, it has been suggested that some species create conditions suitable for others that follow them (i.e. facilitate other species). In fact, Sinclair found that zebra did graze at the front of the wildebeest herd, but this was because wildebeest removed 80% of the grass as they moved across an area, leaving zebra no choice for food but to graze ahead of wildebeest, or to avoid them altogether.

Sinclair observed the locations of grazing herds of the various ungulates from a vehicle driven along transects. Figure 2.15 shows how often herds were observed on 'wildebeest habitat' (short green grass) when wildebeest were present and when they were absent. Figure 2.16 shows how far herds of other grazing species were from wildebeest.

❑ Use the data provided in Figures 2.15 and 2.16 to evaluate each of the three hypotheses listed in Table 2.4. Which species fit which hypothesis best?

■ Figure 2.15 shows that zebra and topi used wildebeest habitat less often than expected by chance when wildebeest were absent and that this did not change *significantly* when wildebeest were present. Both these species were found > 1000 m from wildebeest herds significantly more often than expected (Figure 2.16). On balance, then, although the data do not fit the predictions of hypothesis 1 perfectly, they are consistent with competition for food.

Thompson's gazelle significantly increased its use of wildebeest habitat when this species was present (Figure 2.15), and was found more often than expected 300 m from wildebeest and less often than expected > 1000 m away (Figure 2.16). These data are consistent with hypothesis 2.

Impala significantly decreased its use of wildebeest habitat when this species was present (Figure 2.15) and was found less often than expected < 100 m from wildebeest and more often than expected 300 m away (Figure 2.16). These observations are consistent with hypothesis 1, although it is noteworthy that impala did not move as far away from wildebeest as did zebra and topi (Figure 2.16). Shelter from predation (hypothesis 2) may therefore have played a role in their distribution, too.

The observations in Figures 2.15 and 2.16 are by no means conclusive evidence of the kinds of interactions occurring between wildebeest and other grazers, but they do suggest that, particularly in Thompson's gazelle, shelter from predation may play a role in the interactions among grazing animals in this ecosystem.

Summary of Section 2.2

Predators are animals that eat and kill other animals and include **parasitoids** whose larvae develop in the bodies of other insects and **carnivores** that hunt or trap their prey. **Top carnivores** feed in the highest trophic position, sometimes consuming other carnivores. Animals that feed on a single species (or other limited group) of prey are described as **monophagous**, those that feed on a few species are described as

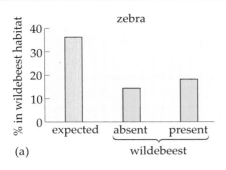

(a)

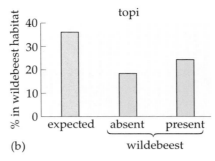

(b)

Figure 2.15 The frequency with which four ungulate species used wildebeest habitat compared to random expectation shown by the grey column. Use of the habitat in the presence and in the absence of wildebeest is also shown.

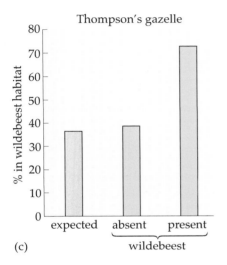

(c)

KEY control   treatment

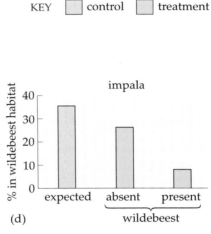

(d)

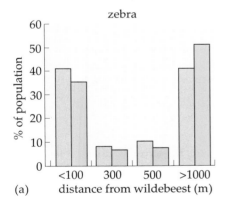

(a)

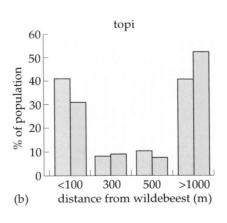

(b)

Figure 2.16 The distance of herds of four grazing species from herds of wildebeest (green columns) compared to distances expected by chance (grey columns). The distributions of all four species are significantly different from random.

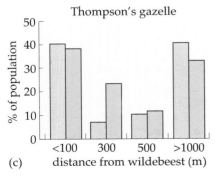

(c)

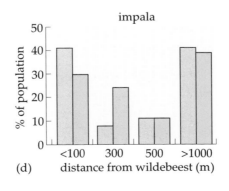

(d)

KEY   expected %   zebra

oligophagous and those that feed on many are **polyphagous**. **Biological control** uses the manipulation of a pest's natural enemies, often predators and parasitoids, to reduce the pest population.

Large carnivores such as birds of prey (**raptors**) have relatively low population densities and are vulnerable to extinction through hunting or the accumulation of pesticide poisons from their prey. Predators may have a significant impact on prey population densities, especially when prey are at moderate densities. Numbers of a predator and its prey may **cycle**, particularly if predators are relatively monophagous. Predators with broader diets **switch** from rare prey species to more common ones as prey densities vary.

Question 2.4   *(Objective 2.1)*

(a)   What are (i) parasitoids, (ii) top carnivores, (iii) raptors? Name one example of each.

(b)   What distinguishes monophagous, oligophagous and polyphagous animals from each other?

Question 2.5   *(Objective 2.2)*

On the island of Newfoundland, Canada, ecologists have studied the population dynamics of lynx (*Lynx canadensis*) and three of its prey: the arctic hare (*Lepus arcticus*) the snowshoe hare (*Lepus americanus*) and the caribou (*Rangifer tarandus*). Lynx are now the main predator of these three herbivores, but before European settlement, the timber wolf (*Canis lupus*) was the major predator on caribou (as elsewhere now in Canada); lynx were very rare and snowshoe hares were absent. Snowshoe hares were introduced in 1864 and by 1900 had reached very large numbers. The lynx population grew and was also very large by 1900 but the wolf became extinct in 1911. In about 1915, the snowshoe hare population of Newfoundland fell dramatically; soon afterwards, the arctic hare population also fell and caribou numbers began to fall. (*Note:* you may find it helpful to draw diagrams as an aid to interpreting these data.)

Ecologists have made the following observations and experiments:

In mainland Canada

(i)   Lynx and snowshoe hare populations show coupled cycles with a period of about ten years.

(ii)   Arctic hares are virtually confined to northern regions where lynx and snowshoe hares are absent.

(iii)   Wolves are the main predator on caribou, and lynx are not important.

In Newfoundland

(iv)   Lynx prey upon young caribou calves and are this species' main predator.

(v)   Arctic hares are now very rare but snowshoe hares and lynx populations in recent decades have shown coupled cycles.

In experiments on small islands

(vi)   When lynx were removed from a caribou breeding area, more calves survived to join the herd. When lynx were placed on a small island where only caribou had been present before, the mortality of caribou calves increased from zero to 65%.

(vii)   Two pairs of arctic hares were introduced onto a small island where there were no snowshoe hares, and no mammalian predators. Six breeding seasons later, their numbers increased to over 1000.

(viii)   When experiment (vii) was repeated on an island with snowshoe hares but no lynx, the arctic hares increased, but less dramatically.

(ix)   All attempts to introduce arctic hares onto islands where lynx and snowshoe hares are both present have failed.

Using your knowledge of predator–prey relationships and of interspecific competition, suggest hypotheses to explain the events in Newfoundland listed as (a)–(g) below. Cite any relevant observations and experimental evidence (i)–(ix) to support them.

(a)   The rapid increase in snowshoe hares after their initial introduction.

(b)   The increase in lynx numbers in the late 19th century.

(c)   The decrease in snowshoe hare numbers about 1915.

(d)   The rarity of arctic hares since about 1915.

(e)   The fall in caribou numbers in about 1920.

(f)   The cycling of lynx and snowshoe hares in the 20th century.

(g)   Variation from year to year in the survival rate of caribou calves.

2.3 Plant–herbivore interactions

There are some clear similarities between the relationships of predator and prey and those of herbivore and plant. Besides the obvious nutritional dependence of higher trophic levels on lower ones, both predators and herbivores tend to be selective to some degree in their choice of food and must therefore search for it. Particularly for highly specialized plant-eaters such as monophagous, seed-infesting insects, the population dynamics of a herbivore may be as tightly coupled to the abundance of its food as that of a parasitoid and its host. Such parallels should be borne in mind, although there are unique features of plant–herbivore interactions, too. Most of these stem from the fact that plants are rooted to the spot and cannot run from a herbivore. Instead, plants defend themselves with an arsenal of physical and chemical deterrents.

2.3.1 Plants as food

To the human eye, a world full of greenery may seem like a well-stocked supermarket for herbivores but, as in a supermarket, that food is packaged and protected from the hungry who must pay a cost to get it. Furthermore, if plants carried nutritional information printed on their sides, few would seem very appetizing to the discerning herbivore. The green tissues of the average plant are mostly water and the bulk of the rest is indigestible cellulose containing almost nothing but carbon. Rich food is found only in storage organs like potatoes hidden underground and in fruits and seeds that are available only seasonally. As if this were not enough to deter the average herbivore, most of the 'goods' are laced with poisons and chemicals that interfere with herbivore digestion or are protected from would-be consumers by an assortment of spines, stinging hairs, tough bark, waxes and latex that clog insect jaws and shiny surfaces that are difficult for small insects to cling to. It is even known for plants to hide from herbivores! A group of Australian mistletoes that live in the branches of eucalyptus trees have leaves that closely resemble the various shapes of their different hosts' leaves. This mimicry is effective because butterflies that lay their eggs on mistletoe search for their larval food plants by sight and eucalyptus leaves are inedible to their caterpillars. Succulent plants in the genus *Lithops* that grow in the Namib desert are known as 'living stones' because of their strong resemblance to the stony ground where they grow. You can easily see (or even grow!) this example of mimicry as a defence against herbivores for yourself, as living stones are common houseplants on sale at most garden centres.

The most important **non-chemical defences** against herbivores are the sheer toughness and indigestibility of leaves which is caused by the presence of thick cuticles, lignified fibres and, particularly in the grasses, by the presence of silica in cell walls. Hairs are another kind of deterrent. These kinds of defence do not prevent herbivores attacking a plant but they do limit the rate at which they can consume it.

Chemicals whose only known function in plants is to deter herbivores and resist pathogens are called **secondary compounds** and fall into three broad chemical groups. **Tannins** are **phenolic compounds** of which several others, particularly **lignin**, are also very widespread amongst plants. Tannins and lignins may also offer plants protection from fungal attack. The name 'tannin' comes from the use of these plant extracts in tanning leather in which process tannins combine with skin proteins.

Another group of secondary compounds is the **terpenoids** which include pine resin and essential oils that give culinary herbs like thyme their characteristic odours. The resin produced by conifers helps repel attack by bark-beetles whose larvae bore tunnels beneath the bark. Bark-beetles infect the trees they attack with pathogenic fungi (the fungus causing Dutch elm disease being an example). The earliest appearance of resin-like material in fossil trees is older than the first fossil evidence of bark-beetles, so terpenoids may have originally evolved as a defence against pathogenic fungi with which bark-beetles later evolved a symbiotic relationship (Langenheim, 1994).

The third major group are nitrogenous compounds which include the **alkaloids**, many of which are highly toxic in very low concentrations. The poisons in hemlock (*Conium maculatum*) and deadly nightshade (*Atropa belladonna*) are both alkaloids. Other **nitrogen-containing toxins** include unusual amino acids (not belonging to the kinds that make up proteins) which are found particularly in certain legume seeds, **cyanogenic glycosides** which liberate hydrogen cyanide when the plant is damaged, and **glucosinolates** which give plants in the cabbage family (Lamiaceae) their bitter flavour. These nitrogenous compounds tend to be produced in greater concentrations when nitrogen is in plentiful supply to the plant.

The wild ancestor of cultivated cabbages (*Brassica oleracea*) contains about five times the concentration of glucosinolates found in domestic varieties because the bitter flavour these compounds produce has been deliberately reduced by selective breeding of the crop. This may be partly responsible for the fact that domestic cabbages are much more susceptible to insect attack than wild ones.

❑ The concentration of secondary compounds has been reduced during the selective breeding of many commercial crops, but an exception to this is cultivated tobacco. Why should this be?

■ Tobacco is cultivated for the commercial exploitation of nicotine, which is one of its secondary compounds.

With exceptions, grasses generally lack chemical defences unless infected with a symbiotic fungus called an **endophyte** which lives inside the stems and leaves of the plant. For example, the fungal endophyte *Acremonium lolii* infects perennial ryegrass *Lolium perenne* and is transmitted between generations in the seed. Fungal endophytes of grasses cause no symptoms of disease in their hosts although some endophytes prevent the plant flowering and cause an increase in its vegetative growth. The endophyte produces alkaloids toxic to insect and mammalian herbivores. As a consequence of these two benefits to the grass, infected individuals have been observed to spread at the expense of uninfected individuals when grazing is heavy. Fungal endophytes are of economic importance because they can make a permanent pasture toxic to sheep.

Secondary compounds have two modes of action. Tannins and other phenolic compounds are called **digestibility-reducers** because they have the effect of reducing the proportion of food assimilated by herbivores. The mode of action may be by inhibition of digestive enzymes, but is not fully understood. These kinds of compounds occur in relatively high concentrations in the leaves of trees and long-lived plants which are attacked by many species of insects. Alkaloids and the other nitrogenous secondary compounds are actually **toxic** to herbivores and exert their effect at much lower concentrations than the digestibility-reducers by interfering with the insect's metabolism. These compounds tend to be more common in herbaceous, relatively short-lived plants which are attacked by fewer insect species.

The distinction between chemicals that reduce the ratio of food consumed to food assimilated (the assimilation ratio) and toxins is a relative rather than an absolute one, especially as plants may contain both types of compound. Even when we know what kinds of chemicals are present in all the plants of a particular habitat and what their effect is upon herbivores in controlled conditions, we may still miss their ecological significance if plant and herbivore species are not studied in detail in the field. A study of the Mexican bean beetle which attacks soybean (*Glycine max*) in North America provides an example.

This beetle was used in an experiment with two varieties of soybean (Price *et al.*, 1988). One was a common variety and the other a variety which produces an unidentified compound that reduces digestibility. Equal numbers of beetle larvae were caged with plants of each type. The survival of larvae to pupation and the amount of leaf area eaten by the end of the experiment were measured. The experiment was done (a) with an hemipteran (bug) predator of the beetle present, and (b) with no predators present. The results are shown in Table 2.5.

Table 2.5   The survival of Mexican bean beetles and area of leaf damaged on a variety of soybean producing a digestibility-reducing compound, when predators of the beetle were present and absent. All values are expressed as % of the normal variety (= 100%).

Predator	Beetle survival	Leaf damage
absent	88%	111%
present	23%	29%

❏ Did the variety containing a digestibility-reducer reduce beetle survival in the absence of predators?

■ Yes, but only by 12%.

❏ Did this result in less leaf damage?

■ No. Though fewer larvae survived, 11% *more* damage was done because each surviving larva ate more, perhaps to compensate for the poor digestibility of its food.

❏ Did the presence of a predator alter the amount of leaf damage?

■ Yes. Far fewer beetle larvae survived (23%) and leaf damage was considerably reduced (from 111% to 29% of the damage on normal plants).

The relationship between herbivore and plant is not a simple one. Each Mexican bean beetle that survived on the soybean variety containing a digestibility-reducer consumed 25% more leaf than a surviving beetle on the control. The growth of herbivores is limited by the amounts of certain amino acids so if assimilation is reduced, consumption must increase if growth is to be maintained. A digestibility-reducing compound will only protect a plant from herbivores if these do not compensate by increasing their consumption, causing more damage.

Some plants apparently increase their defences against herbivores when attacked. Two of many known examples of such **induced defence** are stinging nettle *Urtica dioica* in which the regrowth after damage has a higher density of stinging hairs than the original tissue, and downy birch *Betula pubescens* in which the concentration of phenolic compounds in undamaged leaves arises within two days of damage to other foliage. Fungal and bacterial infection in plants also commonly induces the production of defensive chemicals called **phytoalexins**. An increase in the quantity of terpenoids and a change in types produced occurs when trees are infected by the fungi carried by bark-beetles (Langenheim, 1994). Terpenoids and other secondary compounds may often serve in dual defence against insects and fungi. While many experimental studies have shown that induced defences have a negative impact on the herbivore, it has still to be shown experimentally that induced defences actually benefit a plant over and above the benefit derived from its normal (i.e. non-induced) level of defences.

No plant's defences are ever totally effective against all herbivores, so when herbivory cannot be **avoided** it must be **tolerated**. The chief means by which plants tolerate herbivory is through **compensatory regrowth** which replaces the lost tissue. Grasses are the most tolerant to being eaten of all plants because their growing points (meristems) are placed very low down, in or below the soil surface, where grazing animals cannot damage them. Indeed, compensatory regrowth can be so effective in grasses that some people have argued that herbivory benefits them because they grow better after being eaten! It should come as no surprise that in fact the evidence for such a benefit does not stand up to examination (Belsky, 1986).

❏ Why should it be no surprise that plants do not benefit from being eaten?

■ Because if plants benefited from being eaten, it would be difficult to explain why most plants have evolved so many ways of defending themselves from herbivory. The evolution by natural selection of such defences requires that better-defended individuals have an advantage over more poorly defended ones. A plant's fitness is not determined by how much it grows (or regrows), but by how many descendants it has. This depends upon reproduction and survival, so herbivory must increase these to produce a real benefit.

2.3.2 Herbivores

The diversity and extent of plant defences necessitates some food specialization among herbivores. Mammalian herbivores tend to be less specialized than insects. The giant panda, which feeds only on a few species of bamboo, is a rare exception to this but most **grazing mammals** such as rabbits, sheep and cattle will feed on a wide range of herbaceous species, although they do exhibit preferences among them. Sheep grazing mixtures of ryegrass (*Lolium perenne*) and white clover (*Trifolium repens*) alter their preference between these species depending on the time of day, how much they have already eaten and the relative amounts available. **Browsing mammals** such as mountain hare, deer, moose, giraffe and goats feed on woody plants but include a variable proportion of non-woody material in their diet.

A plant diet presents considerable chemical challenges to any herbivore's digestion. This is particularly true for browsers whose food contains a lot of tannins. Mountain hares, moose and mule deer all produce proteins in their saliva that bind to tannins and neutralize them, but no herbivores produce the enzymes needed to digest cellulose.

❏ How, then, do herbivores digest plant tissue?

■ Cellulose is broken down by enzymes produced by symbiotic micro-organisms in the herbivore's gut (Chapter 1).

The guts of all herbivores are like large fermentation vessels where the chemical work of digestion is done by micro-organisms. Cows have five stomachs to accommodate the process!

Herbivorous insects feed in a variety of ways (Figure 2.17) and on all types of plant organs including those we cannot see below ground. In some orders, e.g. Lepidoptera (butterflies and moths), Diptera (flies) and Hymenoptera (bees, wasps and ants), larvae and adults often differ in their feeding habits. Plant bugs (Hemiptera) such as aphids feed by inserting their needle-like mouthparts through a leaf or stem into the **phloem** which they tap for sap. Adults and juvenile instars may feed on different species of host. The black bean aphid *Aphis fabae* (Figure 2.17c) spends the summer

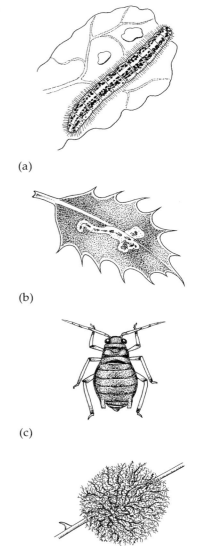

(a)

(b)

(c)

(d)

Figure 2.17 A variety of feeding modes used by insect herbivores. (a) A caterpillar of *Pieris brassicae* chewing a cabbage leaf (× 1.5). (b) A leaf mine made by the larva of the holly leaf miner (× 1). (c) A bean aphid that feeds on sap through tubular mouthparts (× 10). (d) A pincushion gall induced in a rose stem by *Diplolepis rosae* whose larvae feed on gall tissue (× 0.5).

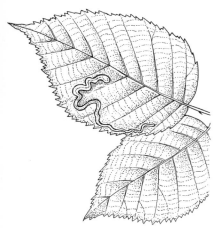

Figure 2.18 Leaf mine of the golden pygmy moth in a bramble leaf (× 1).

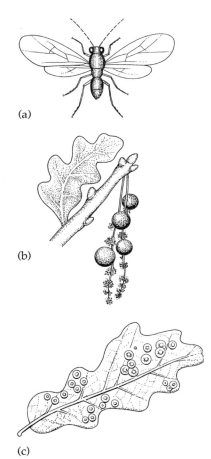

(a)

(b)

(c)

Figure 2.19 (a) *Neuroterus quercusbaccarum* (× 3). Galls of (b) the sexual and (c) asexual generations (× 1).

proliferating on broad bean (*Vicia faba*) and overwinters on woody plants such as spindle (*Euonymus europaeus*). Damage to a plant from aphids and other sucking insects can be severe but difficult to measure.

❑ How would you measure the damage done to a wheat plant as a result of aphids feeding upon it?

■ Experimentally exclude aphids from some plants and compare their growth rate or yield with that of infested plants.

The other main modes of feeding are chewing, mining and gall-forming (Figure 2.17a, b and d respectively). The caterpillars of the cabbage white butterfly *Pieris brassicae* chew the leaves of their food plants, but the adults have sucking mouthparts and need some liquid food (nectar). Mining takes place inside plant parts and can affect roots, stems, leaves (e.g. the golden pygmy moth caterpillar, Figure 2.18), seeds and flowers. The Agromyzidae is an important family of two-winged flies (order Diptera) whose larvae mostly mine leaves. The celery fly is one which attacks wild and cultivated plants of the carrot family. Another is the holly **leaf miner** *Phytomyza ilicis*, whose mines appear as blotches on the leaves of most holly bushes *Ilex aquifolium* (Figure 2.17b).

Galls are proliferations of plant cells produced by the plant host in response to the presence of particular insects. Larvae emerge from eggs, which the female injects into the plant with her ovipositor, and feed inside the gall on the proliferating plant tissue. The Cynipidae (**gall wasps**) are a family of the order Hymenoptera (wasps and bees) and mostly attack oaks in Britain. The gall wasp *Neuroterus quercusbaccarum* (Figure 2.19a) is responsible for two kinds of gall which appear on oaks. *Neuroterus* has an asexual generation at the beginning of the year; this consists only of females which lay parthenogenetic (unfertilized) eggs in male flower buds. The larvae which emerge induce the formation of small spherical galls called currant galls on the catkins (Figure 2.19b). The sexual generation which emerges from the galls, mates and lays eggs on the leaves which produce disc-like spangle galls in late summer (Figure 2.19c). The galls fall off the leaves and the asexual generation hatches from them in spring.

The gall wasp *Andricus kollari* which produces the common, round, hard, marble galls on oaks is thought to have been inadvertently introduced into Britain around 1830 with a load of galls imported into Devon from abroad for industrial use. Marble galls contain about 17% tannin which was used for tanning leather and in the dyeing of cloth. When extracted from oak leaves and added to an artificial diet, a concentration of only 1% tannin by weight produced a significant reduction in the growth of winter moth larvae which normally feed on oak.

❑ If 1% tannin is enough to affect significantly the growth of winter moth caterpillars, is the 17% concentration of tannin found in marble galls surprising?

■ Yes, but the larvae of *Andricus kollari* are able to tolerate this very high concentration.

2.3.3 Dietary specialization by insect herbivores

All herbivores specialize on or prefer particular food plants to some extent, but there is a considerable range in the degree of **dietary specialization** found in different species. Among British butterflies for example, there is the black hairstreak *Strymonidia pruni* for whose caterpillars the only wild food plant is blackthorn *Prunus spinosa* (Figure 2.20). Such extreme specialization on one species is quite rare. Oligophagous species with diets containing a few species are more common; for example, the green-veined white butterfly *Pieris napi* (Figure 2.21) eats many species of the cabbage family (Lamiaceae). Polyphagous herbivores like the green hairstreak (*Callophrys rubi*, Figure 2.22), which has been found on plants in six different families, feed on a wide diversity of food plants but will generally show a preference when offered a variety of edible species.

Specialization on plants within a single family is common in the diet of herbivorous insects and occurs in 75% of British species as a whole. Gall-forming insects such as agromyzid and tephritid flies have narrower diets than the average (>95% of species feed within one family) while some families of larger moths have broader diets than the average (<50% of species feed within one family – Ward and Spalding, 1993).

The mechanisms by which dietary specialization is achieved are complicated. Specialized feeders, such as the cabbage whites, are attracted to food plants suitable for their caterpillars by the very same glucosinolate secondary compounds which make the plants taste bitter and unpalatable to other less-specialized herbivores.

Figure 2.20 A caterpillar of the black hairstreak butterfly on blackthorn *Prunus spinosa* (× 0.5).

❑   If some specialized insects seem to have evolved the capacity to feed upon Lamiaceae, why have not Lamiaceae evolved changes in their secondary compounds in defence?

■   This is a question with no clear answer. In the first place, the plants may indeed have evolved some other defences against *Pieris* caterpillars. Secondly, Lamiaceae – like most plants – are attacked by many different herbivores and their secondary compounds are not protective mechanisms solely related to herbivory by *Pieris*.

The effectiveness of secondary compounds must be judged in the light of all the plant's natural enemies, fungi included. Plant breeders who produced a strain of oilseed rape (*Brassica napus*) low in glucosinolates, so that the seed meal remaining after oil pressing could be used to feed cattle, found that these plants failed to survive in the field because of heavy fungal infection (Bennett and Wallsgrove, 1994). Other strains with normal levels of glucosinolates in the leaves but reduced levels in the seed are now grown successfully.

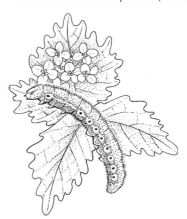

Figure 2.21 Green-veined white *Pieris napi* (× 1.2).

A secondary compound which deters some herbivores may actually benefit or attract others. For example, a compound found in cocklebur *Xanthium strumarium,* which gives the plant some protection against a seed-eating moth, increases its susceptibility to a fly which also consumes its seeds. The compounds produced by plants in the cucumber family (Cucurbitaceae) inhibit one beetle, *Epilachna*, which feeds upon them, but attract another, *Acalymma*. Once it is a specialist feeder upon a plant, the plant's defences may be of assistance to the insect. Herbivores commonly locate their food by responding to the characteristic odour of a plant's secondary compounds. These compounds can also benefit the specialized insect in other ways.

Bark-beetles use conifer terpenes to manufacture substances called pheromones that attract other beetles. The North American monarch butterfly *Danaus plexippus* feeds upon milkweeds (*Asclepias* spp.), some species of which produce compounds called cardiac glycosides which are

Figure 2.22 Green hairstreak *Callophrys rubi* on gorse (× 1).

toxic to vertebrates. Monarchs which have fed upon these species of milkweed as caterpillars are extremely distasteful to bird predators which avoid eating such a butterfly after only a single encounter. Monarchs fed upon milkweeds that lack cardiac glycosides or upon an artificial diet of cabbage do not deter bird predators except those birds that have already tasted a toxic monarch. These birds recognize the coloration and pattern of the monarch and avoid it.

Close relationships such as that between milkweeds and the monarch butterfly led Ehrlich and Raven (1964) to suggest that insects and their food plants have each influenced the other's evolution, or in other words that they have coevolved (Chapter 1, Section 1.8).

❑ How might the hypothesis of coevolution be tested?

■ By phylogenetic analysis (e.g. Chapter 1, Figures 2.58 and 2.59) of the associated insects and plants.

Figure 2.23 shows phylogenetic trees for leaf beetles in the genus *Hoplasoma* and for their food plants in the mint family, the Labiatae. The two trees show a remarkable degree of congruence which strongly suggests a long history of association between the herbivores and their plants, with little transfer of herbivores from one host species to another.

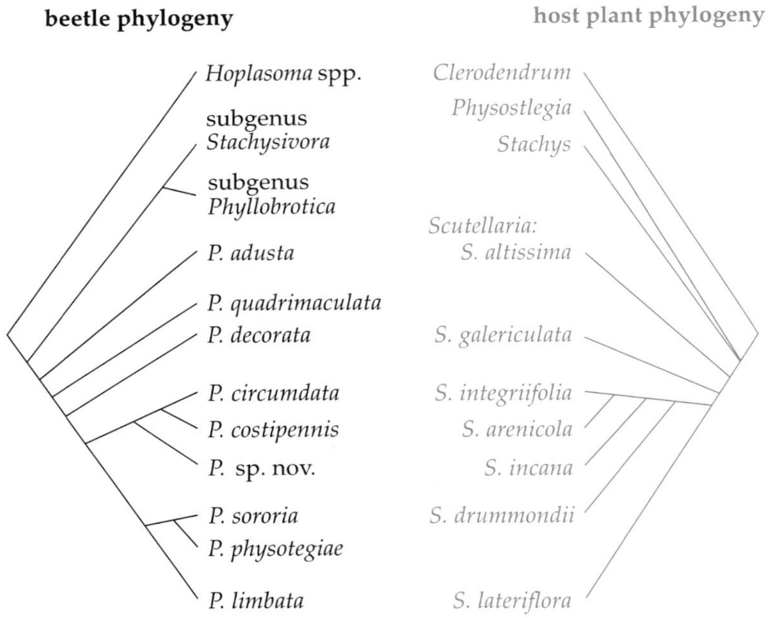

beetle phylogeny

Hoplasoma spp.
subgenus *Stachysivora*
subgenus *Phyllobrotica*
P. adusta
P. quadrimaculata
P. decorata
P. circumdata
P. costipennis
P. sp. nov.
P. sororia
P. physotegiae
P. limbata

host plant phylogeny

Clerodendrum
Physostlegia
Stachys
Scutellaria:
S. altissima
S. galericulata
S. integriifolia
S. arenicola
S. incana
S. drummondii
S. lateriflora

Figure 2.23 Estimated phylogenetic trees for leaf beetles in the subgenus *Phyllobrotica* and for their food plants in the mint family.

Futuyma and McCafferty (1990) carried out another test of coevolution on North American leaf beetles, this time in the genus *Ophraella* which feed upon a range of species in the daisy family (Asteraceae). The plants belong to four different tribes (subgroups) of the Asteraceae: the Astereae, the Eupatorieae, the Heliantheae and the Anthemideae. The *Ophraella* phylogeny in Figure 2.24 shows that the ancestor of the genus fed upon species in the Astereae and that *Ophraella* species later spread to the other three families.

❑ Compare the phylogenetic tree for *Ophraella* spp. with the phylogenetic tree for the plants on which they feed (Figure 2.23). How many shifts to new hosts are indicated by the beetle phylogeny? Are these shifts congruent with the evolution of the hosts?

■ Four shifts to new hosts are shown: two independent shifts to species
in the Heliantheae, and one to each of the other tribes. This pattern is
not congruent with the evolution of the four tribes of Asteraceae.

The lack of congruence between the phylogenies of *Ophraella* and their
food plants suggests that they have not coevolved in this case. In addition
to the four host-shifts that the beetles made between tribes shown in Figure
2.24, there were an additional five shifts (not shown in Figure 2.24) made
between plant genera (within tribes) and two between species in the same
genus. The conclusion is that *Ophraella* spp. have quite often changed hosts
and that this colonization occurred well after the evolutionary appearance
of the tribes on which they feed. Futuyma and McCafferty (1990) argue that
this is probably the rule for most relationships between insect herbivores
and their food plants because many closely related insects feed on quite
distantly related plants.

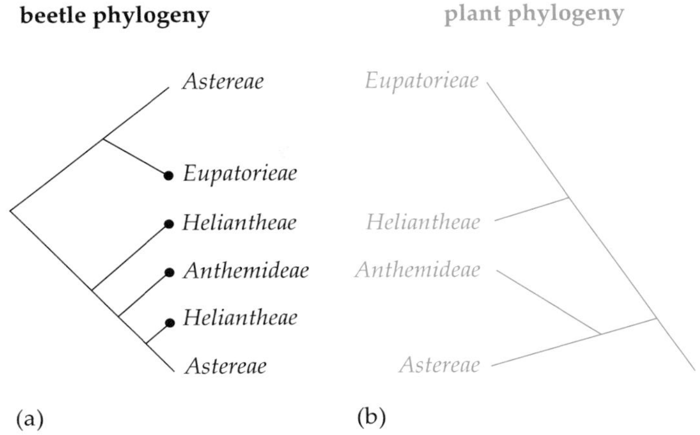

Figure 2.24 Estimated
phylogenetic trees (a) for
Ophraella spp. showing the tribes
of the Asteraceae on which they
feed and (b) for the four tribes of
Asteraceae eaten by *Ophraella*
spp. Each dot on the branches of
the beetle tree indicates a shift
from the host tribe of an ancestor
to another tribe.

Futuyma and McCafferty (1990) obtained a final insight into how
herbivorous insects colonize new hosts by comparing the types of
secondary compounds found in the different plants on which *Ophraella*
spp. feed. They found that the shifts often occurred between chemically
similar hosts. This is to be expected if insects need to be adapted to the
particular secondary chemistry of their host plants in order to feed and
reproduce.

Another important insight into the dietary specialization of herbivorous
insects comes from studies of the insect pests which attack crop plants after
they have been introduced into new areas of cultivation. Studies of the
pests of tea, cacao and sugar cane, for example, have shown that most of
the large numbers of insect species which feed on these plants are from the
native faunas in the area of introduction. In Britain, herbivorous insects
that are polyphagous are much more likely to be pests than species with
more restricted diets. A good example is the winter moth (*Operophtera
brumata*) whose caterpillars feed on most deciduous trees in Britain. Winter
moth is a serious pests of fruit trees and now even feeds on plantations of
the introduced conifer sitka spruce (*Picea sitchensis*).

Some herbivorous insects clearly possess the ability to change their diet
when confronted with novel food. How do they manage this? One
suggestion is that most insects possess rather generalized detoxifying
mechanisms which are capable of dealing with many different (but related)
compounds and that specialization is as much a behavioural phenomenon
as a physiological one (Bernays and Chapman, 1994).

❑   How would you test experimentally whether a dietary specialization
     was the result of (a) behaviour patterns during selection of food by an
     insect, or (b) the result of physiological inability to digest alternative
     food? (Of course, it might be both.)

■   (a)   To test for food selection behaviour, cage the insect with a choice
     of food plants and record which are eaten.

     (b)   To test for differences in digestibility between food plants, give
     insects different plant species in their diet and compare their rates of
     growth, survival and reproduction.

Such experiments were done by Smiley (1978) with three species of
Heliconius butterfly whose caterpillars feed on passion flower vines
(*Passiflora* spp.). The butterflies were caged with a choice of five *Passiflora*
species. One *Heliconius* species deposited eggs on four species equally and
laid some, but fewer, eggs on the fifth. The other two butterfly species laid
eggs principally on a single *Passiflora* sp. only, but each chose a different
one. Caterpillars feed on the plant where they hatch.

❑   How would you describe the choice of food plants for their caterpillars
     made by *Heliconius* species?

■   The first is non-selective and polyphagous (or oligophagous); the other
     two are selective and monophagous.

When the growth rates of caterpillars were tested on diets of all five
Passiflora spp., only one of the monophagous species showed slightly better
growth on its selected host than on the other plants. The other
monophagous *Heliconius* and the polyphagous species grew equally well
on all diets. There is also some evidence from experiments with other
butterflies that the female laying her eggs is much more selective about the
plants on which she lays than the growth rates of caterpillars on different
plants would seem to warrant. This may be because food quality is not the
only factor determining larval success. Females may avoid laying eggs on
certain species or parts of plants that are edible because they harbour
predators or parasitoids. Larvae may be similarly choosy. For example,
Damman (1987) found that caterpillars of a pyralid moth consistently
preferred older host leaves that could be rolled to produce a shelter
to younger ones that provided better food but poorer protection.

Evidence that the presence of protection from natural enemies can allow
herbivorous insects to adopt a broader diet is found in the lycaenid
butterflies, to which the large blue (Chapter 1) belongs. Although the large
blue is parasitic on ants, there are other species among the lycaenids which
feed on plants. Some of the herbivorous lycaenids seek out food plants
patrolled by ants that protect the butterfly's larvae and pupae from
predators and parasitoids; other species do not. Pierce and Elgar (1985)
compared the number of host plant species utilized by ant-protected
lycaenids with the number used by unprotected lycaenids and found that
the ant-protected butterflies tended to have broader diets than the
unprotected ones.

❑   What does this comparison suggest about the role played by protection
     from natural enemies in host-plant selection?

■ It suggests that protection from natural enemies by ants permits the choice of a broader diet, implying that diet choice in unprotected butterflies is limited by *where* it is safe to feed as well as *what* it is palatable to eat.

Herbivorous insect species often have broad geographical distributions. Studies of their food plants are usually confined to a particular site, but when studies from several sites are compared it is often found that a species specializes on different plants in different places. This rather complicates how we describe the diet of such species.

Heliconius is just one of many genera of insects which contain species that specialize locally on one or a few hosts but which over their entire range can only be described as oligophagous. These kinds of specializations can be behavioural or physiological, or both, and there may be an hereditary component influencing the selection of diet. When a genetic component to diet selection is demonstrated in such a situation, local diet specialization is most probably the result of evolution by natural selection.

2.3.4 Effects of herbivory on plant distribution and abundance

Mammalian herbivores range in size from voles to elephants, but voles can have as great an impact on their food supply as elephants on theirs because small mammals often reach very high densities. As we saw in the case study of the rabbit, grazing by mammalian herbivores can determine the whole character of the vegetation in a habitat. In fact, outside those regions of the world that are too dry or too cold to permit tree growth, all grassland habitats would turn to forest if the mammalian herbivores that graze them were removed.

❑ Why are heavily grazed habitats dominated by grasses?

■ The structure of grasses makes them tolerant of grazing (see Section 2.3.1).

❑ Refer back to Chapter 1, Table 1.2. Do these data suggest that all grasses are equally tolerant of grazing?

■ No. Sweet vernal grass was the least abundant grass under rabbit grazing in 1954 (only 4%) and became more abundant after myxomatosis while the other grasses became less abundant. This suggests sweet vernal grass was much less tolerant of rabbit grazing than the other grasses.

The only notable tall herb in areas heavily grazed by rabbits is ragwort *Senecio jacobaea* which contains alkaloids that are toxic to mammals.

Symbioses between insects and plants may influence plant survival and reproduction and as a consequence of this may influence plant distribution. The most dramatic examples are where herbivorous insects have been deliberately introduced into regions containing an introduced weed. A famous example is the biological control of the prickly pear cacti *Opuntia inermis* and *O. stricta* in Australia in the 1930s by the introduction of a moth *Cactoblastis cactorum* brought from the plant's native region in South America.

❑ Would you expect biological control agents such as *Cactoblastis* to be monophagous or polyphagous?

■ Only monophagous insects can safely be used in biological control because introduced polyphages might damage crops and native plants as well as weeds. Furthermore, it is less likely that an introduced polyphage will succeed where native ones have failed than it is that a monophage will succeed in controlling its major food plant.

Plants such as *Opuntia* when introduced into new areas are generally free of their major herbivores. The explosive increases in their new populations which sometimes occur might suggest that these herbivores are very important in population control in the native region, but this is difficult to confirm. For one thing, geographical factors play a large part in the impact of herbivores on their food plants. *Cactoblastis* introduced to southern Africa did not control an infestation of *Opuntia* there, possibly because the moth larvae were heavily attacked by ants.

The results of biological control programmes show that insect–plant interactions can have quite subtle and unpredictable results. A beetle *Chrysolina quadrigemina* introduced to California from Europe to control the St. John's wort *Hypericum perforatum* drastically reduced infestations of the weed in open habitats but failed to remove it from shady woodland margins where beetles would not deposit their eggs. Any ecologists in California who did not know about the history of biological control of *Hypericum* might suppose that this plant tolerates only woodland conditions. In Britain, where the plant is native, it commonly occurs in open habitats.

Not all feeding relationships between insects and plants are predatory or +/− ones. There is a great variety of mutualistic (+/+) interactions, especially between **ants and plants**.

❑ Recall an example of a mutualistic ant–plant association mentioned in Chapter 1.

■ Defence associations between ants and plants occur in both the temperate zone and in the tropics.

Ants may also aid in seed dispersal. The seeds of a number of violets (*Viola* spp.) have a small attached structure called an **elaiosome** which stimulates ants to collect the seeds and carry them off. The elaiosome is removed by the ants and the seed is then dumped in a viable condition on the ants' refuse heap. As well as transporting seeds, this ant behaviour can also hide the seeds from bird predators. A number of different species of plants produce seeds with elaiosomes, whose only known function is to attract ants.

Summary of Section 2.3

Plants possess both **chemical** and **non-chemical defences** against herbivores. Defensive chemicals are known as **secondary compounds** and include **phenolic compounds** such as **tannins** and **lignin**, the **terpenoids** such as pine resin, and **nitrogen-containing toxins** that include **alkaloids**, **cyanogenic glycosides** and **glucosinolates**. Grasses lack chemical defences, but may be infected by **endophytic fungi** that produce alkaloids. Phenolic compounds are **digestibility-reducers**, interfering with herbivore digestion. By contrast, the nitrogenous secondary compounds tend to be **toxic** at relatively low concentrations. Attack by a herbivore may produce an **induced defence** in a plant, while attack by fungi and bacteria commonly stimulates the production of **phytoalexins** that protect the plant against them. Often, herbivory cannot be **avoided**, but it may be **tolerated** through **compensatory regrowth** if a plant's meristems have been protected.

Most mammalian herbivores eat a range of plant species. **Grazing mammals** feed mostly on herbaceous species such as grasses while **browsing mammals** feed on a mixed diet that includes woody material. **Herbivorous insects** feed in a great variety of ways, both inside and outside plants. Two modes of feeding inside plants are **leaf-mining** and **gall-forming**. Dietary specialization differs greatly between insect species, from strict monophagy to extreme polyphagy. However, the diet of most herbivorous insects tends to be confined to the plants within a single family. The more specialized insect herbivores often use the secondary compounds of their host plants to their own benefit in a variety of ways. Some close relationships between insect herbivores and their food plants prove to be **coevolved**, others do not. Dietary specialization may be influenced by the protection some plants provide from a herbivore's natural enemies.

Question 2.6   (Objective 2.1)

(a)   Name three examples of non-chemical defences against herbivores that are found in plants.

(b)   Name two ways in which a plant's chemical defences (secondary compounds) deter herbivores and give an example of each.

(c)   What role do the following play in plant defence: endophytes of grasses; induced defences; phytoalexins; compensatory regrowth?

(d)   Name three ways in which a plant's secondary compounds may be used by specialist herbivores to their own benefit.

Question 2.7   (Objective 2.3)

Ecologists in Japan studied the diet breadth of three species of white butterflies; *Pieris napi* which is locally monophagous (feeding only on one species of Lamiaceae), *P. rapae* which is oligophagous (feeding on four species) and *P. melete* which is polyphagous (feeding on five species). Two alternative hypotheses for the differences in dietary specialization between species were tested:

(i)   That differences in diet breadth were determined by the range of plants on which larvae could grow well.

(ii)   That differences in diet breadth were determined by rates of parasitism on different host plant species.

The weight of pupae from caterpillars of the three species fed on six different species is shown in Table 2.6 and parasitism rates on different host plants are given in Table 2.7. Use these data to answer the following questions.

Table 2.6   The weight (mg) of *Pieris napi*, *P. rapae* and *P. melete* pupae formed after caterpillars were fed on six different species of crucifer in an experiment. Values in the same column with *different* superscript letters are significantly different from each other ($P < 0.05$). The species on which the butterflies laid eggs in the wild are indicated by values in **bold**.

Food plant	Pieris napi	P. rapae	P. melete
Brassica oleracea	180[de]	**198[a]**	**268[a]**
B. pekinensis	215[b]	**189[a]**	**267[a]**
Raphanus sativus	197[c]	**189[a]**	**250[ab]**
Rorippa indica	232[a]	**200[a]**	**236[b]**
Cardamine appendiculata	196[c]	171[b]	**252[ab]**
Arabis gemmifera	**192[cd]**	112[c]	208[c]

Table 2.7   Rates of parasitism (%) of the larvae of *Pieris napi*, *P. rapae* and *P. melete* occurring in the wild on different species of crucifer. A dash (–) indicates no data.

Food plant	Pieris napi	P. rapae	P. melete
Brassica oleracea	–	85%	40%
B. pekinensis	–	73%	13%
Raphanus sativus	–	68%	37%
Rorippa indica	–	61%	27%
Cardamine appendiculata	–	–	63%
Arabis gemmifera	5%	–	–

(a)   Is hypothesis (i) supported or refuted by the data? Cite the evidence for your answer.

(b)   Is hypothesis (ii) supported or refuted by the data? Cite the evidence for your answer.

(c)   What is your overall conclusion about the causes of differences in dietary specialization between the *Pieris* spp? What further experiment would be helpful in deciding between the two hypotheses?

Objectives for Chapter 2

After completing Chapter 2, you should be able to:

2.1   Recall and use in their correct context the terms shown in **bold** in the text. (*Questions 2.1, 2.4, 2.6*)

2.2   Interpret data on the interaction of competing (or apparently competing) species and on predators and prey. (*Questions 2.2, 2.3, 2.5*)

2.3   Interpret data on the dietary specialization of herbivores. (*Question 2.7*)

References for Chapter 2

Alatalo, R. V., Eriksson, D. and Gustafsson, L. (1987) Exploitation competition influences the use of foraging sites by tits: experimental evidence, *Ecology*, **68**, 284–90.

Alatalo, R. V., Gustafsson, L., Lindén, M. and Lundberg, A. (1985) Interspecific competition and niche shifts in tits and the goldcrest: an experiment, *Journal of Animal Ecology*, **54**, 977–84.

Alatalo, R. V., Gustafsson, L. and Lundberg, A. (1986) Interspecific competition and niche changes in tits (*Parus* spp.): evaluation of non-experimental data, *American Naturalist*, **127**, 819–34.

Askew, R. R. (1971) *Parasitic Insects*, Heinemann.

Austin, M. P. (1985) Continuum concept, ordination methods and niche theory, *Annual Review of Ecology and Systematics*, **16**, 39–61.

Bartholomew, B. (1970) Bare zone between Californian shrub and grassland communities: the role of animals, *Science*, **170**, 1210–12.

Belsky, A. J. (1986) Does herbivory benefit plants? A review of the evidence, *American Naturalist*, **127**, 870–92.

Bennett, R. N. and Wallsgrove, R. M. (1994) Secondary metabolites in plant defence mechanisms, *New Phytologist*, **127**, 617–33.

Bernays, E. A. and Chapman, R. F. (1994) *Host–Plant Selection by Phytophagous Insects*, Chapman & Hall.

Bernays, E. A., Cooper Driver, G. and Bilgener, M. (1989) Herbivores and plant tannins, *Adv. Ecol. Res*, **19**, 263–302.

Betts, M. M. (1955) The food of titmice in oak woodland, *Journal of Animal Ecology*, **24**, 282–323.

Bronstein, J. L. (1992) Seed predators as mutualists – ecology and evolution of the fig pollinator interaction, *Insect–Plant Interactions*, Vol. 4 (ed. E. Bernays), pp. 1–44, CRC Press Inc, 2000 Corporate Blvd NW, Boca Raton, FL 33431.

Buss, L. W. and Jackson, J. B. C. (1979) Competitive networks: nontransitive competitive relationships in cryptic coral reef environments, *American Naturalist*, **113**, 223–34.

Clausen, C. P. (1940) *Entomophagous Insects*, Hafner Publishing Co., NY.

Connell, J. H. (1983) On the prevalence and relative importance of interspecific competition: evidence from field experiments, *American Naturalist*, **122**, 661–96.

Damman, H. (1987) Leaf-quality and enemy avoidance by larvae of a pyralid moth, *Ecology*, **68**, 87–97.

Denno, R. F., McClure, M. S. and Ott, J. R. (1995) Interspecific interactions in phytophagous insects: competition re-examined and resurrected, *Annual Review of Entomology*, **40**, 297–331.

Ehrlich, P. R. and Raven, P. H. (1964) Butterflies and plants: a study in coevolution, *Evolution*, **18**, 586–608.

Futuyma, D. J. and McCafferty, S. S. (1990) Phylogeny and the evolution of host plant associations in the leaf beetle genus *Ophraella* (Coleoptera, Chyrsomelidae), *Evolution*, **44**, 1885–1913.

Goldberg, D. E. and Barton, A. M. (1992) Patterns and consequences of interspecific competition in natural communities: a review of field experiments with plants, *American Naturalist*, **139**, 771–801.

Gurevitch, J., Morrow, L. L., Wallace, A. and Walsh, J. S. (1992) A meta-analysis of competition in field experiments, *American Naturalist*, **140**, 539–72.

Harper, J. E. and Webster, J. (1964) An experimental analysis of the coprophilous fungus succession, *Transactions of the British Mycological Society*, **47**, 511–30.

Hassall, M. and Dangerfield, J. M. (1990) Density-dependent processes in the population dynamics of *Armadillidium vulgare* (Isopoda: Oniscidae), *Journal of Animal Ecology*, **59**, 941–58.

Ikediugwu, F. E. O. and Webster, J. (1970) Antagonism between *Coprinus heptemerus* and other coprophilous fungi, *Transactions of the British Mycological Society*, **54**, 181–204.

Imms, A. D. (1971) *Insect Natural History*, 3rd edn, Collins.

Jackson, J. B. C. and Buss, L. (1975) Allelopathy and spatial competition among coral reef invertebrates, *Proceedings of the National Academy of Sciences USA*, **72**, 5160–63.

Lack, D. (1971) *Ecological Isolation in Birds*, Blackwell.

Langenheim, J. H. (1994) Higher plant terpenoids – a phytocentric overview of their ecological roles, *Journal of Chemical Ecology*, **20**, 1223–80.

Lawton, J. H. and Hassell, M. P. (1981) Asymmetrical competition in insects, *Nature*, **289**, 793–5.

Mahdi, R., Law, R. and Willis, A. J. (1989) Large niche overlaps among coexisting plant species in a limestone grassland community, *Journal of Ecology*, **77**, 386–400.

Mitter, C. and Farrell, B. (1991) Macroevolutionary aspects of insect–plant interactions, in *Insect–Plant Interactions* (ed. E. A. Bernays), Vol III, CRC Press, Boca Raton, pp. 35–75.

Muller, C. H., Muller, W. H. and Haines, B. L. (1964) Volatile growth inhibitors produced by aromatic shrubs, *Science*, **143**, 471–3.

Ohsaki, N. and Sato, Y. (1994) Food plant choice of *Pieris* butterflies as a trade-off between parasitoid avoidance and quality of plants, *Ecology*, **75**, 59–68.

Perrins, C. (1979) *British Tits*, Collins.

Pierce, N. E. and Elgar, M. A. (1985) The influence of ants on host plant selection by *Jalmenus evagoras*, a myrmecophilous lycaenid butterfly, *Behavioural Ecology and Sociobiology*, **16**, 209–22.

Price, P. W., Westoby, M. and Rice, B. (1988) Parasite-mediated competition: some predictions and tests, *American Naturalist*, **131**, 544–55.

Price, P. W., Westoby, M., Rice, B., Atsatt, R. S., Fritz, R. S., Thompson, J. N. and Mobley, K. (1986) Parasite mediation in ecological interactions, *Annual Review of Ecology and Systematics*, **17**, 487–505.

Pyke, G. H. (1991) What does it cost a plant to produce floral nectar?, *Nature*, **350**, 58–9.

Rice, B. and Westoby, M. (1982) Heteroecious rusts as agents of interference competition, *Evolutionary Theory*, **6**, 43–52.

Schmitt, R. (1987) Indirect interactions between prey: apparent competition, predator aggregation and habitat segregation, *Ecology*, **68**, 1887–97,

Schoener, T. W. (1983) Field experiments on interspecific competition, *American Naturalist*, **122**, 240–85.

Settle, W. H. and Wilson, L. T. (1990) Invasion by the variegated leafhopper and biotic interactions – parasitism, competition, and apparent competition, *Ecology*, **71**, 1461–70.

Sinclair, A. R. E. (1985) Does interspecific competition or predation shape the African ungulate community?, *Journal of Animal Ecology*, **54**, 899–918.

Smiley, J. (1978) Plant chemistry and the evolution of host specificity: new evidence for *Heliconius* and *Passiflora*, *Science*, **201**, 745–7.

Snoad, B. (1981) Plant form, growth rate and relative growth rate compared in conventional, semi-leafless and leafless peas, *Scient. Hort.*, **14**, 9–18.

Springett, B. P. (1968) Aspects of the relationship between burying beetles *Necrophorus* spp. and the mite *Poecilochirus necrophori* Vitz., *Journal of Animal Ecology*, 417–24.

Trout, R. C. and Tittensor, A. M. (1989) Can predators regulate wild rabbit *Oryctolagus cuniculus* population density in England and Wales? *Mammal Review*, **19**, 153–73.

Underwood, A. J. (1986) The analysis of competition by field experiments, in *Community Ecology* (eds J. Kikkawa and D. J. Anderson), Blackwell Scientific Publications, Victoria.

Vance, T. R. (1978) A mutualistic interaction between a sessile marine clam and its epibionts, *Ecology*, **59**, 679–85.

Vökl, W. (1992) Aphids or their parasitoids: who actually benefits from ant-attendance?, *Journal of Animal Ecology*, **61**, 273–81.

Ward, L. K. and Spalding, D. F. (1993) Phytophagous British insects and mites and their food-plant families – total numbers and polyphagy, *Biological Journal of the Linnean Society*, **49**, 257–76.

Ward, P. S. (1991) Phylogenetic analysis of pseudomyrmecine ants associated with domatia-bearing plants, in *Ant–Plant Interactions* (eds C. R. Huxley and D. F. Cutler), pp. 335–52, Oxford University Press.

Williams, A. H. (1981) Analysis of competitive interactions in a patchy back-reef environment, *Ecology*, **62**, 1107–20.

PHYSICAL FACTORS AND CLIMATE   CHAPTER 3

Prepared for the Course Team by Irene Ridge

3.1   Introduction

Organisms interact not only with *each other* – the subject of the first two Chapters – but also with *non-living components* of their environment. Such interactions between organisms and the **abiotic** environment are the subject of this and the following Chapter.

Amongst the wide range of relevant factors are climate (temperature, rainfall, light, etc.), supplies of mineral nutrients and essential gases (O_2, CO_2), and chemical conditions such as pH in soil or water. The question we ask for each abiotic factor is: how does it affect the ability of different organisms to grow, survive, reproduce and disperse and thus affect the distribution or range of species? But although it is convenient to study one factor at a time, bear in mind that this is nothing like the 'real world': abiotic factors always *interact* with each other and with biotic factors. At high temperatures, water evaporates more rapidly from terrestrial organisms and the risk of damage from desiccation increases, so the effects of water supply and temperature are closely interlinked. In deep shade, plants are short of light (for photosynthesis) and may die or develop in a weak, spindly fashion, or survive if they have a specialized physiology which adapts them to these conditions. The weak plants in deep shade are often more susceptible to attack by pathogens and frequently are also unable to flower. So abiotic factors can affect the ability to survive and reproduce *directly* and also, by influencing biotic interactions, *indirectly*. Increased susceptibility to pathogens or predators, inability to form mutualistic (+/+) associations and, perhaps most commonly of all, reduced competitive ability, are all examples of these indirect effects. To give you an overview of abiotic/biotic interactions, Section 3.2 examines the distribution of organisms in one of the harshest physical environments – rocky shores.

3.2   Rocky shores: a case study

Anyone who has watched big waves pounding a rocky shore will appreciate the tremendous physical stress imposed on shore organisms. Strong forces – pounding and suction – operate so that any organisms not firmly attached or within a sheltered crevice become detached and in danger of being dashed against rocks. **Wave action**, therefore, is one of the dominant abiotic factors acting on rocky shore organisms. The severity of wave action varies greatly between different shores (see Section 3.2.1) and other abiotic factors vary both between and *within* shores, particularly from the top (the highest point reached by high tides) to the bottom (the lowest point exposed by low tides). Given this top-to-bottom variation of conditions on shores, it is perhaps not surprising to find that resident organisms occupy distinct *zones*. All these organisms are aquatic, the dominant plants being attached algae (seaweeds) and the dominant animals heavily armoured molluscs or crustaceans that can attach themselves firmly to rocks. The main question we address in this Section is whether the zonation of these shore organisms is indeed *caused* by varying physical conditions or results primarily from biotic interactions such as competition.

3.2.1 The shore environment

From a shore organism's point of view, the abiotic conditions which have the strongest effects on survival are (1) the severity of wave action and (2) the amount of time that must be spent out of water when the tide is out and physical conditions (such as temperature) during this period.

1   *Wave action* Rocky shores experiencing severe wave action are commonly described as **exposed** for the very good reason that they usually face open ocean (without shielding from offshore islands) and face into prevailing winds. *Local geography*, *aspect* and *wind strength* (which is affected by climate) are thus three factors influencing the severity of wave action. A fourth factor is the *slope* of the shore – its physical gradient – because waves generally break with greater force on a steeply sloping shore.

2   *Time out of water* Around the upper boundary of a rocky shore, organisms may be submerged in seawater for, on average, only a few hours on two occasions per lunar (28-day) month; at the lower boundary, they may be exposed to air on average for a few hours twice a month; and in the middle of the shore, they spend approximately equal amounts of time submerged or in air. So the higher up the shore an organism lives, the greater the time spent out of water. These varying conditions from top to bottom of a rocky shore depend on the *timing and amplitude of tidal cycles* which are the basis for dividing the shore into zones reflecting the vertical extent of tides (Figure 3.1).

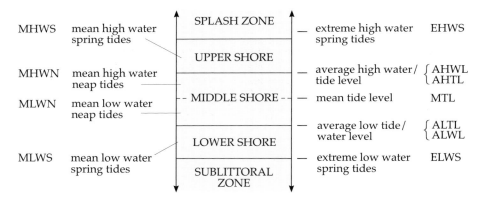

Figure 3.1 The main zones on the seashore defined in relation to vertical tidal range. Box 1 explains the system for naming zones and levels.

Box 1 explains more about tides, the tidal terminology on the right of Figure 3.1, and the reasons why amplitude varies. But consider first the problems facing shore organisms when they are in air. First, there is the risk of *desiccation*, bearing in mind that these are all basically aquatic organisms.

❑   Think of a second problem which is most severe for organisms on exposed rocks in summer.

■   Overheating, i.e. damage from high temperatures: exposed rocks can become very hot and the longer they are in direct sunlight, the hotter they get.

The risk of freezing when exposed in cold conditions is the other side of this temperature problem. Overall, the severity of temperature/desiccation problems increases from lower to upper shore and organisms permanently resident on the upper shore must be able to tolerate very extreme conditions. This is shown in Figure 3.2, which also illustrates how two other factors influence temperature variation and risk of desiccation.

❑   What are these two factors and what are their effects?

■   (1) *Aspect*: north-facing shores remain cooler and less prone to drying out than south-facing shores. (2) The marked difference between shores in the tropics and temperate zone shows the importance of *climate*: high temperatures in the tropics mean that few organisms can survive even in the middle shore.

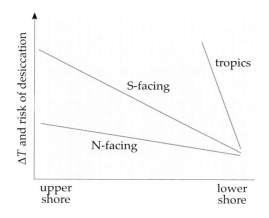

Figure 3.2   Variation from lower to upper shore of temperature and desiccation stress, i.e. daily temperature range ΔT, and risk of desiccation, for south- and north-facing shores in the temperate zone and a rocky shore in the tropics.

Look at Box 1 now.

Box 1   Tides

Tides – the regular rise and fall of sea-level – are the result of two forces acting on large masses of water (the oceans): the gravitational attraction of the Moon (and to a lesser extent the Sun) causes water-level to rise on the side of the Earth nearest to the Moon (or Sun) and the centrifugal force which keeps Earth and Moon apart causes a rise on the opposite side of the Earth. Typically, there are two tides a day with an average interval of about 12 h 25 min between high water. **Tidal range** or **amplitude** is the vertical distance between high water and low water.

During a lunar (28-day) month, tidal range varies as the relative positions of Sun and Moon vary with respect to the Earth. At new Moon and full Moon, the Moon, Earth and Sun are all in line (maximum gravitational pull) and tides with the largest range occur; these are called **spring tides**. At half-Moon, when the gravitational forces of Sun and Moon act at right angles to each other, **neap tides** with the smallest tidal range occur (Figure 3.3).

The range of spring and neap tides is not constant throughout the year but varies with the gravitational influence of the Sun. This is greatest at the time of the equinoxes (March and September), when spring tides with the maximum annual range occur. We can now provide a rationale for the naming of zones and levels on the shore (Figure 3.1) in relation to vertical tidal range and daily tidal cycles. This is the system you should use if you do an ecological project on seashores, and the letters are:

E = extreme

H = high

L = either low (if 1st or 2nd letter) or level (if last letter)

M = mean (for one or more years)

N = neap tide

S = spring tide

W = water

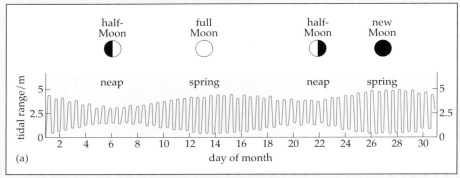

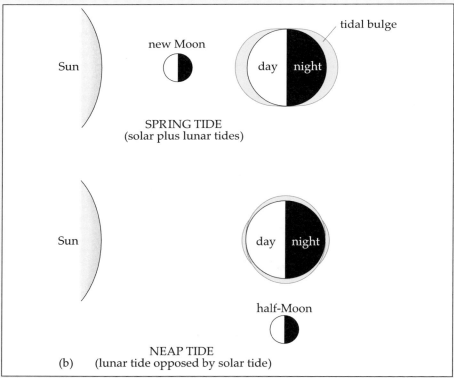

Figure 3.3(a) Example of a tidal record over the lunar month showing the variation in tidal range with neap and spring tides occurring every 14 days. (b) Relative positions of Earth, Moon and Sun for the spring tide at new Moon and a neap tide. Shading around the Earth represents water level, with the tidal bulge corresponding to high tide.

Considering first the levels shown on the right of Figure 3.1, extreme high water and low water for spring tides, EHWS and ELWS, are the points reached at high and low water for the biggest spring tides of the year (March and September). These levels define the boundary of the rocky shore habitat and, more generally, the **littoral**, which links fully aquatic and fully terrestrial habitats on any shore. Below ELWS lies the **sublittoral**, which is never exposed to the air and is inhabited by true marine organisms. Above EHWS is the **splash zone**, which is never fully covered by the sea but is often drenched by spray at high tide. In between are the three zones of the shore. The **lower shore** is between ELWS and the average low water level ALWL (i.e. the average for all 730 tides in a year of the low water point).

❑ So how many times, on average will (a) the upper and (b) the lower part of the lower shore be exposed to air?

■ (a) 365 times; this is the number of tides (half the total) whose low water level is below ALWL. (b) Twice, at low water for the extreme spring tides in March and September.

Clearly, most of the lower shore is under water for most of the time. The **middle shore**, lying between average low and high tide levels, is both exposed to air and covered by the sea during the average tidal cycle. And the **upper shore**, between AHWL and EHWS, is submerged only about twice a year at its upper boundary and for about half the tides in the year at its lower boundary.

Tidal range and the slope of the shore determine how much living space is available to organisms on a rocky shore and you have seen that tidal range varies during the year. However, it also varies geographically (Figure 3.4), mainly because land masses can interfere with the tides. In estuaries or bays that face the open ocean and are very long but shallow, for example, amplitude may be extraordinarily high (see the left side of Figure 3.4). This is because water in such places naturally sloshes to and fro, and when the periodicity of this wave matches the tidal period, very big tides result. Equally, currents and local topography may interact to reduce tidal range and, for 'enclosed' seas such as the Mediterranean and Red Sea, which are effectively cut off from oceans, tidal range is practically zero (Figure 3.4).

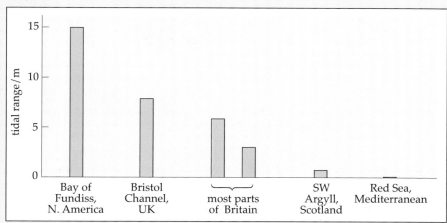

Figure 3.4   Tidal range in Britain and extremes in other parts of the world.

The tidal zones defined in Box 1 correspond closely to the zonation of organisms, which is a universal feature of rocky shores: different groups of species occur mainly in particular zones. This immediately suggests that zonation of the shore defined by tidal range and linked to the increasing risk of desiccation and temperature extremes up the shore *causes* the zonation of organisms, a question explored more fully in Section 3.2.3. However, organism zonation can be influenced strongly by wave action (exposure) as illustrated in Figure 3.5 for the brown seaweed *Pelvetia canaliculata* (channelled wrack).

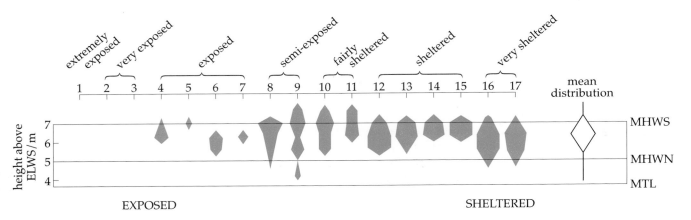

Figure 3.5   The distribution of *Pelvetia canaliculata* on rocky shores of the Dale peninsula, Pembrokeshire, at sites of different exposure (1–17). The width of kite diagrams indicates relative abundance.

❑   By reference to Figure 3.1 and information in Box 1, describe the distribution of *Pelvetia* in Figure 3.5.

■   The seaweed occurs mainly in the upper shore, sometimes extending to the upper part of the middle shore[1]. It extends furthest up the shore at semi-exposed or fairly sheltered sites, which are the most exposed of the sites where *Pelvetia* is abundant (sites 8–17)[2]. *Pelvetia* is absent from extremely and very exposed shores[3], present in small numbers on exposed shores and abundant on all the more sheltered sites. The seaweed is more abundant lower down the shore on very sheltered and semi-exposed sites[4].

❑   For each point marked with a numbered asterisk in the answer above, suggest an explanation based on *Pelvetia*'s tolerance of abiotic factors.

■   [1] *Pelvetia* must have high resistance to desiccation and temperature extremes. [2] Where spray from wave action helps to keep the upper shore moist, *Pelvetia* extends higher up the shore, but [3] it is apparently intolerant of severe wave action. [4] There is no obvious explanation for variations in the lower limit of the seaweed.

In Section 3.2.3, we examine in more detail explanations for the distribution of *Pelvetia*, but before this consider the general types of organisms that occur on rocky shores. You can see examples of these organisms in the TV programme 'Rocky shores'.

3.2.2 The organisms of rocky shores

1   Plants

All the plants on rocky shores are algae that can glue themselves to rock surfaces by sticky secretions. The large **perennial** algae (i.e. persisting from year to year) are all brown or red seaweeds (divisions Phaeophyta and Rhodophyta respectively). Short-lived species of green algae (division Chlorophyta) such as sea lettuce (*Ulva lactuca*) may be found at any level, often in pools or attached to limpet shells or to stones on the lower shore; and unicellular algae, mostly diatoms, also occur on rocks throughout the shore. All the larger algae colonize new sites by means of free-floating propagules (spores or zygotes) which germinate to produce sporelings after settling on a hard surface. Various crustose lichens (Chapter 1) occur in the splash zone, for example species of *Xanthoria* visible from a distance as a bright orange band, and the black species *Verrucaria maura*, may extend upwards from the middle shore.

On sheltered to moderately exposed rocky shores in temperate regions, the most prominent organisms are **fucoid algae** (belonging to the Order Fucales), of which *Pelvetia* is an example. They are all brown algae with a flexible stem or *stipe*, a basal *holdfast* and a flattened leaf-like *lamina*. Figure 3.6 shows the zonation of fucoids at both a very sheltered and a very exposed site on the Pembrokeshire coast. Other brown algae are the **kelps**, which are abundant on the lower shore and sometimes form a kind of submarine forest in the sub-littoral. Also shown in Figure 3.6 (overleaf) is the distribution of the most common molluscan grazers, species of limpets (*Patella*) and winkles (*Littorina*); this allows direct comparison of grazers and fucoids. The main point to note is that both types of grazer have a much wider vertical distribution than any fucoid, particularly on the very exposed shore: here they extend upwards to areas where there are (apparently) no large algae at all! Figure 3.6 emphasizes the point illustrated earlier for *Pelvetia* that the zonation and the *presence* of algae is strongly influenced by shore exposure. On the very exposed shore, only one fucoid occurs in the middle shore with other algae confined largely to the lower shore, and it is tempting to conclude that wave action is *the* factor influencing algal distribution on exposed shores. When you read Section 3.2.4, you will see that such 'obvious' conclusions are not always justified.

2   Animals

Relatively few animal species are permanent residents of rocky shores and the most abundant and ecologically significant of these are either molluscs or crustaceans. Figure 3.7 (on p. 115) shows the mean distribution of the most abundant species on the Dale peninsula, Pembrokeshire. The *grazing herbivores* – limpets, periwinkles and topshells – are all gastropod molluscs that move about when the tide is in, scraping diatoms and algal sporelings from the surface of rocks or large algae. When the tide goes out, common limpets (*Patella vulgata*) return to a home site, clamping down tightly on the rock surface where a groove created by friction exactly fits the shell. The other species retreat to humid, sheltered sites on large seaweeds, under stones or in cracks in the rock.

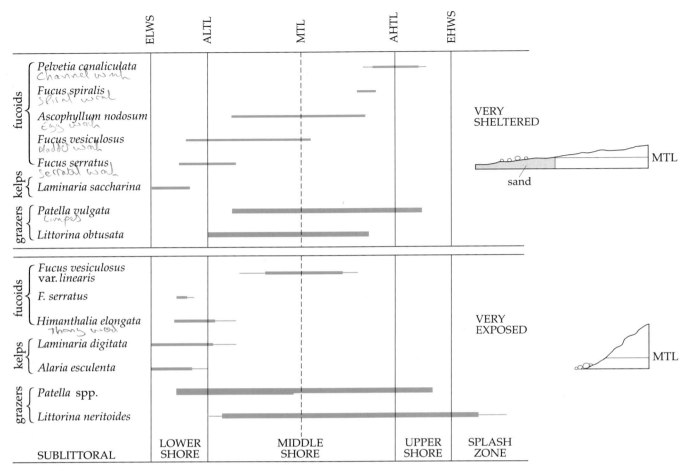

Figure 3.6 The distribution of fucoid algae, kelps and the most common molluscan grazers at two sites on the Pembrokeshire coast. Thick lines indicate higher densities. Common names of seaweeds are: *P. canaliculata*, channelled wrack; *F. spiralis*, spiral wrack; *A. nodosum*, egg wrack; *F. vesiculosus*, bladder wrack; *F. serratus*, serrated wrack; *H. elongata*, thong-weed. The kelps have no common names.

The most abundant *plankton feeders* on the Dale peninsula are **barnacles**, which are attached to rock throughout their adult lives (Figure 3.8). Attachment is by an outer circle of tough, calcareous plates and there are four smaller plates that can be moved and form a hinged cover (operculum) to the space in which the animal's body lies. Long, bristly thoracic limbs are used like a net to collect plankton and the animal retreats behind tightly closed shell-plates when the tide is out. Like limpets, barnacles have planktonic larvae (commonly called **spat**) which drift around in sea currents before settling onto rocks.

Dog-whelks (*Nucella lapillus*, Figure 3.9) are predatory snails (gastropod molluscs) and the commonest carnivore on the Dale coast (Figure 3.7). They feed mainly on barnacles or mussels but also eat limpets and a wide range of other molluscs by sitting on top of the shell and boring a hole through it, using enzyme secretions and their rasping tongue (*radula*). The prey is then paralysed by injection of a narcotic, attacked by digestive enzymes and finally sucked up through the dog-whelk proboscis in the form of a nutritious liquid. Other predators include crabs and starfish, which migrate with the tide, and birds such as oystercatchers, which usually eat mussels. The most abundant shore animals, however, are either sessile (barnacles) or have very limited mobility (limpets, topshells, periwinkles and dog-whelks), although all have planktonic larval stages that are readily dispersed by currents. Like seaweeds, all these animals show zonation on rocky shores.

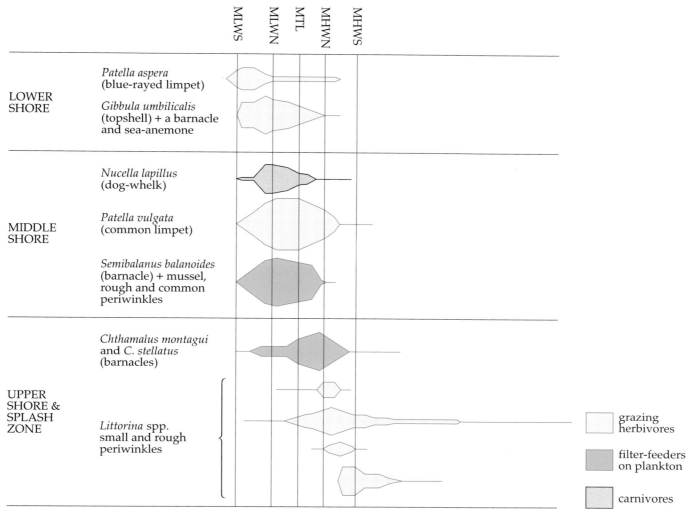

Figure 3.7   Mean distribution of animals on rocky shores of the Dale peninsula, Pembrokeshire. The width of kite diagrams indicates relative abundance.

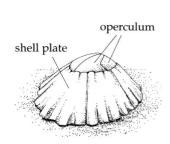

Figure 3.8   The general form of sessile barnacles (*Semibalanus* and *Chthamalus*). (~ ×2)

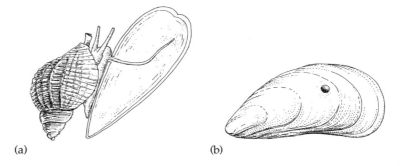

Figure 3.9   Dog-whelk (*Nucella lapillus*) (×1): (a) with proboscis extended through a hole into a living mussel and (b) hole bored in a mussel shell.

3.2.3 Causes of zonation

The question that now needs to be resolved is: why do these rocky shore organisms show zonation? There are two extreme sorts of answer. Zonation on a particular shore may occur because:

1 organisms have a certain, limited tolerance of desiccation and/or temperature extremes and this determines where they occur on the shore (i.e. abiotic factors dominate); or

2 organisms can survive at all levels on the shore but are restricted to particular zones and shores because of competition, grazing or predation (i.e. biotic factors dominate).

In practice (as you might guess), both kinds of factors operate on all organisms and may interact to determine zone boundaries. Often, but not always, the upper boundary is determined mainly by abiotic factors and the lower boundary by biotic factors, and experiments are needed to sort out the web of interactions. Some examples are described in this Section. There is also the question of what determines whether an organism is present or absent from a particular shore: factors such as availability of propagules (spores or planktonic larvae) or suitable habitat (e.g. bare rock or crevices), climate and the severity of wave action are relevant here.

1   Seaweeds

The fucoid algae (brown seaweeds) show clear zonation on most temperate rocky shores but, as Figure 3.6 shows, zones commonly overlap and are strongly influenced by shore exposure. Algae from different zones vary in their physiology. For example, lower shore algae photosynthesize better when submerged than in air whereas mid- to upper shore algae have highest rates when partially dried in air (Table 3.1).

❑ From earlier discussions, what other kinds of physiological differences would you expect between algae from different zones?

■ Differences in tolerance of desiccation and of high or low temperatures.

As suggested for *Pelvetia* (Section 3.1.1), such differences are found. The longer the period of desiccation which an algal species can tolerate, the further up the shore it grows (Table 3.1): *Pelvetia* may lose 96% of its initial water content and still recover after resubmergence. But both algal sporelings which develop above the normal zone of the species and adult plants which are experimentally transplanted on rock chips to higher zones do not survive for more than a few weeks, even if protected from grazers. So it does indeed seem that the upper boundary of seaweed zones – at least in the upper part of the shore – is determined mainly by algal tolerance of abiotic conditions.

For the lower boundary of seaweed zones, however, the situation is different. Schonbeck and Norton (1980) studied this in detail for *Fucus spiralis* and *Pelvetia canaliculata* on the Isle of Cumbrae, Firth of Clyde, Scotland. They carried out three major experiments:

1 measuring growth rates of algae both on the shore and in laboratory cultures of sporelings grown in a range of light levels;

2 transplanting algae on rock chips to lower zones (and, as a control, transplanting them within their normal zone) and monitoring growth and survival;

3   clearing areas of algae in the *F. spiralis* zone and observing the growth
of *Pelvetia* sporelings here (no grazers were present).

Pelvetia was found to have a very low growth rate, both within its own
zone on the shore and in culture (Table 3.2). Growth was also strongly
depressed (more so than for any other alga tested) at light levels
comparable to those underneath a canopy of fucoid fronds. The results of
the transplant experiment are shown in Figure 3.10 (overleaf) and the
clearance experiment (3) showed that *Pelvetia* zygotes settled and
germinated readily in the *F. spiralis* zone (along with those of *F. spiralis*) but,
over a period of eight months, were overshadowed by the faster-growing
Fucus and died out. If *F. spiralis* was removed from the settlement areas,
Pelvetia survived and grew well.

Table 3.1   Tolerance of desiccation for four species of fucoid algae. Plants were
exposed for varying periods in air at 25.5 °C with 50–59% relative humidity. Data
for growth are means $\pm$ 1 S.E. and – means that plants were not measured.

Fucoid species	Duration of exposure to desiccation/h	Subsequent linear growth/ (mm 10/days)	Relative wt. gain %/10 days	Condition of algae after 10 days in culture
P. canaliculata	12	3.1 ± 0.31	25.0	healthy
	24	3.1 ± 0.22	28.7	healthy
	48	2.5 ± 0.31	22.4	healthy
	70		−6.1	damaged but alive
F. spiralis	12	4.3 ± 0.58	16.2	slight damage
	24	−0.9 ± 1.25	14.9	damaged but alive
	48	–	–	dead
	70	–	–	dead
F. vesiculosus	6	0.0 ± 2.20	−6.3	damaged but alive
	12	–	–	dead
	24	–	–	dead
	48	–	–	dead
A. nodosum	7	0.3 ± 0.15	0.7	4 plants healthy, 6 somewhat damaged

Table 3.2   Comparison of growth rates of *Fucus spiralis* and *Pelvetia canaliculata* in
different situations. Data as mm/30 days or, for embryos, μm/30 days.
Comparisons between species are significant, $P < 0.05$.

Growth conditions	*F. spiralis*	*P. canaliculata*
own zone on shore	4.7	1.7
in culture: embryos	750	203
juvenile plants	15.3	4.2

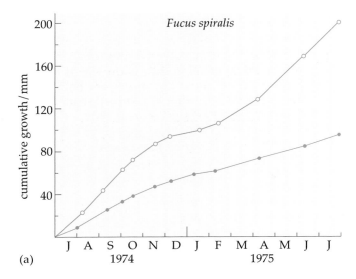

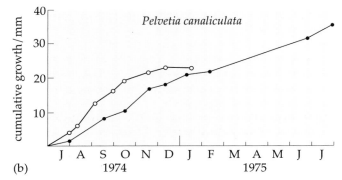

Figure 3.10 Growth in length of (a) *Fucus spiralis* and (b) *Pelvetia canaliculata* when transplanted as young plants lower down the shore (to the mid–middle shore, open symbols) and within their own zones on the shore (solid symbols) in July 1974. *Pelvetia* transplanted to the middle shore began to decay in December 1974.

❏ What do you conclude (a) from Figure 3.10 and (b) from the results of the clearance experiment?

■ (a) Both algae grew well when transplanted lower down the shore – in fact, much better than in their own zones. However, whereas *Fucus spiralis* maintained a high growth rate for at least 18 months, *Pelvetia* showed a decline in growth after 5–6 months and died out after 7 months: it appears unable to survive for long periods on the middle shore. (b) The clearance experiment shows that *Pelvetia* can colonize the adjacent *F. spiralis* zone but did not persist there because of competition with the faster-growing *Fucus*.

The mysterious disappearance of *Pelvetia* in the transplant experiment (Figure 3.10) was investigated further and found to be caused by a fungal infection. It appears that if this seaweed is submerged for abnormally long periods (as it is lower down the shore) it succumbs to fungal decay, leading Rugg and Norton (1987) to describe it as 'a seaweed that shuns the sea'. For *Pelvetia canaliculata*, therefore, the lower zone boundary is determined by two kinds of biotic factors: inability to compete with other fucoids because of its low growth rate, and a high probability of fungal decay when subjected to prolonged tidal submersion. Similar studies indicate that competition is a major factor determining the lower boundary of other seaweeds.

2 Barnacles

In Figure 3.7, you can see that different species of barnacles dominate the middle and upper shore in Pembrokeshire. *Semibalanus balanoides* (previously *Balanus balanoides*) is the middle shore species and two species of *Chthamalus*, *C. stellatus* and *C. montagui*, are more common on the upper shore. Until 1976, these latter species were lumped together as *C. stellatus*; they are very similar and their distributions overlap, but *C. montagui* is usually more common higher up the shore and especially in sheltered sites. Planktonic larvae of both *Chthamalus* and *Semibalanus* settle throughout most of the shore so that selective mortality of spat or adults must occur in order to explain the observed zonation.

Semibalanus is a northern species, close to its southern limit in British waters; the average time at which death occurs from desiccation in air at 18 °C is 6 h for spat and 45 h for adults. *Chthamalus* (lumping the two species together) is a southern species and reaches its northern limits along the west coast of Britain; death of spat occurs on average after 48 h in air at 18 °C and after 165 h for adults.

❏ What does this information suggest about factors causing barnacle zonation?

■ It is likely (and generally accepted) that risk of desiccation (i.e. physical factors) determines the upper zone boundaries, with the more desiccation-tolerant *Chthamalus* occupying the upper zone.

Connell (1961a,b) investigated the factors determining lower zone boundaries in a classic study on the Isle of Cumbrae.

❏ Using the seaweed studies as a guide, suggest the kinds of experiments that would be useful in solving this question.

■ Transplant and exclusion experiments.

Connell did both. He moved rock chips containing *C. montagui* (now known to be the only species present at this site) into the *Semibalanus* zone and monitored the survival of young and old individuals when *Semibalanus*, whose spat settled and developed on the chips, were or were not cleared off: some results are shown in Table 3.3. He also cleared a mid-shore strip of barnacles and removed continuously any *Semibalanus* that settled, the result being that *C. montagui* settled and grew there quite happily.

Table 3.3 Percentage mortality of *C. montagui* on transplanted rock chips with and without *Semibalanus* surrounding them.

| | Percentage mortality: | | | |
| | young *Chthamalus* | | older *Chthamalus* | |
Level on the shore	*Semibalanus* present	*Semibalanus* absent	*Semibalanus* present	*Semibalanus* absent
(1) just below MTL	90	35	31	0
(2) at about MLWN*	86	40	75	36

(1) Middle of *Semibalanus* zone.   (2) Lower quarter of *Semibalanus* zone.

* Stones at this level were under a cage.

❑ Do the data in Table 3.3 support the hypothesis that *C. montagui* is competitively excluded from the *Semibalanus* zone?

■ Yes, they do. Comparing pairs of measurements with and without *Semibalanus*, there is consistently higher mortality of both young and old *C. montagui* when *Semibalanus* is present.

Careful observation revealed the nature of the competition: it was basically for space, with the faster-growing *Semibalanus* elbowing out *C. montagui* by one of the mechanisms shown in Figure 3.11. This is an example of **interference competition** (see Chapter 2).

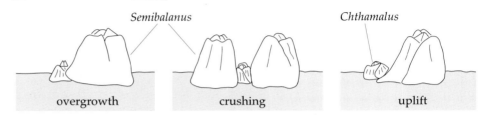

Semibalanus   *Chthamalus*

overgrowth   crushing   uplift

Figure 3.11 Interspecific competition for space between dominant *Semibalanus balanoides* and *Chthamalus montagui* (not to scale).

Note in Table 3.3 that transplanted stones at MLWN were under cages. This was done to exclude the barnacle predator *Nucella lapillus* (dog-whelk), which was common on this shore. Connell considered the possibility that *Nucella* might affect competition between the two barnacle species and might also influence the lower boundary of the *Semibalanus* zone. *Nucella* is absent from the *Chthamalus* zone because it cannot tolerate the degree of desiccation there and it also shows a preference for *Semibalanus* over other barnacles and for larger individuals of either species. From comparisons of the effect of *Semibalanus* on transplanted *Chthamalus* inside and outside cages that excluded *Nucella* (i.e. minus and plus predation), Connell concluded that dog-whelk predation reduced by 60% the *Chthamalus* mortality due to *Semibalanus* crowding at low shore levels. He explained this by observing that *Semibalanus* numbers were reduced by predation *before* they had crowded out *Chthamalus*. Higher up the shore, *Semibalanus* individuals grow more slowly and were not subject to predation until *after* they had succeeded in crowding out *Chthamalus*. Dog-whelk predation, therefore, almost certainly influences the degree to which the *Chthamalus* zone overlaps that of *Semibalanus*: whether it influences the lower zone boundary of *Semibalanus* is less certain, but seems likely.

This elegant study by Connell provides an (almost) complete explanation for the distribution of two species of barnacles on the Isle of Cumbrae. But it does not explain the distribution of these or similar barnacles at all other sites because Connell's explanation depends on certain conditions which do not necessarily apply elsewhere. The most important of these necessary conditions is that *rates of larval settlement must be high and roughly equal for both species.*

❑ What might happen if settlement rates for the two species of barnacles were equal but very low?

■ Interference competition and, therefore, exclusion of *Chthamalus* from the *Semibalanus* zone would be much less likely to happen.

This is indeed observed on coasts of central California, where the zones of balanid and chthamalid barnacles usually overlap and there is vacant space on the rocks as well. Three main factors were found to influence the number of larvae reaching the shore in California (Roughgarden *et al.*, 1987):

1   The direction and strength of surface currents, which may be offshore or onshore in central California.

2   The size of kelp (Section 3.2.2) 'forests' which fringe the shore and provide a habitat for fish and invertebrate predators of planktonic barnacle larvae; larval abundance was 70 times greater on the seaward side of the kelp forest than on the landward side during summer when larvae were approaching the shore prior to settlement.

3   The size of rockfish stocks, *Sebastes* spp., the major predator of barnacle spat in the kelp forests.

On a global scale, factor (1) is of most significance so that regions with predominantly offshore currents tend to have low rates of larval settlement and those with predominantly onshore currents have high rates, modified by local factors such as predator densities. Whether or not larval densities of the two species are similar depends on the geographical distribution of the species, which usually depends on tolerance of climatic factors such as temperature and desiccation. It so happens that on the Isle of Cumbrae, the distributions of *S. balanoides* and *C. montagui* overlap and both are equally abundant. Furthermore, water transport is mainly onshore and the confined circulation of water in the narrow strait between Scotland and Ireland seems to concentrate larvae in this area.

3.2.4   Exposure and more complex interactions that affect zonation

You saw in Figures 3.5 and 3.6 that exposure may affect both algal and animal zonation and the presence or absence of species on rocky shores. Can this be explained solely in terms of varying tolerance of wave action or does exposure also affect biotic interactions? This can be answered only from experimental evidence, but before examining such evidence look at Figure 3.12, which is based on another study of the Dale area, Pembrokeshire (Ballantine, 1961). Figure 3.12 illustrates the strikingly different appearance of shores with different degrees of exposures. In fact, Ballantine devised a system of assessing the exposure of a shore based on the presence and zonation of certain indicator species (a **biological exposure scale**) and his seven grades from extremely exposed to extremely sheltered are shown in Figure 3.5. On Ballantine's *extremely exposed shores* (Figure 3.12a) where there is continuous heavy surf, barnacles (nearly all *Chthamalus*) and limpets (mostly very small) are abundant in the middle and upper shore. Dense algal growth occurs only in the lower shore and the splash zone, where the red alga *Porphyra* (which is used to make laver bread) is found in summer. The only fucoid alga is a short, erect, bladderless form of bladder wrack (*Fucus vesiculosus* var. *linearis*). All these algae are very tough with high tolerance of exposure, and the absence of other species is largely because of their lower tolerance: they are torn from rocks or battered to pieces by strong waves.

extremely exposed shore   semi-exposed shore   very sheltered shore

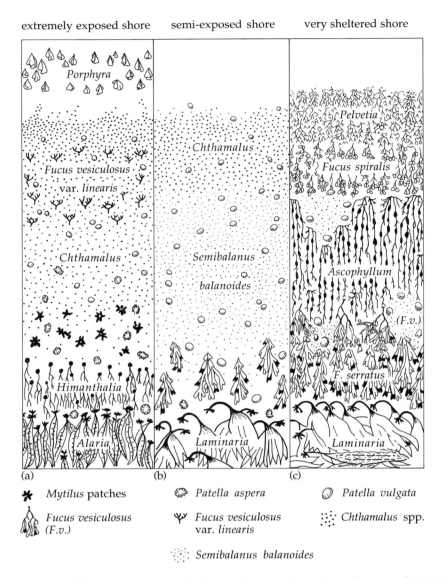

Figure 3.12 The main zonation of algae and animals on three shores in the Dale area, Pembrokeshire.

As exposure decreased (very exposed and exposed shores), Ballantine found that limpets and barnacles remained abundant but with *Semibalanus* replacing *Chthamalus* on the lower middle shore. *Porphyra* gradually disappeared and the only fucoids in the middle and upper shore were scattered *Pelvetia* and bladderless bladder wrack.

❑   Given this information and the appearance of the very sheltered shore (Figure 3.12c), what appears odd about the semi-exposed shore (Figure 3.12b)?

■   The absence of fucoid algae in the middle and upper shore. Some fucoids were present on the more exposed shores and one might expect increasing numbers to occur as exposure decreased, culminating in the luxuriant growth of very sheltered shores.

In fact, Ballantine found some *Pelvetia* on semi-exposed shores (so they *can* survive there) but only as scattered individuals that did not form a distinct zone. This strongly suggests that some factor *other than* intolerance of wave action is responsible for the absence of algae, and experimental evidence is needed to find out what. Figure 3.13a is a diagram of the mid-shore at a semi-exposed site on Holy Island, which lies off the west coast of Anglesey. Algal cover here was low and there was a high density of limpets (more than 90 per m²). Figure 3.13b shows an adjacent strip of shore from which all limpets had been removed one year previously (1974) and any recolonizing limpets subsequently removed.

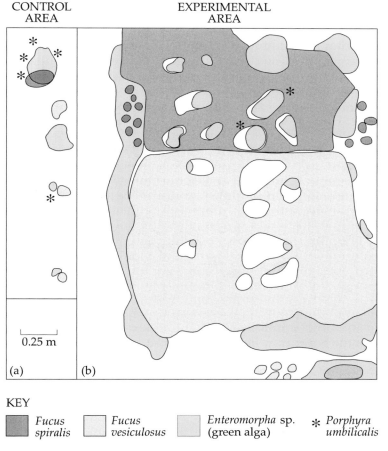

CONTROL AREA

EXPERIMENTAL AREA

0.25 m

(a)   (b)

KEY

■ *Fucus spiralis*   □ *Fucus vesiculosus*   ▨ *Enteromorpha* sp. (green alga)   * *Porphyra umbilicalis*

Figure 3.13 Diagrams showing algal cover in two adjacent strips of a semi-exposed mid-shore (mid-way between MHWN and MTL) at Holy Island, Anglesey, in summer 1975. (a) Control site; (b) experimental site from which all limpets were removed in summer 1974.

❑ Compare Figure 3.13a and b. What conclusions do you draw from this experiment?

■ On the experimental site (limpets removed), there is 80–90% algal cover comprising mainly *Fucus spiralis* in the upper part and *Fucus vesiculosus* in the lower part. There are still a few areas of bare rock and areas dominated by the short-lived green alga, *Enteromorpha*. Since limpets are grazers, the data suggest that the main reason for low algal cover in the control site is grazing by limpets, i.e. a *biotic factor,* and not wave action.

The influence of grazers on algal cover was demonstrated dramatically following the wreck of the oil tanker *Torrey Canyon* off the north Cornish coast in March 1967. In areas where heavy detergent spraying was used to remove oil from rocks, virtually all organisms were killed, but by July rocks which before the oil spill had shown only a sparse covering of fucoid algae were covered by dense growths of green algae. By September, the short-lived green algae were beginning to disappear and were followed by dense growths of young fucoid algae. Wherever the fucoids became well-established *before* recolonization by limpets and other grazers, the shore retained a dense algal cover for some years because the grazers eat only sporelings or very young algae. It is also difficult for limpet larvae to settle where there is dense algal cover and they usually do so only after seaweeds have been torn from rocks by storms, creating gaps. Interestingly, in areas which were moderately fouled by oil but were *not* sprayed with detergent, grazers persisted and a dense algal cover did not develop. Shores recovered quite quickly from moderate oil pollution provided they were not treated with detergent, which is now widely recognized as the main cause of long-term damage after oil spills.

In summary, there is a graded interaction between algae, grazers and exposure (wave action) at sites such as the Dale peninsula and Anglesey:

- On the most exposed shores, only the toughest algae can survive but grazers are also suppressed by wave action, allowing a sparse algal cover to develop in the middle to upper shore alongside abundant barnacles (*Chthamalus* spp.).

- On semi-exposed shores, the middle to upper shore typically has a dense barnacle cover but may be almost bare of algae largely because of grazers. Extra-heavy predation by dog-whelks or disturbances (such as oil spills) which kill grazers may allow algal cover to develop.

- On the most sheltered shores, conditions favour very rapid algal growth and gaps created by disturbance are rare, so that few limpets (and barnacles) become established. Dense algal cover throughout the shore is the norm with limpets and barnacles restricted to scattered bare patches.

Other studies have shown that even more complex biotic interactions can influence the abundance and zonation of algae on rocky shores. For example, Farrell (1991), working on a semi-exposed shore in Oregon, north-western USA, found that a species of barnacle (*Balanus* sp.) greatly facilitated the growth of fucoid algae on cleared areas of rocks; Question 3.3 at the end of this Section is about this study. When predators such as birds or dog-whelks exert significant effects on barnacles and/or limpets, the web of interactions becomes even more complex. So it is certainly not safe to assume that sparse algal cover on a rocky shore reflects solely the effects of abiotic factors such as wave action and desiccation. Nor is interference competition the only type of biotic interaction influencing numbers and distribution of shore organisms.

Summary of Section 3.2

- Rocky shores are a harsh environment occupied by marine organisms. Wave action (determined by the degree of exposure) and time out of water (determined by position up the shore and the timing and amplitude of tides) are the most important abiotic factors influencing the survival and distribution of shore organisms.

- The main tidal zones are the sub-littoral, lower shore, middle shore, upper shore and splash zone. Apart from lichens in the splash zone and upper shore, all other plants on rocky shores are algae, the most obvious being large seaweeds. The most abundant types of larger animals are molluscs (e.g. herbivorous limpets and carnivorous dog-whelks) and crustaceans (e.g. filter-feeding barnacles).

- All shore organisms except those occurring only in rockpools show some degree of zonation. Commonly, the upper zone boundary is determined mainly by tolerance of abiotic conditions and the lower zone boundary by biotic interactions such as competition. Zonation and abundance may be influenced by rates of larval or spore/zygote settlement and intensity of predation and grazing.

Question 3.1   (Objectives 3.1–3.3)

Where on a rocky shore and under what conditions of exposure are you likely to find: (a) common limpet *Patella vulgata*; (b) kelps; (c) *Pelvetia canaliculata*; (d) barnacles?

Question 3.2   (Objectives 3.2 & 3.3)

For each of the following boundaries (a) and (b), identify the main types of interactions that determine its position on rocky shores: (a) lower boundary of *Pelvetia canaliculata*; (b) upper boundary of *Semibalanus balanoides*.

Question 3.3   (Objectives 3.3 & 3.10)

Farrell (1991) studied the middle and upper shore of moderately exposed sites on the coast of Oregon, north-western USA. The sites were dominated by a barnacle B (*Balanus glandula*) and a brown alga A (*Pelvetiopsis limitata*), with small numbers of a second barnacle C (*Chthamalus dalli*) also present. Grazers were abundant, mainly limpets L (*Lottia* spp.) and some periwinkles (*Littorina* spp.). Wave action frequently removed barnacles and algae and Farrell tried to discover what factors influenced the sequence and rate of recolonization. Figure 3.14a (overleaf) shows the typical pattern of recolonization in experimental plots at Site 1 that were scraped bare of animals and algae. Note that settlement rates for both species of barnacle were high in this area.

Figure 3.14 For use with Question 3.3. Experiments on a semi-exposed rocky shore in Oregon. (a) Recolonization of plots at Site 1 scraped bare of all organisms in April 1983; A, brown (perennial) algae; B, barnacle *Balanus*; C, barnacle *Chthamalus*. (b)(i) Control; (ii) the effect on recolonization by A, B and C of excluding limpets in plots at Site 2, scraped bare in April 1984. Analysis of variance was used to compare cover in control and limpet-removal plots: NS, not significant, * and ** $P < 0.05$ and 0.01. (c) Density of algae in plots at Site 2 with and without limpet removal (− and +L) and with and without barnacle removal (− and +B). The control plot is +L+B. Results were analysed by analysis of variance, and a significance test and treatments that did not differ significantly are linked by vertical lines. ** $P < 0.01$.

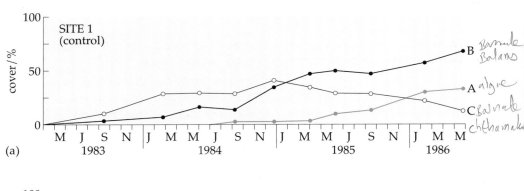

(a)

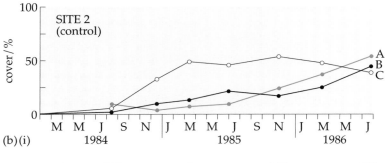

(b)(i)

algae	NS	*	NS	NS
Semibalanus	NS	*	*	**
Chthamalus	NS	**	*	*

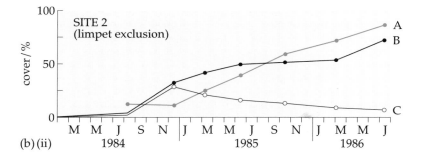

(b)(ii)

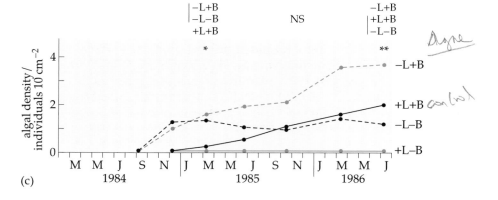

(c)

(a) From Figure 3.14a, describe the pattern of barnacle recolonization and suggest an explanation for the sequence observed.

In two other sets of plots at Site 1, Farrell observed the effect on algal recolonization of removing newly settled barnacles. He found:

   (i)   if plots were kept free of *Chthamalus,* there was no effect on algal density compared with control plots with both barnacles present;

   (ii)   if plots were kept free of *Balanus,* algal density was reduced by 90%.

(b) How do (i) and (ii) help to explain the observed pattern of algal recolonization in Figure 3.14a?

Farrell then carried out experiments to try to distinguish between two mechanisms by which barnacles might facilitate the development of algal cover:

Hypothesis 1   The rough plates of *Balanus* provided a better (e.g. more sheltered) settlement site for algal spores than the bare rock (*substrate-alteration hypothesis*).

Hypothesis 2   Algal sporelings growing in crevices on and between barnacles were protected from herbivores such as limpets (*herbivore-protection hypothesis*).

Study Figure 3.14b (a limpet-exclusion experiment carried out at a second site, Site 2) and Figure 3.14c (an experiment involving limpet and/or barnacle removal during recolonization at Site 2).

(c) Which of hypotheses 1 and 2 is better supported by the data?

(d) Describe overall the effect of grazers (limpets) on the rate and pattern of recolonization of bare rock.

3.3 Classifying abiotic factors

The range of abiotic factors is so wide and the ways in which they act so variable that it is helpful, before considering particular factors in detail, to group and classify them. Think first about the abiotic factors mentioned in Section 3.2.

❏   Make a list of abiotic factors that affect rocky shore organisms.

■   (1) *temperature*; (2) *lack of water* (desiccation when in air); (3) *wave action.* These were the main factors discussed but you might also have mentioned (4) *light supply* (affecting algal photosynthesis and varying with depth of submergence); availability of bare rock or suitable crevices in which to live, i.e. (5) *space*; (6) *water currents* (affecting transport of planktonic larvae or algal propagules to the shore); and (7) *aspect* which influences both (1) and (2).

All but one of these factors can be regarded as affecting the *physical environment* of organisms but (2), water supply, relates to the *chemical environment.* Here is one way of classifying factors and it is shown in the columns of Table 3.4; this Chapter is largely concerned with physical factors and, in particular, with the interlinked set (light, temperature, precipitation, humidity and wind) that are all aspects of *climate.* Chapter 4 is concerned mainly with chemical factors.

Table 3.4   A classification of abiotic factors.

	Physical factors	**Chemical factors**
Resources	light, space,	mineral nutrients, O_2, CO_2, H_2O
Conditions	light, temperature, humidity, precipitation, wind, fire, water movement/wave action, soil (physical state)	salinity, toxins, pH, soil (chemical state)

Table 3.4 also shows that factors can be classified in a second way, as resources or conditions. A **resource** is something which organisms use and which can be in short supply and the focus of competition between individuals. Apart from living space, resources are all absorbed or taken up and the majority are required for body maintenance or growth. **Conditions**, on the other hand, are not used but simply influence the external environment in ways that affect how organisms function. Conditions may affect the outcome of competition but they are not competed *for*. This resource/conditions classification, therefore, looks at abiotic factors from the *organisms'* point of view.

There are some factors which appear in more than one category in Table 3.4, reflecting their complex nature. Light, for example, is a physical resource when used by plants for photosynthesis; but the periodicity of light (daylength), the ratio of red to far-red wavelengths and the proportion of ultraviolet wavelengths present all influence organisms and relate to effects of light as a condition. Water and soil are other factors which can be considered from several points of view. Perhaps the first question you should ask when trying to determine the nature of a factor is: 'is it *used*?' If the answer is 'yes', then this is a chemical resource if molecules or ions are absorbed and a physical resource if not. If the factor is not used, it must be a condition and you have to ask if it affects the physical or chemical environment of the organism. Secondary factors can affect or modify the primary factors listed in Table 3.4; for example, snow cover persisting in sheltered hollows will affect light supply and temperature conditions for herbaceous plants; and you saw in Section 3.2 that *aspect* of a rocky shore (or any sloping surface) influences temperature and humidity. There are also portmanteau terms in Table 3.4 which embrace a wide spectrum of meanings, space being a good example: this can mean simply space on a rock on which to settle and grow (as described for barnacles in Section 3.2) or it can mean special types of space, such as singing posts for songbirds or holes in which to nest or rear young. A whole range of factors that involve other organisms (and so are biotic factors) are not included in Table 3.4. Food is an obvious example.

❏   How would you classify food in Table 3.4?

■   It is a chemical resource that is provided by other organisms, living or dead.

Large organisms such as trees can provide resources (living space and food) and act as conditions (providing shelter and influencing the chemical state of surrounding soil through deposition of leaf litter). Despite these limitations, however, Table 3.4 provides a basic framework that should be helpful as you read the rest of this and the next Chapter.

3.4 Temperature

Temperature is quite unambiguously a condition (Table 3.4); it is the climatic condition with the most widespread and fundamental effects on organisms. An individual can survive and reproduce within only a limited range of temperatures and this range varies between species and, often, between populations within species. Clearly, therefore, temperature is an important axis of the fundamental niche (Chapter 2). In this Section, we consider first the different scales at which temperature varies and may affect distribution (Section 3.4.1). Then we discuss the ways in which temperature affects vital activities (3.4.2), variations between organisms in their temperature tolerance or requirements (3.4.3) and mechanisms by which organisms escape the thermal constraints of a particular geographical area – often by their behaviour or life cycle and especially important in 'extreme' environments such as hot deserts (3.4.4). Finally, we review the evidence that temperature affects distribution and consider briefly the possible effects of global warming (3.4.5).

3.4.1 Temperature variation and scales of distribution

If you travel from the North Pole to the Equator at sea-level, air temperatures will increase; if you climb a high mountain, they will decrease; if you compare summer and winter temperatures in the UK with those of central Russia at the same latitude, they are higher and lower respectively in Russia; and anywhere in the temperate or polar regions, winter temperatures will be lower than in summer.

❑ From these examples, list the four general ways in which temperature varies globally.

■ *Latitude*; *altitude* (temperature falls, on average, 5.5 °C for an increase in altitude of 1000 m); *continentality* (distance from large bodies of water, which heat up and cool down much more slowly than land and hence moderate land temperatures: the further from the Equator and the nearer the centre of a continent, the greater the annual range of temperature); and *season*.

These broad variations in temperature determine global distribution patterns. Together with rainfall (an important proviso), temperature is the main determinant of major vegetation types (**biomes** – Figure 3.15 overleaf). These indicate the kind of plant community that would naturally dominate in a region (**climax community**) – *if* there were no human interference.

❑ From Figure 3.15a: (a) what biome covers most of the UK and (b) where does Mediterranean scrub and woodland occur?

■ (a) Temperate deciduous forest (although, of course, forest clearance has removed most of this vegetation). (b) Around much of the Mediterranean (where the names *maquis* and *garrigue* are applied to taller and shorter scrub, respectively); on the east coast of America in parts of California and Chile; at the southern tip of Africa and in southern Australia. All these areas have a 'Mediterranean' climate with hot, dry summers and warm, wet winters.

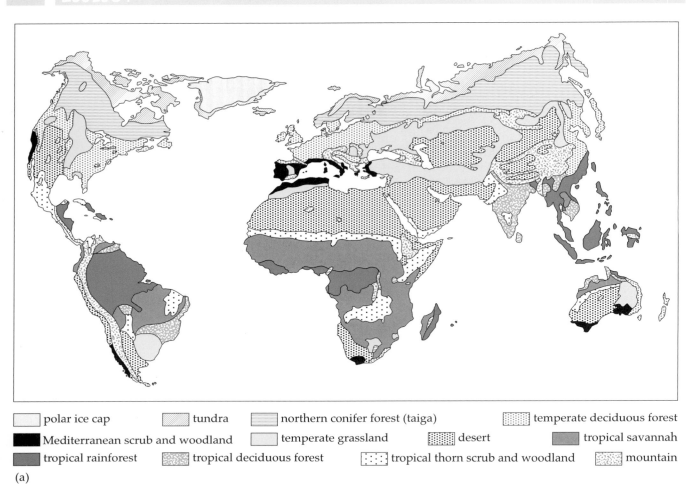

polar ice cap   tundra   northern conifer forest (taiga)   temperate deciduous forest

Mediterranean scrub and woodland   temperate grassland   desert   tropical savannah

tropical rainforest   tropical deciduous forest   tropical thorn scrub and woodland   mountain

(a)

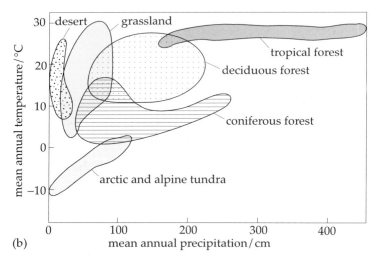

(b)

Figure 3.15 (a) Distribution of the major types of world vegetation (biomes). (b) Six major biomes related to their mean annual temperatures and total precipitation.

Numerous modifying factors affect temperature on a smaller, regional scale. Even quite small hills may have a significant effect, so altitude works at this scale, and the direction in which a hillside faces (*aspect*) influences local temperature via exposure to sun and/or wind. Warm or cold ocean currents are also powerful modifiers: the reason why sub-tropical plants may be grown in some extreme western parts of Britain is because of the warm Gulf Stream that passes close to the coast.

Finally, at an even finer scale, microhabitats only a few centimetres apart may show considerable temperature difference and influence the distribution and behaviour of small organisms. This is particularly marked on land surfaces exposed to direct sunlight because these may heat up to well above air temperature. But soil and rock are poor conductors of heat, so the temperature under a stone or just under the soil surface can be many degrees cooler: noon air temperatures at Luxor (Egypt) in April were 30 °C, surface temperatures 55 °C and temperatures 7.5 cm below the sand surface only 27 °C. Similar considerations apply in cold conditions when temperatures under snow (which is an excellent insulator) are much warmer than above it (see Figure 3.16).

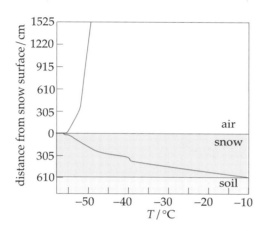

Figure 3.16 A generalized temperature gradient through air above snow and the snow cover to the soil surface.

3.4.2 The ecologically relevant effects of temperature

Temperature affects the rate of nearly all chemical processes – chemical reactions (*except* photochemical reactions) and enzyme activity – and many physical processes such as rates of diffusion. It also affects the properties of membranes and the solubility of gases in water (less oxygen is present in warm than in cold water). These are the main reasons why every aspect of an organism's life is affected by temperature (general effects are discussed further below). However, particular temperatures may also act as triggers or cues for specific activities. Seeds of many temperate species will not germinate unless they have been exposed for a few days or weeks to temperatures below 0–5 °C, for example, and some species will not flower unless seeds or flower buds have experienced a short period of freezing cold. The role of temperature as a cue is discussed at the end of this Section.

General effects of temperature on metabolism

Since metabolism is basically a web of chemical reactions integrated by feedback and various diffusion and other transport processes, temperature inevitably affects metabolism. The rates at which an organism uses energy, grows and develops all tend to increase with temperature. Complications arise when different metabolic processes have different sensitivities to temperature and the best-known example of this is respiration and net photosynthesis (i.e. net gain in carbon, the difference between total carbon fixed and carbon lost through respiration) in green plants (Figure 3.17).

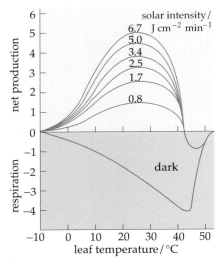

Figure 3.17 The rates of net production (net carbon fixation in photosynthesis) at various light intensities and respiration in darkness as related to leaf temperature for a typical leaf of a temperate climate plant. Both processes are measured in the same arbitrary units which are positive for net production (carbon gain) and negative for respiration (carbon loss).

❑   From Figure 3.17, at which temperatures are the rates of net production and respiration greatest? Is the growth rate (rate of gain in dry weight) for this leaf positive, negative or zero between 45 and 50 °C?

■   Rates are greatest at 25–30 °C for net production and 45 °C for respiration. Between 45 and 50 °C the leaf would have a negative growth rate (i.e. it would be *losing* dry weight) of approximately 1 unit.

Since leaves exposed to strong sunlight may easily heat up to 40–50 °C, this 'imbalance' between net production and respiration at high temperatures can be significant.

High temperatures inactivate enzymes and therefore inhibit metabolic processes, although large differences in **thermostability** exist between different enzymes and the same enzyme in different species or populations. At one extreme, certain bacteria live in hot volcanic springs at temperatures of 85 to over 100 °C! More subtle is the variation in thermostability of enzymes such as malate dehydrogenase (MDH, an important enzyme in the Krebs or TCA cycle) illustrated in Table 3.5.

Table 3.5   Relative thermostability of malate dehydrogenase (MDH) for two species of violet (*Viola*) from northern and southern populations. All plants were collected from the field and grown in the laboratory under identical conditions before extracting the enzyme. Thermostability was tested at 50 °C.

Species	Relative thermostability of MDH:	
	northern	southern
V. blanda	1	5.7
V. fimbriata	4.4	10.3

❑　How does the heat stability of MDH differ within and between the two violet species?

■　For each species, heat stability of MDH is lower for the northern compared with the southern populations; and *V. blanda* has lower heat stability than *V. fimbriata* when comparing populations at the same latitude.

The interspecific differences correlate with the habitats where the two violet species occur. *V. blanda* is found in shady woodlands and *V. fimbriata* in open, disturbed sites where it is often exposed to full sunlight. This illustrates the general principle that, commonly (but not always), organisms 'perform' best at the temperatures encountered in their natural habitat; this applies at the molecular and the whole organism level. Often, however, organisms die at high temperatures which are still well below those at which most enzymes show thermal inactivation and the usual reasons are dehydration (on land) and lack of oxygen (in water). Recall that oxygen solubility decreases with increasing water temperature.

Under cold conditions, metabolism slows down and membrane properties change, either of which can be lethal for organisms adapted to warm conditions. But the factor most likely to kill organisms is the formation of ice crystals inside cells. With few exceptions, living cells cannot survive sub-zero temperatures unless they are highly dehydrated or have mechanisms, such as possession of antifreeze substances, that prevent cells from freezing. Among plants, seeds and spores typically have a low water content and so are well suited to surviving extreme cold.

Temperature as a cue

Many biennial plants, such as henbane (*Hyoscyamus niger*) and autumn-sown crops such as winter wheat will not flower unless the shoot tip is exposed to a temperature of 0–5 °C for at least a week. Seeds of many temperate-region plants will not germinate (e.g. wayfaring tree *Viburnum lantana*) and dormant buds cannot develop ('break', e.g. sycamore *Acer pseudoplatanus*) unless exposed to near-freezing temperatures for periods ranging from a few hours to a few weeks. In all these examples, temperature acts not on metabolic rates in general but on some particular set of reactions which often leads to the synthesis of a growth regulator. The end result is information about the *season* – it's winter.

❑　Why do you think this cold requirement for flowering, germination, etc., has evolved in so many temperate region plants?

■　It reduces the risk of flowering, germinating, etc., in autumn, when there is no chance of setting seed and new shoots would be killed by winter cold.

A declining temperature, and sometimes a threshold temperature, may serve as a signal that winter is approaching and trigger appropriate behaviour in both plants and animals. Examples include dormancy in buds and tubers, hibernation in vertebrates, and dormancy in insect eggs and larvae or pupae. Decreasing daylength is an alternative trigger and sometimes both daylength and temperature signals are required. Temperature can also provide information about *location*. The small seeds of some wetland plants which commonly lie buried in mud germinate best if they experience alternating warm and cooler temperatures – a signal that they are close to the mud surface and not too deeply buried for successful emergence. Finally, there are a number of animals (several types of invertebrates and some fish and reptiles among vertebrates) in which temperature during development influences the *sex* of offspring and so acts as a developmental cue.

Particular temperatures or patterns of temperature change can thus influence development and provide information to organisms about their environment, resulting in appropriately timed behaviour. One of the concerns expressed about global warming is that progressively milder winters may eliminate the winter cold signal for some species so that they no longer flower, germinate or break dormancy.

❏   If this happened, predict the likely effect on the *distribution* of the affected organisms.

■   Distribution might shift or become restricted to higher latitudes or altitudes where appropriate temperature signals still occurred. In the longer term, new genotypes might evolve which did not require winter cold or required less severe cold.

3.4.3 Temperature tolerance and thermal requirements

The extreme temperatures *tolerated* by an organism and the temperatures or amount of heat *required* for persistence in the wild (i.e. for survival and reproduction) set temperature-related boundaries that influence where, in theory, an organism might occur (its *fundamental niche*) and possibly where it actually does occur (its *realized* niche). Before exploring the nature of these thermal limits, it is necessary to distinguish two broad types of organisms: those that generate heat internally and in this way maintain a more-or-less constant body temperature (**endotherms**, i.e. mammals and birds) and those that depend largely on external sources of heat for regulating body temperature or do not regulate it at all (**ectotherms**, i.e. all other animals and all plants). This distinction draws attention to the fact that body temperature (or, more precisely, *core body temperature*, excluding the surface and extremities) is not necessarily the same as the external or ambient temperature. It complicates the relationship between body and external temperature but the basic principles of thermal tolerance and requirements are, nevertheless, similar for ecto- and endotherms.

Temperature tolerance

Figure 3.18 illustrates several thermal responses in pea plants which are typical of thermal characteristics in general.

The upper and lower **lethal temperatures** are those above and below which peas cannot survive because of irreversible damage: they mark the extremes of thermal tolerance and all organisms have such limits. Temperatures that allow peas to grow are shown by the lines for the *upper and lower threshold temperatures,* with the central shaded area indicating optimum temperatures for growth. Clearly from Figure 3.18, the seedling and germination stages have lower temperature tolerance and require higher temperatures than other stages and such *variation during the life cycle* with young stages showing highest sensitivity is widespread among plants and animals. Notice also that only about 7 °C separates the upper lethal

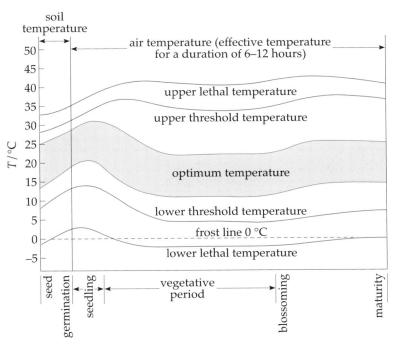

Figure 3.18 The thermal responses of pea plants, *Pisum sativum*, over a period of 64 days. Threshold and optimum temperatures relate to growth processes.

from the upper optimum temperature for seedlings: organisms generally operate closer to their upper temperature limits than to their lower limits.

Even for the same stage of the life cycle, however, temperature limits for members of the same population are not necessarily fixed. Figure 3.19 illustrates one reason for such variation: temperature tolerance may vary depending on *length of time of exposure and on other environmental conditions* (e.g. relative humidity in Figure 3.19).

The woodlouse *Oniscus asellus* has an upper lethal temperature of 36 °C for a one-hour exposure at 50% relative humidity (RH) but only 15 °C for a 24-hour exposure. The explanation for this difference is that woodlice readily lose water by evaporation and die of desiccation during the long exposure period.

❑ Given the above information, why does the upper lethal temperature decrease with relative humidity for a one-hour exposure but increase for a 24-hour exposure?

■ Evaporation of water will reduce body temperature, so for a short exposure when no woodlice die of desiccation, the lower the RH the greater the evaporation and the higher the temperature tolerated. Over 24 hours, the critical factor is desiccation, so the lower the RH the more water is lost and the lower the temperature tolerated.

This example underlines the general principles that organisms may survive brief exposures to unfavourable temperatures but not prolonged exposures and that temperature tolerance may be greatly affected by water supply, humidity and resistance to desiccation.

The second reason why temperature tolerance may vary is because organisms may actively adapt or **acclimate**. This depends on the temperature conditions experienced for several days or weeks prior to exposure to the limiting temperature. In Table 3.6, for example, the lower

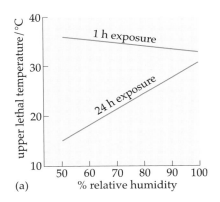

(a)

(b)

Figure 3.19 Relationship between humidity, time of exposure and upper lethal temperature for the woodlouse *Oniscus asellus* (×3.5). The graph shows the highest temperature tolerated during exposure to various relative humidities for 1 h or 24 h.

the **acclimation temperature** for young goldfish, the lower their upper lethal temperature – they had adapted to cooler conditions. In the wild, seasonal changes of temperature and daylength at higher latitudes induce similar acclimation. Woody branches of sycamore *Acer pseudoplatanus* have a lower lethal temperature of –3 to –4 °C in summer but are triggered by the falling temperatures and shortening daylengths of autumn to develop cold resistance, with lower lethal temperatures in the range –30 to –50 °C. This process is often called **cold hardening** – and any gardener knows how important it is to harden off seedlings raised indoors before planting them outside.

Table 3.6   Upper lethal temperature (defined as the temperature at which 50% died) for young goldfish *Carassius auratus* exposed for 14 h to various temperatures after acclimation at three temperatures.

Acclimation temperature/°C	Temperature at which 50% died/°C
1.2	28
10	31
17	33.5

Genetic considerations

The ability to acclimate is under genetic control and species vary considerably in this character. Genetic factors also determine the range of temperatures tolerated for any given acclimation temperature. When an individual shows wide temperature tolerance, it is described as **eurythermal** (from the Greek *eury*, wide) whereas limited tolerance is described as **stenothermal** (Greek *stenos*, narrow). Figure 3.20 shows temperature tolerance diagrams for eurythermal and stenothermal fish; within the limits shown on these diagrams, it is possible to transfer fish from one temperature to another on the same vertical axis without lethal effects and, by acclimation, the lethal limits can be altered upwards or downwards.

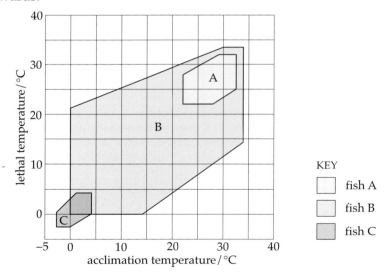

Figure 3.20   Temperature tolerance diagrams for three fishes: a tilapia, *Rutilus* (roach) and *Trematomus*.

❏   In Figure 3.20, which of fish A–C is eurythermal and which is stenothermal? Which of fish A–C comes from British waters (roach), tropical waters (*Tilapia*) or polar waters (*Trematomus*)?

■   Fishes A and C are stenothermal and fish B is eurythermal. This follows from the areas enclosed within the tolerance diagrams, which represent the degrees within the temperature limits. Species A, which cannot survive below 22 °C, is clearly the tropical fish tilapia; species C, which cannot survive above 4 °C, is the polar fish *Trematomus*; and the roach must, therefore, be species B.

These fishes show physiological adaptation to the range of temperatures which are experienced in their normal environment. Sometimes, however, a species may show wide temperature tolerance because it is composed of discrete, local populations each with narrower but different tolerance limits. The northern and southern populations of the two violet species with different thermostability of MDH (Table 3.5) illustrate this point. They can be described as thermal ecotypes, an **ecotype** being an old but still frequently used term for a population with genetically determined and ecologically relevant characteristics. It is now known that most populations of a species differ genetically in at least some ecologically important traits and that these differences are often continuous, rather than abrupt and absolute as the term 'ecotype' tends to imply. Thus, there could well be violet populations with intermediate thermostability of MDH linking the extreme northern and southern populations. The 'heavy metal ecotypes' described in Chapter 4 are a more clear-cut example, where plant populations able to grow on metal-rich soils are sharply and genetically defined.

Temperature requirements: the day-degree concept

So far we have identified temperature cues as being important requirements for the appropriate timing of activities (Section 3.4.2) and temperature thresholds as necessary conditions for growth with an optimum temperature range indicating, rather vaguely, the best growth conditions (Figure 3.18). There is another way of defining temperature requirements for ectotherms, however, which is useful because it takes into account the time dimension: for a given temperature, how much *time* is required in order to grow by a certain amount or attain a particular stage of development? This approach is especially relevant when the growing season, i.e. the time available for growth and development, is limited.

The North American frog *Rana clamitans* requires a minimum average temperature (lower threshold) of 8 °C for eggs and tadpoles to complete development (Figure 3.21a overleaf). At 18.5 °C development is completed in 138 days and at 25 °C (the typical water temperature at the time of breeding) in an average of 85 days (Figure 3.21b). If you multiply together (days required for development) by (number of degrees above threshold), it comes to about the same value: $138 \times (18.5 - 8) = 1449$; $85 \times (25 - 8) = 1445$ and we can say that this frog requires 1445 – 1449 **day-degrees** above threshold for development. It is a combined measure of time and temperature and reflects the varying pace of ectotherm development with temperature. In endotherms, by contrast, the constant body temperature means that only time matters for development: human babies are born after a remarkably standard nine months.

Figure 3.21 The frog *Rana clamitans*. (a) Linear rate of development (egg to metamorphosis) with temperature. The lower threshold temperature is that at which, by extrapolation, development rate is zero. (b) The direct relationship between average time required for development and average water temperature. The grey shaded band indicates the maximum time usually available from egg laying in June to freezing episodes which kill tadpoles (October) at the northern distribution limit.

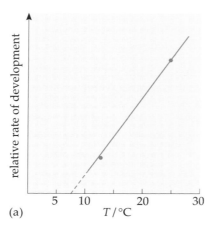

(a)

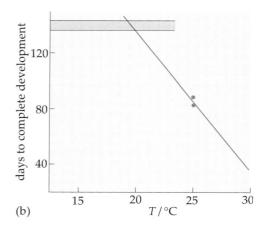

(b)

❑   Use the day-degrees above threshold value of 1445 for *R. clamitans* to calculate the minimum average water temperature necessary to complete development in 140 days (the maximum time usually available). *83 °C*

■ $140 \times$ (temperature above threshold, t) = 1445; so $t = 1445/140 = 10.3$ °C, and the minimum average water temperature is 10.3 + 8 (the threshold) = 18.3 °C.

This is quite a high temperature and probably explains why *R. clamitans* is not found at latitudes above 50 °N.

To summarize the main points in Section 3.4.3:

*   All organisms have thermal limits (lethal and threshold temperatures) which vary during the life cycle, with length of exposure and external conditions, and because of acclimation.

*   Temperature tolerance and the ability to acclimate are under genetic control and may vary widely between and within species; those with wide thermal tolerance are described as eurythermal and those with narrow tolerance as stenothermal.

*   Thermal requirements for particular activities of ectotherms, e.g. development, seed production, germination, can be described in terms of day-degreees which measure time required above a threshold temperature.

3.4.4 Escape from temperature constraints

Unless they live in environments with more-or-less constant temperature – e.g. ocean depths or the floor of a moist tropical rainforest – most organisms experience unfavourable temperatures at some stage in their life cycle. In this Section, we consider the range of mechanisms that allow organisms to survive unfavourable temperatures and the environmental conditions in which such mechanisms can operate. Many plants and small mammals, for example, survive intense winter cold in the tundra only beneath an insulating blanket of snow (Figure 3.16): snow is an essential requirement and these organisms are not found in very exposed areas where snow is blown away.

Endotherms

Maintaining core body temperature in the range 35–42 °C by metabolic activity is an excellent way of not merely surviving cold but of remaining in peak form, able to feed, escape predators and even breed. The 'cost' however, is a high one. Endothermy requires a great deal of energy, especially when external temperatures are low and body heat is lost rapidly, and endotherms survive only where they have a constant supply of food. Whenever food supply is strongly seasonal, as in many polar or cool temperate regions, one or more of the following mechanisms may occur:

(i) Food is stored either as fat deposits in the body or as external caches and the organism **hibernates**, i.e. becomes torpid and reduces metabolic rate and body temperature so that energy is conserved.

(ii) The organism **migrates** to areas of more favourable temperature and food supply and lays down body fat stores if the migration is long.

❑ Think of at least two examples of (i) and (ii).

◼ (i)   Hedgehogs (*Erinaceus europaeus*) and bats (which are insect-eaters in the temperate zone) are examples of hibernators that lay down fat stores before winter. Most species of squirrel hibernate but wake at intervals and eat cached food: North American ground squirrels (*Spermophilus parryi*) spend about 7% of the winter awake and then eat in a few hours as much cached food as they expend during 10 days of hibernation (when fat stored in the body is utilized).

(ii)   Many birds migrate to lower latitudes in winter, notably nearly all those that breed in the Arctic (e.g. many waterfowl) or the strict insect-eaters that breed in cool temperate zones (e.g. swallow *Hirundo rustica*). Large tundra mammals such as caribou or reindeer (*Rangifer tarandus*, Figure 3.22) often migrate hundreds of miles southwards to the forested taiga in winter where the snow is softer and less of an obstacle to grazing.

Figure 3.22 Caribou or reindeer *Rangifer tarandus* (×0.03).

Note that migrations within the tropics are due to the effects on food supply of seasonal rainfall rather than temperature, which is almost constant. Note also that very small mammals such as voles and shrews cannot survive winter cold by hibernation because of their size. Their high surface to volume ratio means that they lose heat more rapidly than larger mammals and would need to maintain quite high metabolic rates even when dormant – which would require impossibly large amounts of stored food. Conversely, the survival of very small tropical hummingbirds actually depends on their ability to become torpid at night: their daytime metabolic rate is so high that it could not be sustained overnight and must be reduced.

All these hibernators and migrators *avoid* unfavourable low temperatures and their life cycles, patterns of movement and large-scale distribution are determined primarily by conditions of temperature and food supply (a biotic resource). Small-scale distribution is quite a different matter and is nearly always determined by a complex set of interacting biotic and abiotic factors – as illustrated earlier for rocky shore animals.

❑ Recall from Section 3.4.1 one way in which temperature can be said to have a direct effect on the local distribution of small mammals in the tundra.

■ Because of very low air temperatures, they are confined during winter months to areas where there is a reasonable covering of snow, which acts as an insulating layer (Figure 3.16). Animals remain active in winter (see above) so there must also be a supply of food in the snow-covered habitat.

(iii) For endotherms that remain active in very cold or hot areas, survival often depends on specialized features of their physiology or anatomy that act to conserve or dissipate heat. In *cold environments*, mammals generally have a compact body shape with short appendages such as tails or ears. Larger mammals also develop extremely good insulation, subcutaneous fat in marine mammals such as seals and whales and two-layered fur – the outer long and shaggy and the undercoat short and dense – in land mammals. The musk ox (Figure 3.23) shows these characteristics clearly. These mechanisms that reduce heat loss mean that the necessity to increase heat production and food intake are reduced as temperatures fall; the arctic fox, for example, does not need to increase body heat production until air temperature falls below −40 °C.

Figure 3.23 Musk ox *Ovibos moschatus* (×0.03) a herbivore from northern Canada and Greenland. Note the shaggy fur and compact body, with short legs and small ears.

In *hot environments*, the ecological effects of temperature and water supply can hardly be separated, particularly if losing heat by the evaporation of water (sweating) is the main cooling mechanism. Bodies with long, slender extremities and large, heat-radiating ears are common, but physiological adaptations are usually related to reducing water loss (discussed further when we consider hot deserts, Section 3.7). An important general point, however, is that the availability of *shelter habitats*, such as underground burrows or shade under trees, is often critical for survival: small mammals may emerge from burrows only at night. Movement into shade or shelter as a means of keeping cool is a type of behavioural thermoregulation (see below) and is of much greater significance to ectotherms.

Ectotherms and plants

Plant shoots and the bodies of ectotherms are by no means always at the same temperature as the ambient air. They can be considerably hotter when in direct sunlight or cooler when in shade but, in the majority of *mobile* animals, systems have evolved that maintain a fairly constant, near-optimum temperature for the longest possible time during the day. In ectothermic land animals, particularly reptiles and larger insects such as dragonflies (Odonata), **behavioural thermoregulation** is by far the most important mechanism for controlling body temperature: it can be defined as *the use of behavioural mechanisms to seek appropriate thermal conditions*. Table 3.7 shows the maximum potential change in body temperature (ΔT_b, where Δ, Greek D or *delta*, means 'difference in') attainable for a 1-kg animal by different behavioural mechanisms, *assuming* that direct sunlight provides variation in air temperature. By contrast, physiological mechanisms such as increasing or decreasing blood flow to the skin, panting or shivering (increasing muscle activity) produce a maximum ΔT_b of only 1–6 °C, with the notable exception of evaporative cooling from a *wet* skin ($\Delta T_b = 20$ °C), which is usually impossible in dry places.

Table 3.7   Estimates of the effects on body temperature T_b of different types of behavioural thermoregulation.

Behavioural mechanisms	Maximum change in T_b for a 1 kg animal/°C	Comments
1 daily range of T_b	55	cools (active at night); warms (active by day)
2 habitat choice	38	cools or warms
3 selection of microhabitat:		
(a) sunlight versus shade	28	warms versus cools
(b) burrowing	45	cools
(c) climbing	35	cools usually
4 postural adjustments:		
(a) orientation to sunlight	18	warms
(b) orientation to wind	1	cools
(c) body-shape changes	5	cools or warms
(d) elevation off substrate	5	cools
(e) conduction to substrate	10	cools

Consider the data of Christian *et al.* (1983) for the Galapagos land iguana *Conolophus pallidus*, a large, herbivorous lizard (4–7 kg) found on an island close to the Equator (Figure 3.24).

❑ From Figure 3.24, describe how average body temperature, T_b, changes over 24 h relative to average ground and air temperatures during the hot season.

■ From about 10 a.m. to 6 p.m., T_b fluctuates around 36–37 °C, which is slightly below average air temperature and far below ground temperature. From 6 p.m. to midnight, T_b falls slowly to about 30 °C, just above air and ground temperature, and then falls more rapidly to a minimum of 25 °C at 6 a.m., which is lower than air and ground temperature and may result from the choice of microhabitat. T_b increases rapidly from 6 to 10 a.m., when lizards bask in warm sites.

❑ *Conolophus* lives in arid habitats with each animal having a territory that comprises a flat scrubby plain area and a steep, north-facing slope. Because of strong south-easterly winds, air and ground temperatures of the exposed plain are up to 5 °C less than those of the sheltered slope (Figure 3.25, overleaf). Given this information, which of the mechanisms in Table 3.7 do you think are used to regulate body temperature?

■ Mechanism 1: *Conolophus* is active during daylight and could not achieve its (presumably) preferred T_b at night; being active at a suitable time of day is the main mechanism by which body temperature is regulated. Mechanism 2, moving between slope and plateau, is probably linked to the rapid warming up in the morning and slow cooling down in the evening. Mechanisms 3a, 4a–b and 4d–e are very likely to be used (given the nature of the terrain) and would be linked to fine temperature adjustments in daytime, especially reduction of T_b in the middle of the day.

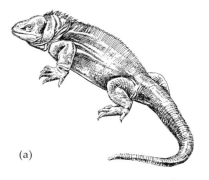

(a)

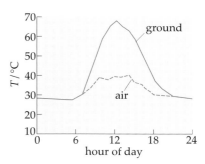

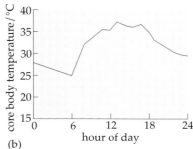

(b)

Figure 3.24   (a) Galapagos land iguana *Conolophus pallidus* (×0.04). (b) Average body temperature of a land iguana and average ground and air temperatures over 24 h during the hot season.

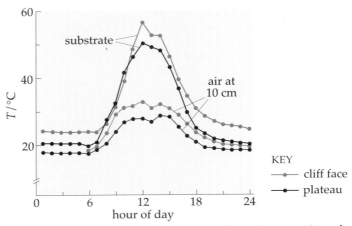

Figure 3.25 Average ground and air temperatures for plateau and north-facing slope habitats during the cool season. The prevailing south-easterly wind results in differences in wind speed, air and ground temperature.

Conolophus does indeed bask in the sunlight in the mornings and move periodically into shade on the cooler plateau in the afternoons; in the cool, cloudy season, it spends proportionately much more time on the warmer cliff habitat. Temperature, therefore, determines directly *Conolophus'* use of space on a daily and seasonal basis. But more importantly, survival for *Conolophus* in this environment requires a range of habitats that permit thermoregulation. For anyone planning a conservation strategy for *Conolophus*, this kind of interaction between temperature and the physical environment would certainly have to be borne in mind.

Many other peculiar features of life cycles and behaviour can be related to the need to warm up, cool down or survive during potentially lethal temperature conditions. Before cold winter seasons, mobile terrestrial ectotherms (e.g. amphibians, reptiles, some insects) either become torpid and retreat to a sheltered site where they are protected from extreme cold and predation by endotherms; or, for insects, eggs, larvae or pupae develop cold resistance (Section 3.4.2) and enter a resting stage known as **diapause**. Some adult flying insects (notably butterflies such as the monarch *Danaus plexippus* in North America) store body fat and migrate to lower latitudes in winter. Herbaceous plants commonly survive winter either as seeds (which have a low water content and very high cold resistance) or underground (as rhizomes, tubers, etc.) where buds are protected. Two examples of a more specific kind are described below.

(i) In the tropical alpine tundra of Mount Kenya in east Africa, air temperatures fall to near freezing every night and rise to around 20 °C on sunny days. A behavioural pattern has evolved in certain insect larvae that allows them to exploit the microclimate within tussock plants to maintain an even body temperature. Figure 3.26 shows the daily temperature variation within a tussock of the grass *Festuca pilgeri*; insulation of the basal region occurs because the fine leaves trap a layer of still air which slows down the rate of heat loss at night and heat gain by day. Larvae of two species of moths (*Gorgopis* sp. and *Metarctia* sp.) build silk tubes from leaf tip to tussock base and move up and down by means of spines on their undersides (Coe, 1967). In this way, they can maintain a reasonably even temperature, moving towards the leaf bases on sunny days and at night and to leaf tips in the early morning, late afternoon or in cloudy conditions.

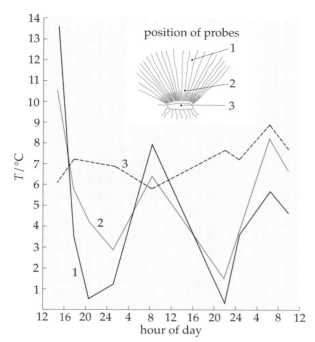

Figure 3.26 Daily temperature variation in a tussock of *Festuca pilgeri* in the alpine zone of Mount Kenya, March 1966.

(ii) For many plants of the arctic or alpine tundra, the short growing season provides barely enough thermal time (day-degrees, Section 3.4.3) to flower and ripen seeds. In one such plant, mountain avens *Dryas octopetala*, Swedish botanists showed that Sun-tracking by the white, dish-shaped flowers (Figure 3.27) resulted in a 0.7 °C increase in average temperature relative to artificially restrained flowers. This increased warmth resulted in significantly heavier seeds which had a greater chance of producing viable seedlings.

❑ For (i) and (ii) above, identify the *habitat requirements* that are necessary in order to obtain suitable temperature conditions.

◼ For (i), presence of tussocks of *Festuca pilgeri* in an open position is required. For (ii), an open, unshaded habitat is necessary for Sun-tracking.

Figure 3.27 Flowers of mountain avens *Dryas octopetala* (×0.6) that maintain a higher temperature by tracking the Sun.

Before leaving the subject of escape from temperature constraints, an interesting 'reverse situation' should be mentioned: actively raising body temperatures to near-lethal values as a mechanism to fight infection. **Fever** in endotherms is well-known as a defence mechanism. Pathogens and internal parasites commonly have a very narrow range of temperature tolerance and although raising body temperature is a risky and (at least for humans) uncomfortable procedure, it can be highly effective in killing pathogens. The same applies to ectotherms. Fish, frogs, lizards, lobsters and cockroaches have all been found to raise body temperatures when infected. Boorstein and Ewald (1987) showed that when grasshoppers were infected with a parasite, they basked for longer in the sunlight and raised body temperatures from their usual 34 °C to a near-lethal 40 °C. Although this grasshopper fever reduced subsequent growth, overall survival was greatly increased compared with insects that were prevented from raising body temperature. You can draw two morals from this example: first, that a habitat that allows ectotherms to attain 'unfavourable' temperatures may sometimes be important for survival; and secondly that the ability of ectotherms to thermoregulate can be a good defence against stenothermal parasites.

3.4.5 Large- and small-scale distribution in relation to temperature

Temperature clearly has a powerful influence on organisms, supporting the earlier point that it probably affects the *potential* distribution of most species. Whether it affects *actual* distribution is more difficult to prove. The relationship is most clear for geographical distribution and least clear for small-scale distribution, as illustrated by the two examples below.

Barnacles

Recall from Section 3.2.3 that the barnacle *Semibalanus balanoides* is at the southern limit of its distribution in Britain whereas *Chthamalus montagui* and *C. stellatus* are at their northern limit.

❑   What factor was the major determinant of these geographical limits for *Semibalanus*?

■   It was resistance to desiccation which, for these sessile animals, relates directly to summer temperature. No explanation was given for *Chthamalus*, but in fact it is low temperature acting mainly on recently settled larvae in autumn and winter.

Information in Section 3.2.3 makes it clear, however, that the vertical distribution of these barnacle species on a given rocky shore is strongly influenced by interacting biotic factors, notably interspecific competition and predation by dog-whelks. Temperature, through its effects on desiccation and larval survival in winter, affects the upper limits on the shore, with shore aspect and exposure exerting a strong modifying effect. But one cannot say that temperature alone determines the local (small-scale) distribution of these barnacles.

British woodlice

Compared with insects, **woodlice** (Crustacea, Isopoda) are regarded as being rather poorly adapted to terrestrial conditions. They occur in humid microhabitats, such as plant litter, under stones or logs and, in one case, inside ants' nests. The availability of these microhabitats, soil conditions and moisture are certainly important factors influencing small-scale distribution. As shown in Figure 3.19 earlier, temperature tolerance depends strongly on humidity. Nevertheless, temperature is thought to be a major factor influencing large-scale distribution. Figure 3.28a and b shows the distribution of two British species and Figure 3.28c and d the average winter and summer temperatures.

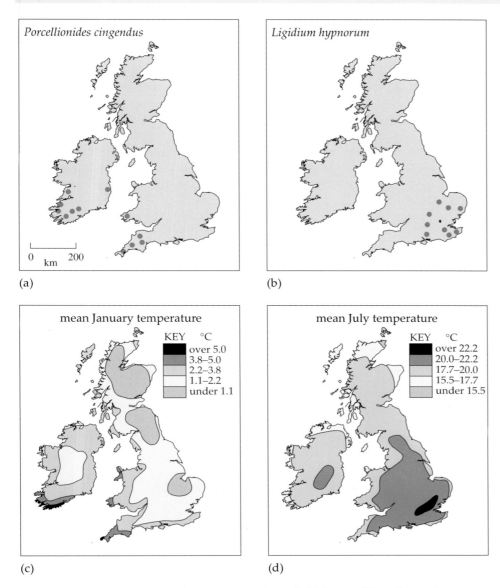

Figure 3.28 (a) and (b) Distribution maps of two British woodlice. Green dots represent vice-counties in which the species have been recorded since 1953. (c) and (d) Average temperatures in January and July respectively for the British Isles.

❑   Suggest explanations for the different distribution patterns in Figure 3.28a and b assuming that temperature is involved.

■   For *Porcellionides* (representative of species confined to the south-west), mild, frost-free winters is the most striking temperature feature of their region (Figure 3.28c): intolerance of severe winter cold is likely to limit distribution. For *Ligidium* (representative of species confined to the south-east), either tolerance of, or a requirement for, high summer temperatures (you cannot distinguish which from these data) is more likely to explain distribution.

For both species, reduced ability to compete or greater susceptibility to disease (i.e. biotic factors) are more likely to explain distribution than direct, lethal effects of temperature outside their ranges. We should also take into account the *history* of these species in the British Isles and their *ability to disperse*. These topics are considered fully in Book 2 but, for example, the two woodlouse species could have spread into Britain from warmer continental areas after the last glaciation; their restriction to the south may then be explained partly by a poor ability to disperse northwards. In the next Section, we examine a final and very striking example of the effects of temperature on distribution.

3.4.6 The treeline

The **treeline** or **timberline** has been described as 'the sharpest temperature-dependent boundary in nature' (Wardle, 1974). It is the boundary between forest and arctic or alpine tundra (Figure 3.29), the latter vegetation comprising herbs, mosses, lichens and low-growing shrubs. Because it is often such a clear boundary, the treeline provides a useful way of reviewing the ways in which temperature affects distribution. We consider first historical evidence which supports the view that temperature determines the position of treelines.

Figure 3.29 Photograph showing a New Zealand alpine treeline.

Historical and palaeoecological evidence

Figure 3.30 illustrates shifts of the treeline in northern Quebec, Canada, where black spruce *Picea mariana* is dominant. Payette *et al.* (1989) studied an area with low rolling hills (the upland sites in Figure 3.30a). By studying tree-ring widths (Figure 3.30b) and the remains of dead trees (which take hundreds of years to decay in this cold area), they were able to reconstruct vegetation history (and therefore where the treeline lay) over a period when average temperatures were known. Black spruce shows great variation in growth form (i.e. **phenotypic plasticity**) in different climatic conditions: it may grow as a single-stemmed tree or, if winter cold is intense and there is little protective snow cover, stem tips die back or branch and dwarf trees develop. This stunted vegetation is called **krummholz** (German for crooked wood) and occurs just beyond the treeline.

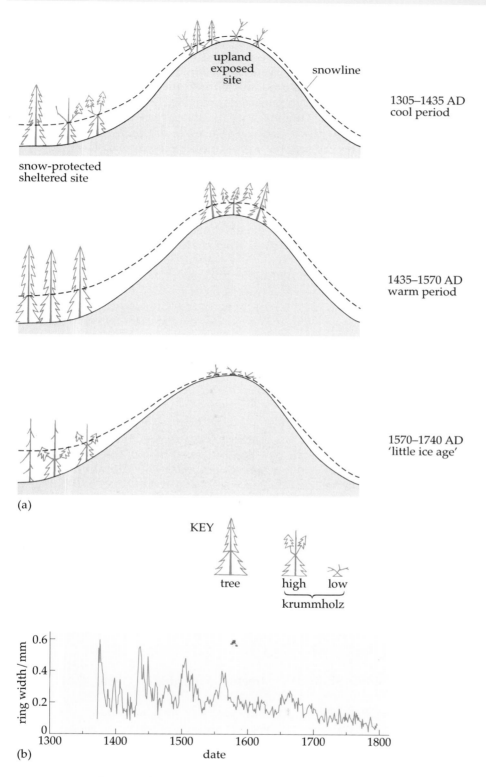

Figure 3.30 (a) The growth form of black spruce typical of three periods at protected and exposed upland sites in northern Quebec. Protected sites were thinly scattered and exposed upland sites are most representative of the general vegetation. (b) Variation in tree-ring widths with time for trees at an exposed upland site.

❑   Use this information to decide where the treeline lay in relation to the study site at each of the three periods shown in Figure 3.30a.

■   For the cool period ((a),top), the study site lay just to the north of the treeline: the vegetation (typified by upland sites) mainly comprised krummholz. For the warm period ((a),middle), the study site lay on or just to the south of the treeline; a good tree cover was present on upland sites. For the little ice age ((a), lower), the study site lay north of the treeline, with very stunted krummholz present on upland sites.

Clearly, the treeline moved north during the late Middle Ages and then retreated south during the little ice age driven by changes in temperature.

Studies of geologically earlier periods (**palaeoecology**) showed that climate warming occurred between 5000 and 4000 years ago and was accompanied by a northward advance of the black spruce treeline, achieved in the remarkably short time of 150 years (MacDonald *et al.*, 1993). In fact, black spruce migrated north at a remarkable 200 km per century after the last ice age, 10 000 years ago, and part of the explanation for this speedy advance is thought to lie with its plastic growth form and life history. Spruce krummholz cannot form seeds and regenerates vegetatively, but once winter temperatures rise above a threshold, so that apical buds are not killed every year, the tree form develops and this *can* produce seeds. Spruce seeds are well dispersed by wind or animals so they provide a rapid invasion potential.

How does temperature act at the treeline?

There is no single answer to this question, because temperature acts in different ways on different tree species, but two kinds of hypothesis can be distinguished. (1) *Winter cold* hypotheses emphasize the impact of winter conditions, either low temperatures *per se* or *winter desiccation* – the drying out of buds and young shoots under the combined effects of low temperature and strong winds. Water lost in winter cannot be replaced because the soil is frozen. (2) *Summer warmth* hypotheses emphasize the lack of warmth (i.e. too few day-degrees above relevant thresholds) during the short, cool growing season.

❑   Which type of hypothesis appears to be most relevant for black spruce at the arctic treeline?

■   Type 1: winter cold or desiccation when apical growing points are exposed above snow in exposed sites.

This explanation is quite convincing but we shall see later that another interpretation of the data is possible.

The strongest evidence against winter cold hypotheses for alpine timberlines is the *continentality effect*. In continental interiors, there is a greater annual temperature range – colder winters and warmer summers – compared with oceanic regions on the edges of continents or on islands.

❑   From the data in Table 3.8: (a) are alpine timberlines at a higher or lower altitude for continental sites compared with oceanic sites at similar latitudes? (b) Why does this support summer warmth (hypothesis 2) rather than winter cold (hypothesis 1) as the primary factor influencing treeline position?

■   (a) Alpine timberlines are at a higher altitude in continental areas, especially in the temperate zone; this is the continentality effect. (b) It cannot be explained by winter cold because *greater* cold is experienced at continental sites (see the temperature data in Table 3.8). Rather, it appears to be higher summer temperatures that allow treelines to rise, i.e. *summer warmth* and not winter cold is the crucial factor.

Table 3.8   Altitudes of treelines for areas at different latitudes with continental or oceanic climates. Temperature data for sites just below the treeline are included for two sites.

Climate	Locality	Latitude	Treeline altitude/m	Average temperature/°C: warmest month	coldest month	annual
Tropics/subtropics:						
± oceanic	New Guinea	6° S	3900–4100			
continental	N. Chilean Andes	19° S	up to 4900			
oceanic	Hawaii, Mauna Kea	22° N	3000			
Temperate:						
continental	US, Colorado Rocky Mts	39° N	3350–3600	13.8	−8	2.3
oceanic	New Zealand, Mt. Ruapehu	39° S	1490	11.7	2.2	7.0
continental	Canada, Alberta, Rocky Mts	50° N	2150–2300			
oceanic	UK, Scotland, Cairngorms	57° N	6–700 (potential) 500 (actual)			

The implication of summer warmth hypotheses is that a certain amount of heat (or day-degrees) is required to complete a vital process – but which process(es)? One possibility is *net production*: growth is possible only if trees fix more carbon during the growing season than is used in respiration by woody, non-photosynthetic tissues over the whole year. As trees get bigger, their respiratory demands increase and, eventually, photosynthesis during the brief, cool summer may not meet the demand. On the whole, the little evidence available does not support this *carbon balance hypothesis*. Some treeline species seem to have plenty of fixed carbon to spare for growth. An alternative hypothesis is that *'ripening' and the development of cold hardiness* (Section 3.3.3) is the critical process. This involves the deposition of lignin in cell walls so that they are stiffened and do not crumple when desiccated and frozen. Interestingly, two exceptionally hardy species, the deciduous conifer larch *Larix gmellini* at the Siberian treeline and *L. lyallii* at the alpine treeline in the northern Rocky Mountains, are distinguished by having woody buds.

❏   Use this ripening hypothesis to suggest an alternative explanation for the limitation of black spruce at the treeline.

■   Inadequate summer warmth prevented the ripening of spruce buds and new shoots so that they succumbed to winter cold and/or desiccation above the snowline.

This explanation combines elements of hypotheses 1 and 2: winter cold or desiccation may cause death but summer warmth determines the level of cold resistance. The same explanation can even be applied to alpine timberlines in the tropics, where there is no winter season but freezing can occur at night, erratically and at any time of year. Treelines would be set wherever the average interval between freezing episodes was too short to allow ripening to occur, although this hypothesis has not yet been tested.

3.4.7 Temperature change: the impact of global warming

Global warming is a rise in global mean temperature that, it is predicted, will occur as a result of the increasing concentrations of atmospheric 'greenhouse' gases, notably CO_2, because of human activities such as burning fossil fuels. The wider effects that this may have on agriculture, human societies and ecosystem processes are discussed in Book 5. Here, we consider simply the impact of a rise in mean temperature on the distribution, survival and reproduction of individual organisms. It is by no means certain how climate would change in specific geographical areas as a result of global warming but the consensus view is that general warming, milder winters and a longer summer season would be most marked for polar and temperate regions and progressively less evident going towards the Equator.

> Before reading on, think back over this Section and make your own list of changes in distribution that might occur as a result of these sorts of climatic changes.

The most obvious effect is an upward shift in latitude or altitude for species where temperature limits distribution. This happened with treelines and there is some evidence that it has happened with the lizard orchid *Himantoglossum hircinum* (Figure 3.31), a chalk grassland species that reaches its northern limit in England. A classic study by Ronald Good in the 1930s indicated that its spread from the extreme south-east from 1900 to the 1930s correlated with warmer, wetter winters; the spread has continued albeit at a slower rate since the 1930s. However, the situation is less predictable when competition between species with similar niches may occur following range extension. Potential distribution may alter but actual distribution may not, because of biotic interactions or because of other factors such as poor dispersal ability, barriers to dispersal (ranging from motorways to mountain ranges), lack of breeding or germination sites or other essential conditions and resources.

Figure 3.31 The lizard orchid *Himantoglossum hircinum* (×0.5).

One of the best approaches to this problem is through palaeoecological studies of past shifts in distribution linked to climate change, although global disturbance by humans (e.g. deforestation) means that the situation today is not strictly comparable. Another approach is through detailed studies of day-degree requirements (Section 3.4.3), which has been done for some crop species: it is reasonably certain that less hardy crops (e.g. French beans *Phaseolus vulgaris* and maize *Zea mays*) could be grown without protection at higher latitudes following a global rise of 2–4 °C.

Another way in which global warming could affect distribution is through interference with *temperature cues* (Section 3.4.2). Species that require a certain period of freezing temperatures before flowering or bud burst might fail to do so if winters became milder. Since species' response to temperature cues is genetically determined, they might adapt to changed conditions, i.e. evolve a different response, but this depends on the rate of global warming: would there be enough time to adapt in this way?

Behavioural patterns such as migration that are linked to temperature and food availability might also be expected to change. There is evidence from bird studies that this can happen quite quickly. For example, the blackcap *Sylvia atricapilla* was a summer migrant bird to Britain but, since 1960, increasing numbers have overwintered here (Berthold *et al.*, 1990). Ringing indicates that these are not British blackcaps staying behind but birds breeding in Continental Europe that reach Britain in autumn via a new migration route. This new route, which was shown to be genetically determined, has evolved in the past 30 years so there must have been strong selection pressure that favoured birds wintering some 1000–1500 km north of their traditional areas in the western Mediterranean. It would clearly be advantageous *not* to undergo a dangerous and energetically expensive migration; milder winters in Britain coupled with increased food supply at bird tables are thought to have allowed this change in blackcaps.

Summary of Section 3.4

- Temperature varies globally with latitude, altitude, continentality and season. Together with rainfall, it determines the nature of biomes. On a regional scale, factors such as aspect and warm or cold ocean currents modify temperatures and microscale variation occurs below the soil surface or under snow, stones, etc. Two general effects of temperature can be distinguished: it affects the rate of nearly all metabolic processes; and it can act as a cue or signal, indicating season or place, triggering events such as germination or flowering, and influencing development (e.g. sex determination). Lethal effects of high and low temperatures are usually caused, respectively, by dehydration (on land) or lack of oxygen (in water) and by the formation of ice crystals in cells.

- Thermal tolerance or requirements can be described in terms of lethal and threshold temperatures and day-degrees above threshold for specific activities. Such characteristics vary during the life cycle, with external conditions and because of acclimation; they are under genetic control, species or populations with wide or narrow thermal tolerance being described as eurythermal or stenothermal, respectively.

- In birds and mammals, endothermy is the primary means of escape from unfavourable temperatures. It requires a large supply of food (energy), especially in cold conditions. Endotherms and ectotherms may avoid unfavourable temperatures by hibernation or dormancy and by migration. For both endotherms and ectotherms, the availability of shelter habitats is often essential for survival in very hot or cold conditions. In addition, specialized features of anatomy, physiology, life cycle and behaviour are commonly involved. Behavioural thermoregulation is the most important mechanism for ectothermic animals.

- The distribution of British woodlice illustrates how temperature interacts with water supply or humidity and the availability of suitable shelter habitats to influence small-scale distribution. The influence of temperature on large-scale (geographical) distribution is illustrated by British barnacles and by arctic or alpine treelines. The current best guess is that treeline position depends mainly on summer warmth (or the interval between freezing episodes), which determines ability to withstand subsequent cold periods. Global warming could influence potential and actual distributions by general effects (e.g. treeline, upward shifts in latitude or altitude) or by altering temperature cues or behavioural patterns (e.g. migration).

Question 3.4   *(Objective 3.1)*

Identify statements (a)–(e) as true or false and, if false, explain why.

(a) A species in which all individuals were stenothermal might still occur over a wide range of temperatures if it contained numerous thermal ecotypes.

(b) Upper lethal temperatures are determined mainly by the thermal stability of enzymes.

(c) Stenothermal organisms can persist only in regions where there is little seasonal variation in temperature.

(d) Endothermy is a special kind of behavioural thermoregulation.

(e) Ability to acclimate may increase the range of temperature over which an organism can survive and reproduce.

Question 3.5   *(Objective 3.5)*

A perennial herbaceous plant *Geum turbinatum* grows in open conditions and has a wide latitudinal distribution (over about 20°) in eastern USA, where the terrain is vary varied and mountains run in a N–S direction. However, within this geographical area, local distribution is strictly limited. Using the information below, predict the most likely habitat for this species, giving reasons for your choice.

- The plant reproduces mainly by seeds and these require five weeks below 0 °C to break dormancy.
- Seedlings have a wide tolerance of soil and moisture conditions.
- Mature plants grow best at temperatures below 15 °C and do not flower at higher temperatures.

Question 3.6   *(Objectives 3.4, 3.5 & 3.10)*

This question is about snow buttercups *Ranunculus adoneus*, which occur only in the alpine tundra of the Rocky Mountains. These plants grow in sheltered areas where snow melts very late in the season (snowbeds); the length of the growing season is determined by the time of snowmelt, which depends on temperature. Galen and Stanton (1993) carried out experiments at two sites in a snowbed, one near the edge and one near the centre (Figure 3.32).

(a) From information in Figure 3.32, identify ways in which conditions and / or resources differ for snow buttercups at these two sites.

Galen and Stanton manipulated growing season length in plots B and D at the two sites (Figure 3.32a) and the effects on plant growth, as measured by the percentage area of ground covered, and on seed weight are shown in Figure 3.32b and c respectively.

(b) From Figure 3.32a–c, what factor(s) appear to influence most strongly (i) the growth (% cover) and (ii) seed weight of snow buttercups? Explain your reasoning.

(c) Other studies showed that larger seeds have a better chance of producing new plants but are less likely to be dispersed away from the parent. Use this and all the other information provided to assess the overall effect of temperature on the distribution of snow buttercups and comment on the likely reasons for their restriction to snowbeds.

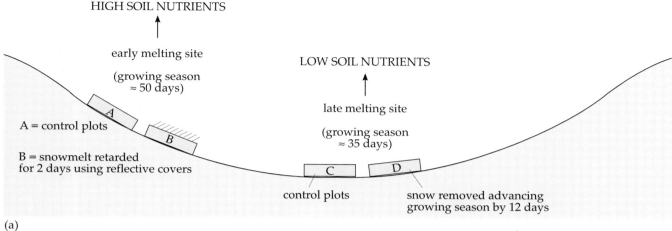

(a)

Figure 3.32 (a) Section through a snowbed showing the two sites studied by Galen and Stanton and the experimental manipulations carried out at these sites. (b) Percentage cover of snow buttercups on plots A–D in (a). (c) Average seed weight for plants on plots A–D. Values are significantly different ($P < 0.05$) where letters *above* histograms differ.

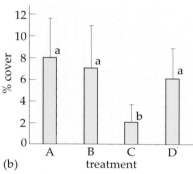

(b)

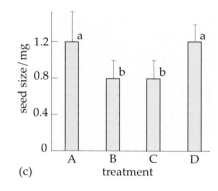

(c)

3.5 Wind

Wind, defined in the broad sense as movement of air masses, is a conditional factor that has significant and somewhat different effects on individuals, communities and ecosystems. Three kinds of effects can be distinguished:

1   Wind acts as an agent of *dispersal* and *transport*.
2   It can be a major *disturbance factor*.
3   It can influence *temperature and water relationships* and, in plants, has a *pruning and dwarfing* effect.

3.5.1 Dispersal and transport

Wind may disperse pollen, seeds, spores and small animals such as insects and spiders (dispersal mechanisms are discussed in Book 2). Such dispersal can occur over great distances so that, inevitably, geographical distribution may be affected. This is especially clear for isolated volcanic islands such as the Hawaiian archipelago, lying in mid-Pacific some 4000 km from the nearest continent (America). Most native plants and insects reached the islands initially by air and the majority (80% of plants and 95% of insects) came from south-east Asia. They were carried by the high altitude, fast-moving jet stream, which slows down over Hawaii with the result that transported material tends to settle out. Wind dispersal may also have a considerable effect on local distribution; for example, weak-flying insects such as aphids spread more rapidly downwind, and the 'rain' of wind-dispersed seeds is much heavier, with a greater chance of seedling occurrence, in the downwind direction. Thus, wind affects the distribution of individuals and species composition and distribution in communities.

At ecosystem level, wind-blown organisms and organic debris provide the major or sole source of food for animals that exist in the most extreme conditions on Earth: the tops of high mountains and the interior of Antarctica and Greenland. No vascular plants or mosses grow in these barren **aeolian zones** which can be described as wind-dependent ecosystems and will be considered further in Book 4. Wind likewise plays a transport role in less extreme ecosystems, carrying materials such as dust, salt spray and airborne mineral nutrients and pollutants. It was wind transport that led to the spread of radioactive material from the nuclear reactor incident at Chernobyl in 1988; deposition was favoured by high winds, large particle size, abrupt deflection of air flow and, above all, heavy rain, which explains the patchiness of contamination. On the plus side, certain ecosystems receive nearly all their inputs of mineral nutrients through wind transport plus deposition in rain, the peat-accumulating areas (*blanket bog*) over large parts of upland Britain being one example. The dominant organisms here are *Sphagnum* mosses and most mosses and lichens (Chapter 1) growing in exposed places depend similarly on aerial nutrients. Another example is the 'green shadow' downwind of penguin rookeries – areas of lush grass on otherwise sparsely vegetated South Atlantic islands where ammonia from penguin droppings provides a nitrogen input. Finally, there are **loess ecosystems** where the soil derives entirely from wind-blown material (loess). Vast areas of central Europe, North America and Asia fall into this category as a result of intense global dust-blowing during past glacial periods. Wind can thus be an important agent determining the nature of soils.

3.5.2 Disturbance

The reverse of loess deposition is *wind erosion,* which is a major disturbance factor where soils are bare of vegetation. Mobile dunes are a feature of coastal and desert sand dune systems and tolerance of repeated burial by sand is essential for plants in these areas. Growth of marram grass *Ammophila arenaria*, the dominant species of young coastal dunes in Britain, is actually stimulated by repeated burial and marram is the chief agent by which dunes are eventually stabilized.

Very strong winds can, of course, snap off or uproot trees and when this occurs on a large scale it is described as **catastrophic windthrow**. Although rare, catastrophic windthrow exerts a strong influence on the species composition and appearance of forests in humid areas (fire – Section 3.9 – has a similar effect in dry areas) – the reasons for these effects on forests will be discussed fully in Book 3, Chapter 3. For example, some species regenerate only if the shading canopy is removed over quite a large area: they are present in a forest only if substantial disturbance occurs above a certain frequency. Canham and Loucks (1984) calculated from historical records and the ages of trees along transects the frequency of catastrophic windthrow in hemlock–hardwood forests of northern mid-west USA (Minnesota and Wisconsin): it occurs somewhere in the forest zone about every 15 years and every 1 ha patch of forest is likely to be blown down at least once during a period of 1210 years. Especially severe winds such as the hurricane that hit the north-east USA in 1938 leave an imprint on the landscape in the form of large, even-aged stands of trees which regenerated where the entire forest was flattened. However, less severe winds cause smaller-scale patchiness because only the most exposed sites are affected and individual trees vary in their susceptibility to windthrow. Variation depends on both tree height and type, conifers being generally more susceptible than hardwoods. Interestingly, old, hollow trees where the heartwood has been destroyed by fungi, are usually *less* likely to blow down than are trees with a solid trunk because hollow cylinders are better able to resist forces at right angles than solid structures containing the same cross-sectional area of solid material; this is why in Windsor Great Park, England, a higher proportion of old trees than young trees escaped destruction in the great storm of October 1987.

3.5.3 Effects on temperature and water relationships and dwarfing effects

The desiccating and cooling effects of wind were mentioned in Section 3.4 (recall the dire effects of winter desiccation on black spruce at treeline, Section 3.4.6). In general, the stronger the wind, the greater the rate of water and heat loss from an organism. This is the explanation of *wind chill* and why shelter from strong wind is important for animal survival in cold conditions. It also explains why animals seek exposed, windy sites in hot conditions.

❑   Recall from Section 3.4.4 an example of an animal that does this.

■   The Galapagos land iguana, *Conolophus.*

Alpine and arctic birds and mammals may behave in a similar way in summer although mammals have an additional reason for doing so: it reduces the attacks of biting insects which are a major nuisance in the tundra. The extent to which wind speed influences water and heat loss by convection depends strongly on the thickness of the insulating **boundary layer** of relatively still air which every solid surface has. Increasing wind speed reduces boundary layer thickness and the smoother the surface of an organism, the greater its sensitivity to wind speed. If the surface is rough or covered with hairs, boundary layer thickness is increased and water and heat losses are reduced for a given wind speed.

In plants, the desiccating effect of wind is particularly important, although you need to bear in mind that the relative humidity of air and differences in temperature between organism and air also affect rates of water loss. Water loss through transpiration can be reduced by closing stomatal pores but since this prevents CO_2 entry for photosynthesis, it is not a viable option in permanently windy sites. Only plants which can restrict wind-induced increases in transpiration *without* closing stomata (e.g. by rolling up leaves as happens in marram grass or having stomata in pits on the leaf surface) are able to grow in very exposed sites. Plants unable to restrict transpiration in such ways are characteristic of sheltered habitats such as below the tree canopy or in ravines; many species of *Rhododendron* are in this category. The boundary layer is also influenced by leaf shape and orientation: it is thinnest at the leaf edge pointing into wind and, in general, small needle-like leaves have a thinner layer compared with large, flat leaves.

Where wind carries a heavy load of abrasive particles, these tend to damage leaf (and also insect) cuticles and reduce their waterproofing qualities. Roadside verges where passing vehicles produce violent gusts of dust-laden wind and upland areas or tundra where winds carry snow crystals are especially prone to this kind of damage.

Figure 3.33 One-sided growth caused by strong and persistent winds. Buds fail to develop on the windward side so all the branches point downwind.

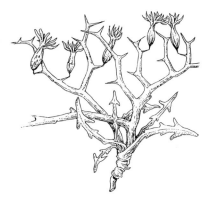

Figure 3.34 *Launea cervicornis*, a spiny cushion plant characteristic of the dwarf community on very exposed coastal sites in Minorca.

It was thought that these desiccating effects of wind coupled with damage from transported material or actual shoot breakage could explain the stunting and deformation so common among plants in exposed sites along coasts and in the uplands and tundra (e.g. krummholz, Section 3.4.6, and see Figure 3.33). Recent work, however, suggests that even when there is ample water supply and no damage, winds above a certain speed have a dwarfing effect (Retuerto and Woodward, 1992). This may be a plastic response to wind, i.e. plants are dwarfed in exposed sites and 'normal' in sheltered sites, as described for black spruce in Section 3.4.6. But in places that are almost permanently windy, species or ecotypes may be genetically dwarf. The result is the same, however: distinctive and often species-rich communities of dwarf plants such as the *dwarf Atlantic heath* of the Scilly Isles, *hedgehog heath* of Mediterranean mountains and the similar community found on the most exposed coastal sites of Minorca and Majorca. These last two communities are so named because the dominant species grow as spiny hummocks resembling hedgehogs (Figure 3.34) although spininess is more likely to have evolved as a defence against grazers than in response to wind.

There are good reasons why the dwarf habit is advantageous in these exposed places. The ground itself has a boundary layer a few cm thick so that wind speed is much reduced close to the ground; thus if wind speed at 2 m above ground is 25 m s^{-1}, then at the canopy level of dwarf shrubs 3 cm tall, it is only 2.5 m s^{-1}.

Summary of Section 3.5

- Wind is a conditional factor which affects dispersal and transport, acts as a disturbance agent, and influences temperature and water relationships. It is a major dispersal agent for many species, can affect distribution patterns in communities and transports material between ecosystems. Aeolian zones depend entirely on energy inputs transported by wind and the soil of loess ecosystems derives wholly from wind-blown dust.

- Wind causes two main types of disturbance: soil erosion and windthrow of trees. Windthrow may cause small-scale patchiness in forest communities or, with hurricane force winds, whole forests may be felled. Species composition is strongly influenced by the frequency of catastrophic windthrow.

- The cooling and desiccating effects of wind restrict some organisms to sheltered microhabitats. In permanently windy places, plants are stunted and deformed and genetically dwarf ecotypes may evolve.

Question 3.7   (Objectives 3.1 & 3.7)

Each of (a)–(c) consists of two statements, the first of which you can assume to be true. Decide whether the second statement is true or false and, if true, whether it adequately explains the first statement.

(a) 1   In some situations, crop yields are increased by an average of 10–30% immediately downwind of a shelter belt.
2   By reducing wind speed, the shelter belt reduces boundary layer thickness and hence reduces water loss (evapotranspiration) from the crop.

(b) 1   Wood ants cannot breed successfully in open habitats; they build high, loose nest mounds and their larvae are very susceptible to desiccation.
   2   Higher wind speeds in open areas would cause high mounds to blow away and increase rates of evaporation and hence the risk of larval desiccation.

(c) 1   The dominant grasses in dry, exposed coastal habitats commonly have very stiff stems and relatively broad, stiff leaves with a very tough cuticle.
   2   The desiccating effects of wind probably limit growth in these habitats and this morphology would minimize water loss and mechanical damage to cuticles caused by wind-buffeting.

Question 3.8   *(Objective 3.5)*

List at least four ways in which wind may influence the distribution and growth of trees and give examples where possible.

3.6 Water in terrestrial habitats

If you look back at Figure 3.15b, it shows that precipitation (rain and snow) interacting with temperature determines the main types of vegetation. Precipitation is clearly an important ecological factor; it is an aspect of climate (like temperature and wind) and one of the chief determinants of water supply for terrestrial organisms. Liquid water is a vital *resource* for all living organisms and the survival and growth of organisms on land depends on water supply exceeding water losses for at least part of their life cycle. What matters for the organism is that tissues should be sufficiently hydrated at some stage to allow growth and reproduction and that some mechanism permits survival during periods of unfavourable water balance (when losses exceed supply). Figure 3.35 summarizes the factors that influence water supply to and loss from organisms and we shall refer to it throughout this Section. At this stage, notice two main points:

1   *Properties of organisms* (the grey arrows in Figure 3.35) affect both water supply and water loss. On the supply side, extensive plant root systems can 'forage' for water deep in the soil and mobile animals can seek water to drink. On the loss side, surface cells may be covered by waterproof layers and activities that lead to water loss through evaporation or transpiration (e.g. opening stomatal pores for plants and moving about in strong sunlight for animals) may be restricted to certain times of day.

2   A range of *physical or climatic factors* (green boxes in Figure 3.35) affect the supply of external liquid water: precipitation is dominant but also significant are soil properties such as compaction and particle size which influence the amount of water entering and held by soil rather than running off the surface or draining through. Evaporative water loss both from external sites and from organisms is affected by the four climatic factors shown. These influence one or more of the three factors that determine rate of evaporation from a surface: the gradient of temperature, the gradient of humidity and the thickness of the boundary layer (Section 3.5).

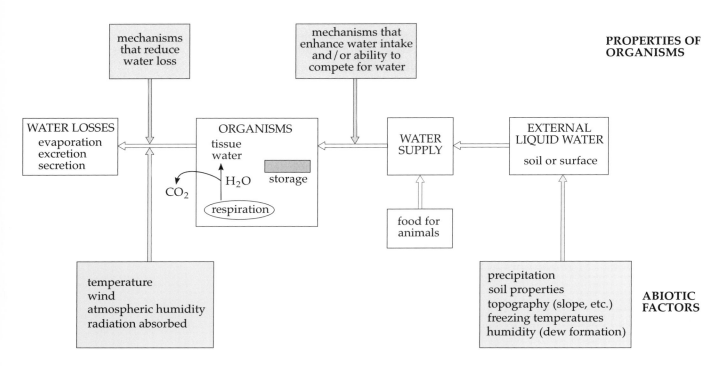

PROPERTIES OF ORGANISMS

ABIOTIC FACTORS

Figure 3.35 Factors influencing water supply to and water losses from organisms.

❑ Identify the factor(s) in Figure 3.35 which have been much influenced by human activities and affect the amount of external liquid water available to organisms.

■ Soil properties is the most obvious factor, particularly drainage. In areas of high rainfall and intensive agriculture (such as parts of the UK), land drainage coupled with the virtual elimination of natural floodplains, culverting of streams and embankment of rivers mean that there are far fewer natural wetlands than formerly. Water does not stay around as long as it used to but passes more rapidly to the sea.

One aspect of water supply in soil is not included in Figure 3.35 and this is the condition of **waterlogging**, when there is so much water present that air is excluded from soil. For many plants, prolonged waterlogging is lethal because their roots die if deprived of oxygen. But plants characteristic of wet soils and floating-leaved plants that grow in fresh water have evolved a way round this problem: roots receive oxygen from shoots via hollow stems and tissue called **aerenchyma** that contain large air-filled spaces (see Figure 3.36). These plants also have protective systems against the toxic mineral ions which are present in anaerobic, waterlogged soils, a subject discussed further in the next Chapter. Clearly, water in soil is both a resource and – in excess – a condition.

Also omitted from Figure 3.35 is an aspect of water relationships that profoundly affects survival and distribution: the tolerance of individuals to water loss and desiccation. We discuss this next.

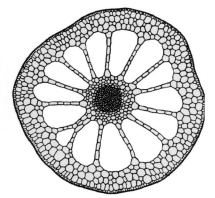

Figure 3.36 Aerenchyma tissue in water milfoil *Myriophyllum spicatum* (×40).

3.6.1 Desiccation tolerance

Very few types of organisms can survive periods of water shortage by simply drying up – losing water from their tissues.

❑ From earlier Sections in this Chapter and from general knowledge, think of some exceptions.

■ Seaweeds from the upper shore (e.g. the brown alga *Pelvetia canaliculata*, Section 3.1), many mosses (Bryophyta), lichens (Chapter 1) and some ferns (Pteridophyta).

The striking characteristic of these organisms is that they can desiccate almost completely *even when actively growing* and then recover on re-wetting. Mosses and lichens that tolerate weeks or months of desiccation are able, therefore, to colonize rocks or tree bark where there is no soil. They grow slowly, absorbing water and dissolved mineral nutrients over their whole surface when rain falls and must often tolerate high temperatures when dry. But their desiccation tolerance means that they have access to prime living space without competition from rooted plants. Most bryophytes, however, tolerate desiccation for hours or days rather than weeks and because they have no mechanisms for restricting water loss, grow in relatively moist, humid or sheltered microhabitats such as the woodland floor or beneath herbaceous vegetation. They also need free water for sexual reproduction. It is the high rainfall, humidity and mild climate of western Britain that makes it probably the richest bryophyte area in Europe.

Among flowering plants and animals, species that tolerate desiccation when metabolically active are very rare. We discuss some examples found in hot deserts in Section 3.7. However, the majority of plants and many invertebrates (particularly insects) have at least one stage of their life cycle when they are *not* metabolically active but are extremely tolerant of desiccation.

❑ From general knowledge, what stages do you think these are?

■ The seed stage in angiosperms, the spore stage in other plants and fungi, the egg or pupal stage in insects and the cyst stage in many other invertebrates (including some parasites, Chapter 1).

Possession of these resting stages, when not only desiccation but also wide extremes of temperature are tolerated, allows organisms to persist in exposed places during 'unfavourable' periods such as summer drought or cold winters (when water is frozen). It is the standard survival mechanism for annual plants and many insects. Perennial plants may have an intermediate type of resting stage that is less desiccation-tolerant than seeds but more so than the growing leafy plant; examples include bulbs, tubers and rootstocks that persist protected underground during periods of drought (and/or severe temperatures). Simply shedding leaves – the organs through which most water is lost – is common among woody plants in areas with very dry seasons (e.g. the seasonal tropical forest on Barro Colorado Island, Panama, shown in TV 'The conundrum of coexistence'). They are described as **drought deciduous**. Soil animals such as earthworms or slugs, which are quite intolerant of desiccation, retreat deeper into the soil during severe drought and reduce activity.

3.6.2 Survival mechanisms during water shortage

For longer-lived terrestrial organisms, survival as a resting stage during drought can be regarded as a last resort. We consider here the types of mechanisms that allow organisms to remain metabolically active when water is in short supply. They relate to three processes: water conservation, foraging or uptake, and storage, which relate to the top two boxes in Figure 3.35 and are discussed briefly below. Some of the most extreme and interesting examples occur in desert organisms and are described in Section 3.7.3.

(a) For effective **water conservation**, the first line of defence is a waterproof outer layer – e.g. the cuticle of ferns, flowering plants and insects and the skin of reptiles, birds and mammals. The snag here is that essential atmospheric gases – oxygen for animal respiration and carbon dioxide for plant photosynthesis – cannot pass through a waterproof layer; there must be entry points for gases and these, inevitably, allow water vapour to escape. In plants, *stomatal pores* on leaves are the entry / exit points. Their aperture can be varied and whenever water is in short supply a conflict arises between the need to conserve water (close pores) and the need to obtain CO_2 for photosynthesis and growth (open pores). Many mechanisms have evolved that act to minimize water loss through open stomata (see Section 3.7.3) but especially noteworthy are the C4 and CAM strategies which involve a change in the biochemistry of carbon fixation.

C4 plants are mostly sedges and grasses (including sugarcane, *Saccharum officinale*) and occur mainly in open tropical habitats such as the African savannah. They are mostly absent from cool or shady habitats but one of the few C4 plants in the temperate zone grows in British estuaries – *Spartina anglica*, cord grass. The name C4 derives from the fact that CO_2 is fixed initially into a four-carbon acid whereas in most species it is fixed into a three-carbon acid (**C3 plants**). The four-carbon acids move from the outer cells of a leaf to rings of special green cells surrounding the vascular bundles (veins) and here the CO_2 is re-released and fixed by the C3 pathway. These green cells are visible if you hold leaves up to the light, and are described as **Kranz anatomy** (*kranz* is German for wreath), which is diagnostic of C4 plants. The affinity for CO_2 of the initial fixation enzyme in C4 plants is much greater than that of the enzyme in C3 plants and this has two important consequences: it allows C4 plants to fix more carbon and grow faster in habitats which are warm and well-lit and where CO_2 limits the rate of photosynthesis; and it allows C4 plants to fix as much carbon as C3 plants but with stomata open less widely or for a shorter time. C4 plants are said to have a greater **water-use efficiency** than C3 plants because less water is lost for every gram of carbon fixed. However, it takes more energy to fix carbon by the C4 pathway and it appears that in cooler climates with lower light levels this disadvantage outweighs the advantages.

CAM plants are mostly succulents and occur worldwide in hot, dry habitats, such as deserts, cracks in exposed rocks and as epiphytes on the upper branches of tropical trees. The name derives from Crassulacean acid metabolism, Crassulaceae being a family of succulents, and like C4 plants they fix CO_2 initially into a four-carbon acid. The unique feature of the group is that initial fixation takes place *at night* when conditions are cooler and more humid and far less water is lost through open stomata than during the day. The four-carbon acid is stored in cell vacuoles and next day

CO_2 is released from it and light energy is used to effect complete fixation to organic matter while stomata remain *closed*. When combined with highly waterproof cuticles and water-storing tissues (see below), the CAM strategy is undoubtedly a Rolls-Royce among water conservation systems. However, it is usually associated with low growth rates, probably because the acid-storing step limits the amount of carbon that can be fixed, so that CAM plants are usually poor competitors when water is *not* in short supply. CAM is nevertheless very widespread, occurring in 20 families of angiosperms, and is the best photosynthetic strategy for extremely dry conditions.

When water shortage is associated with cold and, especially, windy conditions as in moorlands and tundra, effective waterproofing, protection of stomata and a low, bushy or hummocky growth form are the main mechanisms that conserve water in plants. Patterns in plant distribution here often seem to reflect relative tolerance of drought or waterlogging.

❑   *Erica tetralix* (cross-leaved heath) and *Erica cinerea* (bell heather) occur in wetter and drier areas respectively of British moorlands. Which data relate to which species in Figure 3.37 and why?

■   The species on the right is *E. cinerea*: in dry soil, it loses much less water (i.e. it conserves water more efficiently) and therefore grows better than *E. tetralix*.

Among animals, next to waterproofing the most effective means of conserving water are behavioural: retreat to cooler, more humid microhabitats, for example. Generally, foraging for water rather than conserving it is of greater significance for large, mobile animals.

(b) **Foraging for water**: only a minority of animal species, mostly large, terrestrial vertebrates, need to drink water. Other species obtain water entirely from food so that energy, chemical and water requirements are combined. For drinkers, however, the availability of free water may limit distribution and the ability to locate or remember the location of water and travel sometimes long distances to it can be important for survival. The situation on the East African savannah during droughts in the 1980s and 1990s provides a graphic illustration. Increasing numbers of grazing mammals at the few remaining waterholes caused intense disturbance through trampling and overgrazing for many miles around and can be regarded as an environmental impact at the landscape level.

Plants also forage for water by the growth, size and variable geometry of their root systems. Roots may grow towards and branch extensively in moister parts of the soil. On sand dunes and other very free-draining soils, young seedlings develop long, unbranched roots that reach water deep underground (see Figure 3.38).

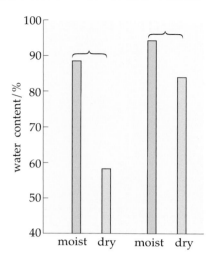

Figure 3.37   Water content relative to that at full turgor of two species of *Erica*, *E. tetralix* and *E. cinerea*, after 2 h at 45% relative humidity, 20 °C and in moist or very dry soil.

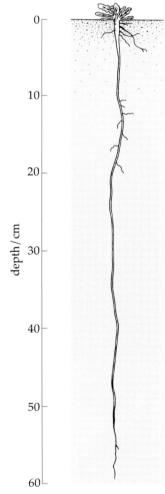

Figure 3.38   Root system of a first-year seedling of cat's ear *Hypochoeris radicata*. The roots of older plants develop side branches but do not attain any greater depth.

Figure 3.39 A 19th century drawing of a landscape showing bottle trees in north-eastern Brazil.

Root physiology is also important in foraging because water becomes more strongly bound by adsorptive and capillary forces as soil dries out. To maintain water uptake, plants must develop a stronger 'suction force', which can be done by increasing the concentration of solutes in root cells and hence their *osmotic pressure*. Species vary greatly in their ability both to vary root geometry and to increase osmotic pressure so that the ability to compete for water also varies.

(c) **Water storage**, in the sense of maintaining within tissues more water than is required for normal metabolism, is rather a rare adaptation to water shortage. The most striking examples occur in deserts (Section 3.7.3) but among plants two systems are more widespread. The first is **succulence** which occurs in CAM plants and involves water storage in special parenchyma cells of stems or leaves. Biting stonecrop *Sedum acre* is a common British succulent of dry walls and grassland. Even more widespread is water storage in tree trunks, where the central heartwood stores water. Water uptake by tree roots commonly fails to keep up with transpiration losses from leaves, especially in hot or windy conditions, and water stored in heartwood makes up the deficit. The result is that trunks shrink during the day and swell at night when stomata close, transpiration ceases and water uptake exceeds losses. In semi-desert conditions, some trees have evolved more specialized water-storing trunks, e.g. the baobab *Adansonia digitata* in African savannah and the barrigudos or bottle tree *Cavanillesia arboria* in north-eastern Brazil: bottle trees may have trunks nearly 5 m across (see Figure 3.39).

3.6.3 Reproduction and establishment

Mechanisms for conserving, obtaining or storing water are often least well developed in juvenile or reproductive phases, which are therefore especially sensitive to water availability. This is a major influence on life cycles, influencing where and when organisms can reproduce sexually. At one end of the scale are organisms that require free water for fertilization and/or egg laying and development. These include most amphibians, many insects (with aquatic larval stages), terrestrial algae, bryophytes and ferns. Highly specialized amphibians do occur in deserts but the general point is that amphibians (and bryophytes and ferns) are more diverse and abundant in wet regions.

Plant seeds and spores take up considerable amounts of water during germination, and water availability at this stage is critical. Look at Table 3.9.

❑ From the data in Table 3.9, (a) in which general habitats would you expect to find *Juncus effusus* in Britain? (b) How would you explain the finding that *J. effusus* germinates better under a sward of grass and clover than on bare soil?

■ (a) *Juncus effusus* clearly germinates best in very wet soils where the water table is close to the surface. As you might expect, the rush normally grows in poorly drained pastures or beside streams. (b) The grass–clover sward traps a layer of moist, still air and prevents drying out of the soil surface. This is a more favourable microhabitat for the germination of soft rush seeds, which, on bare soil, may be prevented from germinating by rapid evaporation of water.

Table 3.9 Percentage germination of soft rush *Juncus effusus* in relation to depth below the surface of water-saturated soil (position of the water table).

Depth of water table/cm	% germination
0	90
5	42
13	2
20	0

Soft rush grows on wet soils because it spreads by seeds that require wet conditions for germination and also because mature plants tolerate waterlogged soils. This is one level of explanation for the distribution of soft rush; however, suppose, for example, that an area of grassland suddenly became waterlogged because of blocked field drains. The habitat is now 'suitable' for soft rush but whether or not it establishes there will depend on at least two additional factors. (i) *Seed availability*: unless there is a bank of viable seeds in the soil or unless seeds are dispersed to this area from mature *Juncus* nearby, no establishment can occur. (ii) *Competition*: any seeds that germinate in the new area must be able to compete with the existing vegetation. In fact, soft rush commonly has large, persistent seed banks in areas that are or were once waterlogged and seedlings are very good competitors, so it often *does* appear in drained sites that become waterlogged again. This example illustrates, however, the range of factors that may influence distribution and the difficulty of attributing distribution patterns to any single factor.

3.6.4 Water supply and distribution

Figure 3.35 illustrates that the water relations of organisms are influenced by water supply and water loss functions. As a broad rule of thumb (with numerous exceptions), the distribution of organisms that cannot tolerate desiccation and are poorly waterproofed is influenced primarily by the risk of water loss and thus by factors such as temperature, wind and humidity that influence water loss. Woodlice are rather poorly waterproofed and you saw in Section 3.4.5 that geographical distribution may be determined partly by temperature. Local or microdistribution, however, is nearly always influenced by the risk of water loss, so that woodlice are mostly active at night and spend the day in soil, litter or humid crevices. Amphibians are likewise very poorly waterproofed; their distribution may be limited not only by the availability of water bodies for breeding but also, for adults, by the availability of cover where they can feed and avoid desiccation.

By contrast, the dominant or climax vegetation in a geographical region (see Figure 3.15) is more commonly influenced by water supply and thus by precipitation and soil conditions that affect external liquid water. In western and central Europe, for example, the natural climax vegetation is mainly deciduous forest which changes in the south to evergreen Mediterranean woodland and scrub. Pigott and Pigott (1993) studied the transition between these two types of vegetation and concluded that 'the southern limit of summer-green deciduous woodlands of western and central Europe is probably largely determined by water supply and in particular by the severity of summer drought in the Mediterranean zone'.

Other studies indicate that the boundary between northern forest (taiga) and steppe grassland is also determined mainly by water supply, with precipitation having the greatest effect and soil properties and topography having small-scale, local effects.

Summary of Section 3.6

- The survival of organisms on land depends on water supply exceeding water losses for some part of the life cycle. Precipitation is the main factor influencing supply, and temperature and wind are the main factors influencing losses.

- Excess water in soil (waterlogging) is a condition that restricts the growth and survival of some plants and soil animals. The distribution of most terrestrial organisms is influenced by water supply, relative tolerance of desiccation and ability to obtain water and restrict its loss. Tolerance of extreme desiccation when actively growing is one mechanism for surviving water shortage and is characteristic of lichens, some mosses and ferns and a few vascular plants and insects. Desiccation-tolerant resting stages (seeds, cysts, etc.) are widespread among invertebrates, plants and fungi.

- When water is in short supply, ability to conserve, store or forage for water whilst remaining metabolically active is critical for survival and reproduction. Mechanisms that help to conserve water such as the C4 and CAM strategies in plants often carry a 'cost' that reduces competitive ability in wetter or cooler habitats.

- Juvenile stages are often the most sensitive to water shortage and this may limit distribution. Some organisms require free water for sexual reproduction (e.g. many amphibians, bryophytes and ferns). Overall, conditions of temperature, wind and humidity that influence the risk of desiccation are most likely to determine the local distribution of organisms that have a poor ability to restrict water loss (e.g. woodlice). Water supply commonly determines the general type of vegetation.

Question 3.9   *(Objectives 3.5 & 3.7)*

Suggest explanations for (a)–(c) and indicate, where relevant, whether water supply or factors affecting water loss are involved.

(a) Woodlands in the north and west of Britain usually contain more species and larger populations of ferns and bryophytes than do similar woodlands in the south and east.

(b) Epiphytes (Chapter 1) in the canopy of tropical rainforests may suffer from water shortage and show characteristics typical of desert plants; epiphytes close to the ground in such forests do not usually experience water shortage and do not show such characteristics.

(c) Plants such as ragged robin *Lychnis flos-cuculi* usually occur naturally in marshy areas but will grow and flower well if transplanted to well-drained sites.

Question 3.10   *(Objectives 3.5 & 3.10)*

In their study of the boundary between deciduous forest and Mediterranean evergreen forest in southern Europe, Pigott and Pigott (1993) carried out experiments in hilly areas of southern France where south-facing slopes are dominated by evergreen holm oak (*Quercus ilex*) and steep, sheltered northern slopes are occupied by deciduous woodland at its southern limit, with abundant lime trees (*Tilia cordata*). They first observed the response to water shortage of the two species in 12-year-old trees that were growing in the University Botanic Gardens, Cambridge: during a prolonged summer drought in 1990, *Quercus* was quite unaffected but *Tilia* stopped growing and many leaves wilted, yellowed and were shed. Figure 3.40 shows changes in the water potential* of shoot tips, a measure of the extent of water loss, for the two species during a day at the height of the drought.

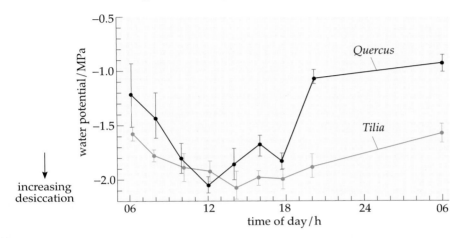

Figure 3.40 Change of water potential in the shoot tips of 12-year-old trees of *Quercus ilex* and *Tilia cordata* over a period of 24 h during a prolonged drought in Cambridge, UK.

(a) How do the data in Figure 3.40 demonstrate the superior drought tolerance of *Quercus*?

Pigott and Pigott found that the critical water potential for *Tilia* below which damage from desiccation occurs is about −2 MPa. They then measured water potential in shoot tips of *Tilia* growing on north-facing slopes in France and, as stunted saplings, on south-facing slopes with *Quercus ilex* (Figure 3.41).

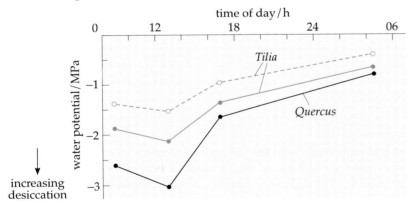

Figure 3.41 Change of water potential in the shoot tips of *Tilia cordata* growing on north-facing slopes (open green symbols) and, as stunted seedlings, on a south-facing slope (solid green symbols) and of *Quercus ilex* growing on south-facing slopes (solid black symbols).

(b) Do the data in Figure 3.41 support the hypothesis that water supply determines the deciduous–Mediterranean evergreen boundary?

(c) Suggest an alternative hypothesis about the way in which water determines this boundary.

* For a full discussion of water potential, see Book 3 of S203 *Biology: Form and Function*. Water potential is measured in units of pressure, pascals (Pa) or, more commonly, megapascals (MPa). Pure water has a potential of zero and the more dried out an organ or organism becomes, the lower (more negative) its water potential.

3.7 The ecology of hot deserts

Hot deserts are a classic example of an extreme environment where abiotic conditions (notably temperature) are exceptionally severe; some resources (notably water) are also in very short supply although others, such as space and light, are abundant. In this Section, as in Section 3.1, we consider the habitat as a whole, rather than focusing on one abiotic factor, and use hot deserts as a case study to illustrate the range of mechanisms by which organisms survive and persist in a harsh environment.

The barren emptiness of deserts is more apparent than real because a great many organisms live there. The majority are dormant for much of the time but the rest remain metabolically active all the year round and provide many striking examples of adaptations that aid survival in extreme conditions of temperature and water shortage. An interesting question which you should bear in mind (we do not answer it in this Section) concerns the relative strength of biotic and abiotic interactions between desert organisms: do they play an equal role in determining the distribution of desert organisms and shaping desert communities? Are biotic interactions *strong* because resources are in such short supply or are they *weak* because conditions are so severe? This question is considered fully in Book 2, Chapter 4.

3.7.1 Conditions and communities

Deserts are places where there is a chronic shortage of water and, in hot deserts, this is brought about by two main conditions: high average temperatures and low rainfall (at most 30 cm per year but often less than 10 cm). High temperatures, often accompanied by strong winds and low relative humidity, mean that water evaporates rapidly from both soil and organisms and this *high potential evapotranspiration* is perhaps the most distinctive feature of **desert climates**. Figure 3.42 and Table 3.10 illustrate two other common features of *temperature conditions* in hot deserts.

❑ What are these common features?

■ (1) Temperature varies greatly, both seasonally and daily (Figure 3.42). (2) The most extreme temperatures and the greatest variation occur at the ground surface; a few cm above or below the surface, conditions are much more equable (Table 3.10).

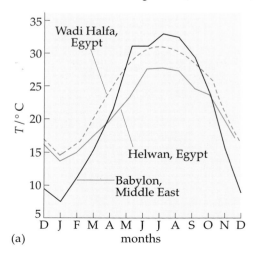
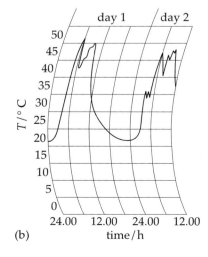

Figure 3.42 (a) Average monthly mean shade temperatures at three desert stations. (b) Daily fluctuations in air temperature at Khartoum, Sudan.

Table 3.10 Air, surface and sub-surface temperatures at desert sites.

Site	Month/time (h)		air		Temperature/°C: surface	below surface
Red Sea Hills (Sudan)	Sept 13.00		40.0		72.5	60.5 (3.0–3.5 cm)
	Sept. 6.00		26.5		26.0	26.0 (3.0–3.5 cm)
Luxor (Egypt)	April 12.00		30.0		55.0	27.0 (7.5 cm)
Mourzook (West Sahara)	April 12.00		–		45.0	25.0 (30.5 cm)
	April 24.00		–		10.0	25.0 (30.5 cm)
		at 10 cm	at 2 cm		at surface	
Sonora (Mexico)	late spring 7.30	25	36		37	–
	12.30	25	30		44	–
	15.00	31	41		46	–

From these data, desert organisms and especially perennial plants must have very wide temperature tolerance; virtually all desert animals show behavioural thermoregulation (Section 3.4.4) with most endotherms and some ectotherms active mainly at night and smaller species spending all or part of the day underground. *Relative humidity* is also much higher at night (Figure 3.43) and this, combined with lower temperatures, means that transpiration losses are lower, providing another advantage for nocturnal animals and the CAM strategy in plants (Section 3.6.2).

Table 3.11 illustrates the highly erratic nature of *desert rainfall*. In North Africa, all rain for one year may fall in a few days as torrential storms, which can cause serious flooding. In the deserts of the USA and Mexico, by contrast, there is a more predictable 'wet' season in winter with occasional storms in summer but the general situation is similar: long rainless periods interspersed with brief comparatively wet conditions. This rainfall pattern has led to the evolution of two different types of desert community. On the one hand are the **persisters**, a community of animals and perennial plants that remain metabolically active throughout dry periods and often show striking adaptations to water shortage. On the other hand are the **avoiders**, a community that is metabolically active only for a short period after rain and completely dormant for the rest of the time; they show few, if any, adaptations to drought.

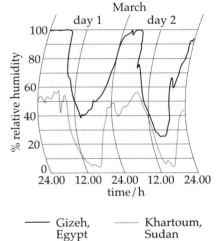

Figure 3.43 Daily fluctuations in relative humidity at two desert stations in North Africa.

Table 3.11 Desert rainfall.

Location	Annual average rainfall/cm	Max. recorded in 1 day/cm
Basra (Middle East desert)	16.7	7.8
Amara (Middle East desert)	20.3	17.1
Cairo (Egypt)	30.0	44.0
Damascus (Syria)	23.4	7.6
California (Sonora desert)	–	8.5

When torrential rain occurs in desert areas with little or no vegetation, a high proportion of the water *runs off*: only the top few cm of soil are moistened and most of the water flows into shallow depressions via runnels or wadis, where it percolates deeper into the soil. By concentrating water in a relatively small proportion of the total desert area, run-off gives rise to two types of habitat. In the lower, **drainage habitats**, water is available deep in the soil for all or most of the time; in the higher, **run-off habitats**, water is available only for a short time after rain and only in the surface layers of the soil. Not surprisingly, the vegetation in these two habitats has rather different characteristics (see Section 3.7.2).

Drainage basins (called *playas* in North America) where temporary lakes form after rain may be associated with the formation of *salt pans*. These arise because the evaporation of water leads to upward movement of water and ions from deeper layers and deposition of salts (including carbonates and sulphates) at or near the surface. Irrigation can have a similar effect and one of the major problems of irrigated desert or semi-desert areas today is the **salinization** of soil, which makes it useless for conventional crops. Huge amounts of soil may be eroded from run-off habitats during rain and in all areas dry surface soil is susceptible to wind erosion with strong, dry winds, which are common in many deserts, giving rise to severe dust storms. The instability of desert soils or, in stony deserts, the virtual absence of soil, is another factor that makes hot deserts a very difficult habitat for perennial plants.

3.7.2 Desert organisms I: the avoiders

Cloudsley-Thompson (1977) described how, a few hours after a violent thunderstorm in the barren wastes of northern Chad in the Sahara, he had to shout to be heard 'above the fantastic chorus of croaking toads and stridulating insects'! A few days later, the area was carpeted with brilliant flowers and the temporary pools teemed with invertebrates and tadpoles. From these examples, it would appear that avoidance is a very successful mechanism for surviving in deserts and, indeed, it is: 50–80% of desert plant species are avoiders. We consider them first.

Plants

Most avoiders survive during dry periods as seeds but, after rain, germination, growth, flowering and seed set occur in rapid succession and plants may die within a few weeks, which justifies their description as **ephemerals**. These plants are not especially resistant to desiccation and their persistence in hot deserts depends on three main features:

1 the ability to complete life cycles in a short period while water is available; this requires reasonable soil fertility to support rapid growth;

2 production of seeds that are very resistant to desiccation and temperature extremes and that remain in the soil as a **seed bank** during dry periods;

3 mechanisms which ensure that germination occurs only when conditions are suitable for completing the life cycle.

Considering each of these features in turn, a requirement for ephemerals that grow after heavy rain in *very* hot conditions is that they fix the maximum possible amount of carbon in their short life so that seed production can be maximized.

❑ What carbon fixation system (C3, C4 or CAM, Section 3.6.2) would you expect to occur in these plants, and why?

■ C4: this mechanism allows high fixation rates and growth rates in hot, high-light conditions. It is also associated with high water use efficiency, which is useful as the soil dries out.

In the Mojave desert of North America, all the ephemerals that appear after summer rain are indeed C4 species (Mulroy and Rundel, 1977). However, in deserts that are less hot and/or have more reliable summer rain (e.g. parts of the Sonora desert), a significant number of C3 species occur, especially in drainage habitats where water is available for longer periods. These species tend to have longer life cycles than the C4 ephemerals and are often described as **summer annuals**. Different species, all of them C3, germinate after autumn rainfall in the Mojave and Sonora deserts. These may persist throughout the relatively cool winter (and are often called **winter annuals**), growing initially as flat rosettes before sending up tall flowering shoots as temperatures rise in spring. Such a growth pattern is analogous to the behavioural thermoregulation of the Galapagos iguana (Section 3.4.4): leaves grow where temperatures are optimal for photosynthesis, i.e. in the warmest position, close to the ground at mid-winter and in cooler positions 10 cm or more above the ground (see Table 3.10) in spring. As a rule, therefore, the hotter the conditions and the more erratic the rainfall, the shorter the life cycle and the higher the proportion of C4 species.

Seeds of desert annuals are produced in vast numbers, possibly in part because of selection pressure exerted by seed-eating animals such as ants and small rodents. In the Sonora desert, Arizona, average soil seed densities ranged from 4000 to 63 800 per m^2, while values of 83 000 per m^2 were reported for the Australian desert in New South Wales. The patchiness of desert seed banks in space and their variability over time result partly from the patchy distribution of parent plants and partly from the activities of seed-eating (**granivorous**) animals. Granivores are often quite choosy about the seeds they gather and this selective foraging very probably (it is difficult to be sure) affects the local abundance and distribution of desert ephemerals.

Regarding seed germination, temperature and soil moisture are the two factors which usually determine that summer annuals germinate only after heavy rain in summer and winter annuals only after heavy rain in autumn–winter. Numerous fail-safe mechanisms have evolved which ensure that the seed bank is never totally depleted. For example, *Salsola volkensii*, a summer annual of the Negev desert, Israel, produces two types of seeds, one able to germinate immediately but losing germinability over two years, and one in which peak germinability is reached only after two years and retained for at least five years (Negbi and Evenari, 1961). Winter annuals from the Chihuahua desert, Arizona, showed annual cycles of dormancy/non-dormancy which ensured that some buried seeds were always retained in the seed bank (Baskin *et al.*, 1993).

A few perennial desert plants have also evolved avoidance mechanisms. They survive dry periods as underground bulbs (notably in the northern Sahara) or as tubers or rootstocks (notably in the Namib desert of southern Africa), which store both food and water. When rain falls, shoot growth and flowering occur very rapidly supported by these underground reserves but instead of allocating virtually all surplus energy to seed production (as ephemerals do), much is used to re-stock the storage organs. These organs are important food items for desert animals (e.g. the blind naked mole rat *Heterocephalus glaber* in the Namib desert) and, as for ephemerals and granivores, the question arises as to what factors are most significant in determining abundance and distribution: is it animal predation of seeds/storage organs, dispersal capacity, or competition for space, suitable germination sites or water?

Animals

The majority of animal avoiders behave in a similar way to ephemeral plants. They are active only for a short period after rain, when abundant food and moisture are available, and they persist during dry periods as desiccation-resistant cysts (Protista), eggs (crustaceans and insects) or even as mature animals enclosed in a cocoon deep underground (amphibians). This last mechanism is called **aestivation** and in an Australian desert toad is accompanied by considerable water storage in the body; some aborigines are expert at locating the toads and tapping their stores for drinking water!

The chironomid (non-biting) midge *Polypedilum vanderplankae* has remarkable *larvae* that tolerate desiccation above ground. They develop in desert rockpools in Nigeria, survive when 93% desiccated for 3–10 years and, in their dry state, tolerate temperatures up to 70 °C (Hinton, 1960).

Generally, however, avoidance in both plants and animals depends primarily on the life cycle, especially on its brevity and/or the timing of different stages. The morphology and physiology of avoiders are usually much the same as those of organisms living in moist climates. By contrast, persisters usually have striking modifications of structure, physiology and/or behaviour.

3.7.3 Desert organisms II: the persisters

To this group belong most of the plants and animals regarded as 'typical' desert species. Many do indeed remain active during long periods without rain, although some become dormant or aestivate if drought is very prolonged.

❑   Recall from Section 3.6.2 three processes that may be important for survival during drought.

■   (1) Water conservation; (2) effective ways of obtaining water by foraging or uptake; (3) water storage.

In addition, (4) some persisters tolerate a higher degree of water loss from tissues than do organisms from wetter climates. Since evaporation of water cools the body, organisms which conserve water efficiently must either tolerate a rise in body temperature (many plants do this) or keep out of direct sunlight (most small animals do this). Larger animals may tolerate both a rise in body temperature and some desiccation.

(a) Plants

Among desert perennials there is a broad distinction between those growing in run-off habitats and those in drainage habitats (Section 3.7.1). The main characteristic of the latter is illustrated in Figure 3.44.

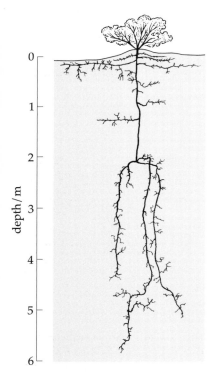

depth/m
0
1
2
3
4
5
6

Figure 3.44 Semi-diagrammatic representation of the deeply penetrating root system of *Acacia socotrana*, a shrub which grows in drainage habitats in deserts.

❑   What is the characteristic and to which of mechanisms (1)–(4) above does it relate?

■   A very deeply penetrating root system, able to reach water deep underground in drainage habitats. This relates to mechanism (2), root foraging.

Plants of drainage habitats, which are mainly shrubs or trees, neither store water to any marked degree nor tolerate much desiccation but many conserve water efficiently and some lose leaves and become dormant during prolonged drought when even their water supply is exhausted. By contrast, plants of run-off habitats are generally more specialized.

One group of run-off species, the **succulents** (Section 3.6.2), shows mechanisms (1)–(3). They store water in stems or leaves and a large barrel cactus (Figure 3.45) may yield over two litres of drinkable water! When rain falls, they absorb water very effectively by an extensive system of shallow roots (Figure 3.46). And by virtue of CAM metabolism and exceptionally waterproof outer layers they also conserve water remarkably well: an uprooted *Echinocactus* remained alive for six years without water supply and without becoming desiccated.

Reduction of leaves to spines, which is such a characteristic feature of cacti, is often cited as a mechanism that reduces water loss: green stems have fewer stomata than leaves and spines also trap a layer of more humid air close to the surface (i.e. increase the boundary layer). However, there are many leaf succulents (e.g. euphorbias in Africa) that also possess spines and, whilst loss of leaves in cacti can be regarded as an adaptation to water shortage, spines are just as likely to have evolved as a protection against grazers.

The other group of plants that grow in run-off habitats are the **xerophytes** – non-succulent perennials which are very tolerant of drought. In fact, there is wide variation in drought tolerance among xerophytes. At one end of the scale are extreme xerophytes such as the creosote bush *Larrea tridentata* (Figure 3.47), an American species that shows three of the drought-survival mechanisms. It has an extensive root system and extraordinary powers of absorbing water from soil, achieved by increasing solute concentration and reducing turgor in cells (i.e. reducing cell water potential). Water is quite well conserved, with leaves covered by a thick resinous layer (that also acts as a defence against herbivores and wind abrasion) but more significant for drought tolerance is *Larrea*'s tolerance of desiccation: mature leaves may lose 30–40% of their water content without wilting and young leaves may lose over 70% but retain some metabolic activity and recover fully when re-wetted. Older leaves are commonly shed during prolonged drought but younger leaves are retained. Similar characteristics are shown by sagebrush *Artemisia herba-alba* of Middle East deserts and this species, like *Larrea*, is dominant in the community. An interesting question, therefore, is whether these extreme xerophytes dominate because of their superior tolerance of abiotic conditions or because of their superior competitive ability: is their battle with the elements or other organisms, or both equally? We return to this question in Book 2, Chapter 4.

Figure 3.45 Barrel cactus *Echinocactus acanthodes* ($\times 0.08$) from California. Note the grooved, water-storing stem and leaves reduced to spines.

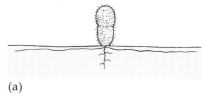

(a)

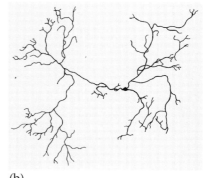

(b)

Figure 3.46 Root system of the cactus *Ferocactus witzleni*: (a) vertical section; (b) viewed from above. Note the shallow but extensive roots.

Figure 3.47 Leafy shoot of the creosote bush *Larrea tridentata* (~×0.5) from Northern and Central America. Young leaves can withstand extreme desiccation.

Less extreme xerophytes do not show such tolerance of desiccation and have varying capacities for extracting water from drying soil and for conserving water. At one end of the scale are species, mostly with C3 photosynthesis, that grow only in cool winter seasons and die back or shed leaves in summer. At the other end are predominantly C4 species that retain leaves in summer, have highly impermeable cuticles and open stomata only in the cooler early morning.

For all these persistent plants, but especially succulents and extreme xerophytes from the hottest and driest areas, *seedling establishment* is difficult and is most likely to limit distribution. The main problem is that seedlings are less waterproof and more easily grazed than larger plants and, unless there is an exceptionally wet year, they do not develop sufficiently impermeable layers or store enough water to survive a long period of drought. Thus, Jordon and Nobel (1981) showed that the length of the first major drought after germination limited the establishment of various succulent species in the Sonora desert (USA) with *Agave desertii* apparently having established in only one year out of the previous 17 (see also Question 3.14). The following example illustrates the importance of biotic interactions for seedling establishment.

In parts of the Chihuahua desert (Texas, USA), the cactus *Opuntia leptocaulis* commonly grows in close association with *Larrea tridentata* (see above). Yeaton (1978) found that the bird- or rodent-dispersed seeds of *Opuntia* tended to be deposited in the fine, wind-blown soil that accumulates below *Larrea* and seedling establishment was much better here than in the hard, compacted soil elsewhere; *Larrea* can be described as a **nurse plant**. What happens subsequently is illustrated in Figure 3.48, which also shows Yeaton's suggested explanations for the sequence of changes. You can see that there appears to be a *cyclical replacement* of *Larrea* and *Opuntia*, a type of relationship that will be discussed more in Book 3.

❑   Use the +/−/0 notation (Chapter 1) to characterize the relationship between *Larrea* and *Opuntia* at points 1 to 3 in Figure 3.48.

■   At (1) and (2), the relationship is 0/+ (*Larrea* promotes seed deposition and seedling establishment, acting as a nurse plant); at (3) there is probably competition for water (−/−) and also root protection from rodents for *Opuntia* (0/+).

(b)   Animals

No desert animals are able to function for long periods *in the open* without water, so there are no animal equivalents to succulent plants or extreme xerophytes. However, many small animals live in places too dry for any perennial plants and in the complete absence of free water or even moist food. Their remarkable tolerance derives mainly from great efficiency in conserving body water and in avoiding temperature extremes, both of which are achieved primarily through *behavioural mechanisms*. Some possess ingenious methods for obtaining water and a few actually store it, but few tolerate tissue desiccation or large rises in body temperature. By contrast, larger desert animals, particularly mammals that remain in the open, must lose water by evaporation as a means of keeping cool. Their survival does not depend primarily on water conservation but on being able to range widely for food and waterholes and to tolerate some degree of tissue desiccation. Examples of different mechanisms are described below.

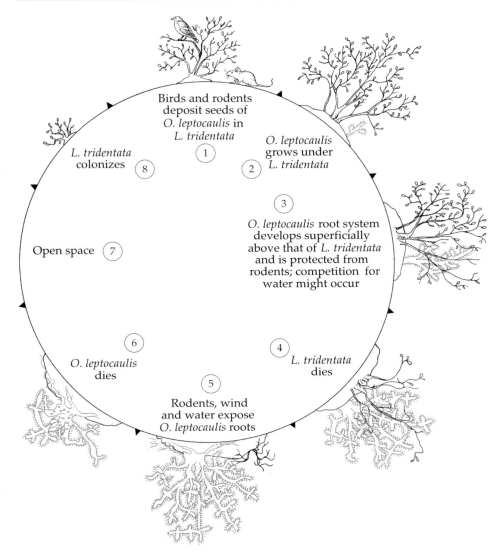

Figure 3.48 Biotic interactions between *Larrea tridentata* and *Opuntia leptocaulis* in the northern Chihuahua desert and the postulated cyclical pattern of change.

Small animals: Obtaining food and, above all, maintaining water balance (losses = gains) are the two basic requirements for the survival of small desert animals. Growth and reproduction require a positive water balance (gains exceed losses) and you can see how difficult this is to achieve if you look at Figure 3.49. This shows the water balance sheet for one of the kangaroo rats, **Dipodomys** spp., which are small rodents from the North American deserts. They live on a diet of plant seeds and are strictly nocturnal, spending their days in cool, humid burrows underground: this is the main way in which water is conserved because evaporation is the chief route of water loss. Dry faeces (with one-fifth the moisture content of those of white rats) and concentrated urine also help the cause of water conservation. Water gain derives largely from **metabolic water** generated during respiration.

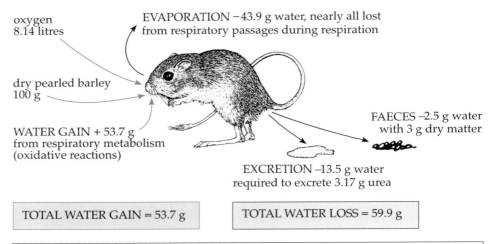

oxygen
8.14 litres

EVAPORATION −43.9 g water, nearly all lost
from respiratory passages during respiration

dry pearled barley
100 g

WATER GAIN + 53.7 g
from respiratory metabolism
(oxidative reactions)

FAECES −2.5 g water
with 3 g dry matter

EXCRETION −13.5 g water
required to excrete 3.17 g urea

TOTAL WATER GAIN = 53.7 g	TOTAL WATER LOSS = 59.9 g

NET WATER LOSS = 6.2 g; this can be made good if the food is not completely dry but
is in equilibrium with air of 20% relative humidity

Figure 3.49 Water relations of the kangaroo rat *Dipodomys merriami* (×0.3).

❑ For each molecule of CO_2 respired, how many molecules of water are
generated?

■ One: the basic equation for respiration is:
$$C_6H_{12}O_6 + 6O_2 = 6CO_2 + 6H_2O$$

However, on a *completely* dry diet, water balance is still negative (Figure
3.49) and *Dipodomys* gets by only through another behavioural mechanism:
seeds are stored underground in equilibrium with moist air.

For *small ectotherms*, cold desert nights mean that nocturnal activity is not
always an option. Many lizards and arthropods are active during the day,
restricting water loss and maintaining a suitable body temperature by
elaborate patterns of behavioural thermoregulation (Section 3.4.4).
Flightless diurnal ground-beetles (*tenebrionids*) from the Namib desert
(south-west Africa) have some of the lowest known rates of water loss
(Louw, 1993), which is achieved by a combination of:

* a very thick waterproof cuticle;
* protecting spiracles (the openings for respiratory gas exchange) by
  fusion of the hard wing cases (elytra) to form a humid cavity over the
  abdomen: this greatly reduces water loss during respiration;
* having long legs that hold the body above the hot ground and allow
  high-speed dashes between sheltered sites;
* having a low metabolic rate and respiring *intermittently* at intervals of
  up to 60 min and only when in shade: this further reduces evaporative
  losses during respiration;
* absorbing almost all water from faeces.

Despite this battery of water-saving mechanisms, the low metabolic rate of
Namib tenebrionids means that metabolic water and their very dry food
still do not compensate fully for water losses. They achieve balance by
interesting water-collecting systems which depend on the fact that the
Namib, a coastal desert, is often foggy: Figure 3.50 illustrates one system.

Figure 3.50 Collection of fog
water by the dune tenebrionid
Onymacris unguicularis (×1) in
the Namib desert. The beetle
climbs to a dune crest, assumes a
head-down position into the fog
wind, and water droplets
condense on the abdomen and
run towards the mouth where
they are taken in.

In contrast to desert beetles, ants, also highly successful desert insects, depend less on being super-waterproof and more on social organization and behaviour. They are commonly active only in the early morning and late afternoon when temperatures are neither too hot nor too cold. They also excavate huge communal nests that may extend up to four or five metres downwards and, like *Dipodomys*, store dry food in humid chambers. Species that feed on plant sap or the honeydew secretions of aphids during wetter periods transfer fluids to special honeypot worker ants (Figure 3.51), which regurgitate fluids to feed and water the colony during dry seasons.

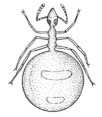

Figure 3.51 Honeypot ant *Myrmecocystus* sp. with fluid stored in its swollen abdomen (×1.7).

Larger desert animals include mammalian grazers such as the oryx and camel (Figure 3.52). They keep cool by sweating and must, therefore, drink water (unless eating vegetation with a high moisture content). But the main reason for their success in deserts is the capacity to go for several days between drinking bouts, which means that they can forage over a much larger area than similar non-desert grazers. Four of the characteristics that contribute to this capacity are listed below.

1   Camels and desert antelopes tolerate **hyperthermia**, a rise in core body temperature of 6–8 °C. For camels, this minimizes the water required for evaporative cooling, while antelopes allow body temperature to rise during high-speed dashes to waterholes after periods of grazing. Both types of animal dissipate excess heat during the cool nights.

2   Much greater desiccation is tolerated than for non-desert species: camels can lose up to 30% and antelopes up to 15% of body water, most of which is withdrawn from *tissues* rather than *blood* so that the blood does not become too viscous for rapid circulation.

3   When the desiccated animals finally drink, they imbibe enough water in a few minutes to rehydrate the body completely. A human who attempted to do this would probably die of osmotic shock caused by too rapid dilution of body fluids but camels and antelopes store water temporarily in their capacious rumens from where it is released gradually.

Figure 3.52 An Arabian camel or dromedary *Camelus dromedarius* (×0.02).

4   Apart from sweating, all other sources of water loss are minimized: faeces are dry and very little urine is produced (urea being secreted into the rumen where it nourishes rumen bacteria).

From these examples, desert animals are seen to be highly specialized, as are desert plants. The reason these organisms do not occur outside deserts is probably because they compete less well or are more susceptible to pathogens. But this still leaves open the question of whether the distribution and abundance of desert organisms *within deserts* depends mainly on abiotic factors (temperature, water supply, etc.) or on biotic interactions (competition, predation, etc.). Strong biotic interactions certainly do occur (as you will see in Book 2) so although it is easy to view all major characteristics of desert plants and animals as adaptations to the harsh abiotic environment, this temptation should be avoided.

Summary of Section 3.7

- The main features of the hot desert environment are low and erratic rainfall; a wide range of seasonal and daily temperature with cooler and relatively humid conditions at night and beneath the ground surface; dry and often dust-laden winds; and shallow or unstable soils. Heavy rain runs off higher **run-off habitats** causing much soil erosion and accumulating underground in lower **drainage habitats**. Salt pans may develop when flooded drainage basins dry out.

- Because of the erratic rainfall, two kinds of communities live in hot deserts: **avoiders** are active for only a brief period after rain and dormant for the rest of the time; **persisters** are metabolically active for all or most of the time. Avoiders are not highly specialized to restrict water loss. Most plant avoiders can be described as ephemerals, summer annuals or winter annuals depending on whether they appear regularly at a particular season. The proportions of C4 and C3 plants depend on temperature and moisture conditions in the habitat and avoiders are characterized generally by:
  - a short life cycle which is completed before soils dry out fully;
  - production of large numbers of highly resistant seeds, which form a seed bank that is a major source of food for granivorous animals;
  - mechanisms which ensure that seeds germinate only under suitable conditions and that the seed bank is never totally depleted.

- A few plant avoiders are perennials that survive dry periods as underground tubers, bulbs or rootstocks. Animal avoiders include amphibians and insects, and crustaceans that spend all or part of their lives in water.

- Tolerance of desert conditions by persisters is determined by their relative ability to obtain water, conserve it and withstand desiccation and/or high body temperatures.
  - The most tolerant plants are succulents (e.g. cacti) and extreme xerophytes (e.g. *Larrea tridentata*); they grow in run-off habitats.
  - Less tolerant plants vary widely in ability to conserve water and some grow in drainage habitats where they tap water supplies deep underground. Many shed leaves during prolonged drought.
  - Desert animals do not tolerate such high body temperatures and extreme desiccation as the extreme xerophytes. Invertebrates, reptiles and small mammals mostly avoid the problem by appropriate behaviour patterns (e.g. nocturnal activity) and access to cool underground microhabitats. They show a range of mechanisms for obtaining water and conserving it and dissipating excess heat.
  - Large mammals (e.g. antelopes, camels) tolerate some degree of hyperthermia and desiccation but must drink and avoid lethal hyperthermia by sweating.

Question 3.11   *(Objective 3.9)*

List at least six reasons (more if you can) why the environment of hot deserts is described as 'extreme'.

Question 3.12   *(Objectives 3.1 & 3.9)*

Three species of desert plant show the following characteristics:

Species A has a deep root system, a high rate of transpiration, and cannot tolerate leaf temperatures above 40 °C.

Species B has C4 photosynthesis, a shallow root system, a high rate of transpiration and produces two kinds of seeds which require different conditions for germination.

Species C has a shallow, spreading root system, a low rate of transpiration, tolerates leaf temperatures of 55 °C or more and keeps stomata closed throughout the daytime.

For each species, select one or more items from (i–viii) and one from (ix) and (x) that can be applied.

(i) Ephemeral; (ii) extreme xerophyte; (iii) succulent; (iv) winter annual; (v) drought tolerant; (vi) resistant to desiccation; (vii) likely to grow mainly in run-off habitats; (viii) likely to grow mainly in drainage habitats; (ix) avoider; and (x) persister.

Question 3.13   *(Objective 3.9)*

Which factor(s) might limit the distribution or abundance of:

(a) small rodents in a hot, rocky desert?

(b) wild camels in any hot desert?

Question 3.14   *(Objective 3.10)*

In a study of seedling establishment in the barrel cactus *Ferocactus acanthodes*, Jordon and Nobel (1981) measured seedling height in a 1 ha area and estimated seedling age on the basis of net production measurements. They concluded (in 1980) that no seedlings had established successfully since the late summer of 1976. Germination requires high temperatures and occurs only in summer. Figure 3.53 shows rainfall and soil moisture data for the Sonora desert, where this study was carried out.

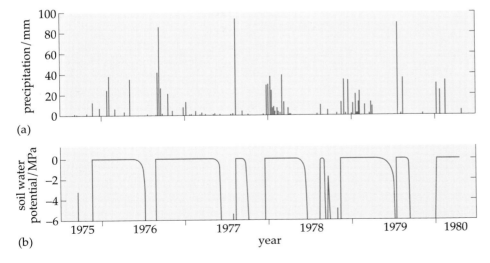

Figure 3.53 Rainfall (a) and soil moisture (b) in the Sonora desert. Note that soil moisture is measured as soil water potential; a moist soil has a value of zero and the more negative the value, the drier the soil.

(a) From Figure 3.53, deduce why 1976 allowed successful seedling establishment whereas 1977–1979 did not, and suggest what is likely to be the critical requirement for seedling establishment.

In further studies, it was found that *Ferocactus* seedlings die if they lose more than 84% of tissue volume through loss of water. Study Figure 3.54 and read the legend carefully.

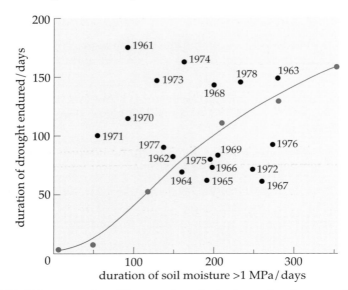

Figure 3.54 Tolerance of seedlings to drought is shown by the green line. The relationship shows that drought tolerance (measured by how long seedlings can survive without water) increases with the length of the growing season (measured as duration of soil moisture > 1 MPa) preceding the drought. The area of the graph below the green line represents all combinations of growing season length and drought duration that permit seedlings to survive. The area above the line represents all combinations of these variables that seedlings cannot tolerate. Black dots show the environmental conditions that occurred in the years shown.

(b) From Figure 3.54: (i) for how many years between 1961 and 1978 was seedling establishment possible, and for what reasons; (ii) do the data in this Figure confirm your answer to (a)?

3.8 Light and daylength

Light is a form of radiant energy usually defined as that part of the electromagnetic spectrum (with wavelengths from roughly 400–700 nm) which is visible to the human eye. Like water, it is an ecological factor that is both a resource and a condition (Table 3.4). For green plants, light is a source of energy which is trapped and used in photosynthesis but for both plants and animals, the relative length of day and night (**photoperiod**) and the amount or ratios of certain wavelengths are important conditions affecting growth and development. Radiant energy with wavelengths just shorter or longer than those of visible light (ultraviolet, UV, and infrared radiation respectively) affects or can be used by certain organisms: UV damages cells and causes mutations, for example, and some photosynthetic bacteria use longer wavelengths between 800 and 900 nm in photosynthesis. We include the effects of these invisible wavelengths in the discussion of light.

3.8.1 Light and photosynthesis

Light and CO_2 are the two resources that may limit carbon fixation in photosynthesis. If plants have adequate supplies of water and mineral nutrients (and many, of course, do *not*), then their growth (i.e. increase in dry weight) will be limited by the amount of carbon they can fix. In these conditions, dominant plants can be expected to have a photosynthetic strategy that maximizes net carbon fixation.

❑   C4 plants occur mainly in open, tropical lowlands. From your knowledge of C3 and C4 plants and information in Section 3.6.2, suggest an explanation for this distribution.

■   C4 plants have higher net carbon fixation than C3 plants when light intensity and temperatures are high – as in tropical lowlands. The flux of light energy is very high in open tropical habitats and so photosynthesis will tend to be CO_2-limited. C4 plants have an internal CO_2-concentrating system that circumvents this problem and allows them to utilize much more of the available light than C3 plants. The same system also reduces carbon loss through *photorespiration* whereas C3 plants lose much carbon by this process when temperatures are high.

These two points, the better performance of C4 plants at high light levels and temperatures, are illustrated in Figure 3.55. 'Performance' here is related to net photosynthesis, NP, which defines the carbon available for growth and is the difference between total CO_2 fixed (gross photosynthesis) and CO_2 lost in respiration, both photorespiration (in the light only) and normal, maintenance respiration (in both dark and light). Bear in mind also (Section 3.5.2) that the better water-use efficiency of C4 plants can be an advantage over C3 plants in very dry conditions.

Overall, the evidence suggests that when photosynthesis is limited by CO_2 supply, and especially in hot, dry conditions, C4 plants have the edge over C3 plants and usually dominate herbaceous vegetation. And yet most plants are C3: virtually all trees and those growing at higher latitudes (temperate/polar zones) and altitudes. This is explained largely by two factors, one relating to temperature and the other to light.

Try to think what these explanations are, and then read on.

The *temperature* argument hinges on the fact that C4 plants usually grow no better and often significantly *less* well than C3 plants if temperatures are low (e.g. in the range 5–20 °C) for much of the growing season (Section 3.6.2). Similarly for the *light* argument, C4 plants have no advantage over C3 if photosynthesis is limited mainly by light rather than by CO_2. This is easy to understand for plants growing in shady habitats such as beneath a tree canopy but less obvious for trees, especially in the tropics. To appreciate why tree photosynthesis is indeed usually limited by light, you need to think about tree structure and the many layers of leaves in a canopy. When a beam of light passes vertically through the canopy of a beech tree (*Fagus sylvatica*), for example, it encounters, on average, four to six leaves between the top of the canopy and the ground; we say that the **leaf area index (LAI)**, the ratio of leaf area to ground occupied by the plant, is 4–6. Thus, while the upper leaves receive ample light and are usually CO_2-limited, the lower leaves are light-limited and light limits photosynthesis for the plant as a whole.

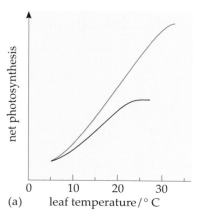

(a)   leaf temperature/° C

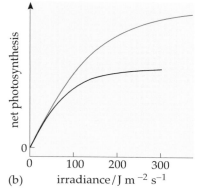

(b)   irradiance/J m $^{-2}$ s^{-1}

Figure 3.55 Rates of net photosynthesis for a C4 species (*Atriplex patula*) (green line) and a C3 species (*Atriplex rosea*) (black line) for: (a) varying leaf temperatures and (b) varying light fluxes (irradiance).

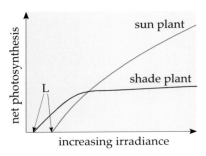

Figure 3.56 Net photosynthesis of plants from sunny and shaded habitats at different light intensities. L indicates the light compensation points (NP = 0).

Plants in shade

The lower leaves on a tree or whole plants growing in shady conditions typically show modifications of leaf structure and photosynthetic machinery which act to maximize photosynthesis in low light conditions. *Shade leaves* are thinner, paler and more delicate and, in shade, have slightly higher rates of gross photosynthesis than *sun leaves*. More importantly, shade-adapted plants or leaves have lower respiration rates and this is the main reason why they achieve higher growth rates than light-adapted plants in shady conditions: remember that NP = GP – respiration (R), and NP equates to growth potential. A useful way of summarizing these photosynthetic adaptations to shade is the **light compensation point**, which is the light intensity at which gross photosynthesis equals respiration (i.e. NP = 0): its value is lower for shade plants (Figure 3.56). Photosynthetic adaptation to shade has a 'cost', however, and this is also shown in Figure 3.56.

❑   What is the cost, i.e. when are shade plants at a disadvantage compared with sun plants?

■   In open conditions with relatively high light intensity, shade plants have much lower NP (and therefore growth) than sun plants.

In fact, shade plants may be damaged by strong light and show growth inhibition (**photoinhibition**). These striking differences in the photosynthetic responses to light of sun and shade plants are sometimes genetically fixed but not always: species such as herb robert (*Geranium robertianum*) will grow well if it develops from seed in the open and equally well if it develops in moderate shade. It *acclimates* (Section 3.4.3) to the light environment experienced and can be described as having a broad niche with respect to light.

Shade is a variable condition, however, and even on the floor of a dense forest, temporary patches of strong light may occur (**sunflecks**). Plants which occupy habitats where the canopy is relatively open and high-intensity sunflecks are frequent commonly have characteristics intermediate between those of sun and shade plants: their light compensation point is higher than that of shade plants but so is their maximal rate of photosynthesis in strong light, i.e. they can *use* the extra energy available in sunflecks.

Aquatic environments

An important additional factor influencing the relationship between light and photosynthesis for aquatic (submerged) plants is that *water absorbs light* and, in fact, different wavelengths are absorbed selectively (Figure 3.57). Both total light penetration and the selectivity of absorption are influenced by the amount of dissolved organic matter (DOM) in the water, as you can see by comparing Figure 3.57a (the Sargasso Sea, with very little DOM) and 3.57b (the Baltic Sea, with very high DOM).

❑   How does DOM influence total light penetration and the selectivity of absorption?

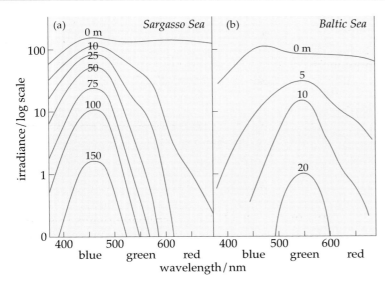

Figure 3.57 Light penetration and its spectral composition at different depths for (a) the blue waters of the Sargasso Sea and (b) the green waters of the Baltic Sea. Total irradiance at different depths is proportional to the area under the curve.

■   It reduces greatly total light penetration so that, at 20 m depth total irradiance is less than that at 150 m in the clear Sargasso Sea. DOM also absorbs selectively the shorter (blue) wavelengths so that the light penetrating most deeply is green–yellow rather than blue–green as in the Sargasso.

Because water absorbs light, photosynthesis is possible only at the surface of a water body and the depth of this **euphotic zone** is influenced also by the numbers of photosynthetic organisms (mostly unicellular algae). It ranges from about 200 m in clear, nutrient-poor water where algae are scarce to only a few cm in nutrient-rich water containing vast numbers of algae. The ways in which light supply interacts with nutrients and other factors and affects the abundance of algae are discussed in Chapter 4, Section 4.8.1.

3.8.2 Shade tolerance in plants

We emphasized in the previous Section the role of light as a resource and the different photosynthetic characteristics of sun and shade plants. This is clearly relevant to the question of **shade tolerance** but it is not the whole story. When plants grow beneath other plants – the commonest type of shading – they experience not only a low supply of light but also a qualitatively different light environment.

❑   From Figure 3.58, how and why is light quality changed in leaf shade?

■   As light passes through leaves, certain wavelengths, particularly those absorbed by photosynthetic pigments, are filtered out. The transmitted light contains a much higher proportion of wavelengths longer than about 660 nm.

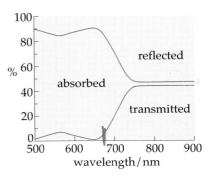

Figure 3.58 The percentage of wavelengths between 500 and 900 nm that are transmitted, absorbed and reflected by wheat leaves.

An important characteristic of leaf shade is the *decreased ratio of red to far-red wavelengths* (R (660 nm) : FR (730 nm)): the ratio may change from 1 : 1 in sunlight to around 1 : 10 after transmission through a single leaf. Light now becomes important as a *condition* because the R : FR ratio influences the activity of the regulatory pigment **phytochrome** which controls a wide spectrum of plant activities. These include stem elongation, leaf expansion and germination: seeds of many plants do not germinate in environments rich in far-red, so these species cannot become established in densely shaded habitats. The variable nature of phytochrome-controlled responses is a major determinant of shade tolerance.

❑   What advantage might there be for a plant whose seeds remain dormant when the R : FR ratio is lower than in sunlight?

■   The seeds will not germinate when competing plants are already established but will germinate only if or when there is no shade from leaves. In these conditions, seedlings are more likely to survive.

Extreme intolerance of shade is most common among plants that grow in bare, disturbed areas, particularly the pioneer species of primary and secondary succession (Book 3) and the species of sparsely vegetated, harsh environments such as sand dunes or deserts. The right-hand side of Table 3.12 lists the main features related to **shade intolerance**, some linked to photosynthesis and production (e.g. 1 to 3) and others to phytochrome responses (e.g. 4 and 6). The list of features related to **shade tolerance** (middle column of Table 3.12) is more complicated because two quite different kinds of plants are involved. On the one hand (Type I) are understorey species – summer-flowering woodland herbs, some shrubs and trees and many mosses – that may reproduce and spend their whole life in shade. On the other hand (Type II) are seedlings of trees that are dominant in climax forest; they establish in shade and persist – sometimes for years – as slow-growing, stunted plants with small leaves until a break in the canopy (e.g. through tree fall) allows them to grow, mature and reproduce. These seedlings are essentially playing a 'waiting game' and Grime (1979) has summarized their strategy in shade as one of survival with minimum expenditure of energy.

By contrast, shade-intolerant species with large seeds and the shade-tolerant species that may complete their life cycle in shade often show *increased* stem elongation and leaf area of seedlings when shaded (item 4, Table 3.12). The response may be interpreted as maximizing light interception or even as foraging for light but usually there are differences between the two types of plants. For shade-intolerant plants, it is essentially a short-term response, effective as a light-seeking response during brief periods of shading by neighbours, for example, but not sustainable for long periods; if shading is deep and prolonged, seedlings show excessive, weak stem growth (**etiolation**) with *reduced* leaf expansion and eventually they collapse and die. For shade-tolerant plants, however, although stems and petioles are usually longer in shade, they do not collapse and leaves are typically larger and thinner. Questions 3.16 and 3.17 at the end of this Section are about responses to shade.

There is in reality much variation in the morphological changes produced by shade, particularly among Type I shade-tolerant species. What matters most, however, is not the precise nature of morphological changes but the overall performance of a plant – its gain in weight and capacity to

Table 3.12 Characteristics of shade-tolerant and shade-intolerant species.

Feature	Shade-tolerant species		Shade-intolerant species
1 respiration rate	low; little increase with temperature		high; increases sharply with temperature
2 light compensation point	low; <1% of full sunlight		high; about 5% of full sunlight
3 growth rate (dry wt. increase)	slow ± shading		fast in light, slow in shade
4 seedling growth:	*whole life in shade (Type I)*	*seedlings of climax trees (Type II)*	
(a) in strong light	elongation and leaf expansion inhibited	elongation and leaf expansion promoted	elongation usually promoted
(b) in shade	elongation increased, plants well supported (no etiolation); leaf area increases	elongation and leaf expansion inhibited in proportion to the degree of shading	(i) if seeds large: excessive weak growth (etiolation), plant often collapses; leaf area may increase (ii) if seeds small: stunted growth, little elongation
5 resistance to fungal attack	high		low (especially if etiolated)
6 seed germination under a leafy canopy	little or no inhibition		inhibition by low ratio of R : FR wavelengths
7 flowering and seed production	inhibited only by deep shade		inhibited by light shade
8 seed weight	relatively high		relatively low
9 evergreen leaves	common for herbs		rare for herbs

reproduce in shade. Some Type I species, such as bluebell *Hyacinthoides non-scriptus*, grow and reproduce better in open habitats *provided* they are not heavily grazed or surrounded by vigorous competitors; they occur commonly in lightly shaded conditions because their shade tolerance is greater than that of competitors from open sites. In contrast, the wood speedwell *Veronica montana* appears to be physiologically adapted to shade and unable to perform well in open habitats: Dale and Causton (1992) found that *V. montana* raised in unshaded sites showed leaf yellowing (*chlorosis*), probably caused by photo-oxidative damage, and produced hardly any flowers.

Certain plants which grow on the floor of lowland tropical rainforest in constant deep shade show a striking *blue–green iridescence* on their leaves, which has been interpreted as a shade adaptation. Lee and Lowry (1975) investigated this in the fern ally *Selaginella willldenovii* and showed that the outer walls of epidermal cells acted like an interference filter, much like the coating on a camera lens. The effect was to increase reflectance of wavelengths least active in photosynthesis (400–500 nm) and increase penetration into the leaf of the most active wavelengths (600–680 nm). In addition, epidermal cells are egg-shaped with a curved outer surface and appear to function as lenses, focusing light onto specially oriented chloroplasts in the cells below.

Such specialized mechanisms are comparatively rare, however, and the ecological significance of the three general strategies related to shade tolerance can be summarized as:

1   Type I shade-tolerant species can all establish, grow and compete well in shade. Moderately shade-tolerant species may reproduce in moderate (but not deep) shade and they are usually excluded from open habitats by competition. Extremely shade-tolerant species can reproduce in deep shade, may show specialized adaptations to shade and may be unable to grow or reproduce in open habitats irrespective of competition.

2   Type II shade-tolerant species are mostly trees which can regenerate under a dense canopy of vegetation. Their *seedlings* are extremely shade tolerant and may persist in shade for years but the full life cycle is completed only if a gap arises in the canopy.

3   Shade-intolerant species cannot establish, usually grow slowly and compete poorly in shade. They grow rapidly and may be strong competitors in open habitats and are typically either colonists of bare, disturbed ground or dominant herbaceous species.

Many species, inevitably, do not fall neatly into one of these groups but it should be clear from this Section that shade tolerance has a very significant effect on the local distribution of plants.

3.8.3 Photoperiod and UV

Two further effects of light as a *condition* are considered in this Section (R/FR ratios were discussed in Section 3.8.2). The first is light periodicity or **photoperiod** – the relative length of day and night – which, for all except the equatorial zone, is the most reliable indicator of season of the year. The second is the amount of short-wave ultraviolet (UV) radiation reaching the Earth's surface, none of which is used in photosynthesis but which can cause biological damage.

Photoperiodic responses

An ability to 'sense' the time of year is important for nearly all plants and animals that experience seasonal climates, whether hot and cold or wet and dry. This is because vital activities such as breeding, flowering, migration or entering or breaking dormancy are carried out at particular times of year. Some species use photoperiod as their sole or major seasonal cue: **short-day species** flower or breed only when nights are longer than a critical length and **long-day species** when they are shorter than a critical length. Common long-day plants in Britain are dog daisy (*Leucanthemum vulgare*), meadow foxtail grass (*Alopecurus pratensis*) and red clover (*Trifolium pratense*).

❑   Many organisms breed or flower only if critical photoperiods are accompanied by suitable temperatures; some long-day plants require a period of winter cold before flowering is initiated in long days. Suggest a reason for this dual control by temperature and daylength.

■   Seasonal climate can be highly variable and a control system involving temperature takes this into account so that, for example, flowering does not occur too early in a very cold spring. The winter cold requirement of long-day plants ensures that seeds germinating in spring produce plants that grow in their first year but flower only in their second, when they have substantial food reserves. This is typical of biennial plants.

When species have a wide latitudinal range, then they are either day-neutral, depending only on temperature or moisture cues, or they show different, genetically determined photoperiodic requirements in different parts of their range, i.e. they are **photoperiodic ecotypes**. Alpine sorrel *Oxyria digyna* (Figure 3.59), for example, grows on British mountains and also in the high arctic tundra, where it requires much longer days to induce flowering than in Britain. Species may thus evolve a response to daylength such that they carry out vital activities at the appropriate times of year and this, presumably, is how ranges have expanded across latitudes in the past. A current concern is whether global warming and climate change could force species to shift to higher latitudes too rapidly to allow the evolution of these responses: we do not know how rapidly climate change might occur nor how long it takes for new photoperiodic responses to evolve.

UV radiation

The same problem of evolved responses to global change occurs again in relation to UV radiation. This time, however, the change is not global warming but thinning of the stratospheric **ozone layer** which filters out most of the more damaging UV radiation (**UV-B**, 280–315 nm) before it reaches the Earth's surface. Thinning is caused by release into the atmosphere of ozone-destroying chemicals such as *chlorofluorocarbons* or CFCs (used as refrigerants and in aerosols and fire extinguishers) and was first detected in the mid-1980s over Antarctica in spring. Despite restrictions on the use of CFCs, ozone layer destruction in spring continues in the Antarctic, has now been detected over the Arctic and appears to be spreading to lower latitudes in both hemispheres.

Figure 3.59 Alpine sorrel *Oxyria digyna* (×1), an arctic alpine plant that shows different photoperiodic requirements for flowering in different parts of its range.

Damaging effects of UV-B include increased rates of DNA mutation (e.g. leading to skin cancer in humans) and reduced photosynthesis, abnormal development and suppression of flowering in plants. Phytoplankton and marine bacteria (*bacterioplankton*) appear to be especially sensitive. Because the atmosphere below the ozone layer absorbs most of the residual UV-B, there is considerable variation in UV-B input for different parts of the Earth.

❑ From general knowledge and common sense, which two types of areas on Earth will receive the highest UV-B inputs?

■ High altitudes and the equatorial zone: in both of these, light traverses a shorter distance before reaching the Earth's surface (because of the angle of light rays and/or atmospheric thickness). Both are well-known as areas with 'strong' sunlight where people are easily sunburnt!

Alpine and tropical plants have evolved protective mechanisms against UV-B. The bluish appearance or *glaucous bloom* of leaves in many high alpine species, especially conifers, is caused by a special layer of cuticular wax, and Clark and Lister (1975) showed that wiping leaves to remove the wax greatly increased UV-B penetration into leaves: the wax acts like a protective suncream. Leaf epidermal cells in other alpine species contain phenolic compounds that absorb UV-B selectively and while some lowland species can synthesize such compounds and acclimate to UV-rich light, others species cannot do this. The ecological worry is whether high latitude species, such as those of the arctic tundra and taiga (the conifer belt), could evolve such protective mechanisms should ozone thinning continue. High latitude phytoplankton are a particular concern because UV-B can penetrate between 10 and 40 m through seawater, depending on clarity, and these sensitive plants support a rich diversity of polar marine life. Currently (1995), there is insufficient knowledge to reach any consensus about whether increased UV-B would or would not cause serious damage to phytoplankton. There is evidence, however, that surface bacterioplankton *are* inhibited by UV-B and appear to have evolved no protective mechanisms (they were just as sensitive as species that live permanently in deep water – Hernd *et al.*, 1993): this could be important because bacterioplankton play a vital role in decomposition and the recycling of mineral nutrients in surface waters (Book 4).

Summary of Section 3.8

- Light as a resource may limit photosynthesis. If light levels and temperature are high and photosynthesis is not light-limited, the C4 strategy is usually optimal (e.g. lowland tropics); in cool, low-light environments and when light-limitation occurs, the C3 strategy is usually optimal.

- Light limitation may occur because of high leaf area indices or shady conditions on land. Plants may adapt to shade by acclimation or show genetically fixed shade tolerance; shade plants have a low light compensation point and, in the open, show photoinhibition and grow less well than sun plants. Plants characteristic of habitats with dappled shade (frequent sunflecks) are intermediate between sun and shade plants. In aquatic environments, light absorption by water is influenced by the DOM content and the numbers of algae present.

- Light acts as a condition because: (a) the R : FR ratio decreases in shade and influences a wide range of phytochrome-controlled processe (e.g. germination); (b) photoperiod influences many processes which are carried out at particular seasons (e.g. flowering, breeding); (c) UV radiation can be damaging and varies with altitude and latitude: some plants show adaptations to high-UV environments. It is uncertain whether high latitude species will adapt to increased UV-B caused by ozone-layer thinning.

- Shade tolerance involves both photosynthetic adaptation and appropriate responses to low R : FR conditions. Characteristics of shade-tolerant and intolerant plants were listed in Table 3.12. Two types of shade-tolerant species were distinguished, those that live permanently in shade and those that tolerate shade only as seedlings (mostly trees dominant in climax forest). The ecological significance of shade tolerance and intolerance was summarized at the end of Section 3.8.2.

Question 3.15   (*Objective 3.1*)

Explain which of statements (a) to (e) are false, partially false or wholly true.

(a) Shade plants that are able to grow in deep shade harvest light for photosynthesis very efficiently and this means that they have very low light compensation points.

(b) The leaf area index of a tree is influenced by leaf arrangement and canopy shape.

(c) It is likely that the proportion of C4 grasses and sedges in tropical grassland decreases from lowland to high altitude sites.

(d) The depth of the euphotic zone depends only on the amounts of dissolved organic matter and inorganic particulate matter.

(e) Sun plants that were unable to germinate on the forest floor may do so in shade cast by rocks or buildings.

Question 3.16 *(Objectives 3.7 & 3.10)*

Figure 3.60 shows for three species, X–Z, how leaf area changed when seedlings were grown for three months under increasing shade. The species were (not in order): *Fagus grandiflora*, a Type II shade-tolerant species of beech; *Lamiastrum galeobdolon*, a moderately shade-tolerant, Type I herb that grows in British woods; and *Liriodendron tulipifera*, the shade-intolerant tulip tree. Match each of X–Z with one of these three species, giving reasons for your choice and, after checking your answer, label the curves in Figure 3.60.

Question 3.17 *(Objectives 3.7 & 3.10)*

Use the information in Table 3.13 and Figure 3.61 to classify species A and B as Type I or Type II shade tolerant or as shade intolerant. Give reasons for your answer.

Table 3.13 Seedling characteristics of the two herbaceous species shown in Figure 3.61. F, full sunlight; S, deep shade.

Species	A	B
seed wt./mg	1.62	0.08
dry weight gain in F/mg plant^{-1}	10.7	10.8
dry weight gain in S as % of F	27	3

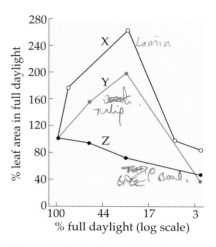

Figure 3.60 Leaf area (relative to that in unshaded conditions) as a function of shade treatment over a period of 13 weeks for three species, X–Z.

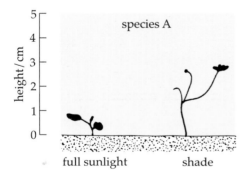 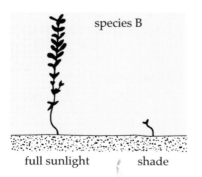

Figure 3.61 Silhouettes (× 0.6) and shoot growth of seedlings of two herbaceous species (A and B) after eight weeks in full sunlight or in shade.

Question 3.18   *(Objectives 3.1, 3.5 & 3.6)*

Comment on or suggest explanations for the following observations:

(a) Although bluebells *Hyacinthoides non-scriptus* are regarded as a typical woodland species, the best stands often occur in woodland clearings.

(b) In one of the Hawaiian Islands, a shrub with C4 photosynthesis grows in the shaded understorey of moist tropical forest.

(c) If significant global warming occurred over about 50 years, it is possible that a sub-tropical long-day plant that normally flowered in midsummer when daylengths were longer than 13 hours could spread to higher latitudes; it is unlikely that a short-day plant which normally flowered in late spring when daylengths were shorter than 12 hours could do so.

3.9 Fire

In a Chapter concerned mainly with climatic factors, which act continuously on organisms, the inclusion of fire, an *intermittent* factor, may seem odd. However, over about half the Earth's land surface the type of community (e.g. whether grassland or forest) and the relative abundance of species present are determined largely by the interaction of fire and climate. If you look at the biome map (Figure 3.15a), the only biomes which never burn naturally are the polar ice caps and tropical rainforests. Some or all parts of the other biomes experience fire with a frequency ranging from once every 1000 years (e.g. some areas of alpine tundra and moist coniferous forest) to every 1–25 years (mid-continental grasslands and savannah) (Chandler *et al.*, 1983). In Figure 3.15b, the biomes with a mean annual precipitation of between about 25 and 250 cm and mean annual temperature above −5 °C are those where fires occur.

❑   From the information above, what two characteristics of vegetation influence most strongly the chance of fire?

■   (i) Its flammability, which usually equates with dryness; hence the forests in very wet areas do not burn. (ii) The biomass – there must be sufficient material to sustain a fire; hence the driest deserts and coldest tundra areas do not burn.

Fire acts primarily as an agent of *disturbance* but it also causes longer-term changes in local conditions of soil, climate and mineral nutrient supply, so it affects both conditions and resources. For the past few thousand years, fires have been started deliberately or accidentally by humans: on the African savannahs, the North American prairies and the Australian grasslands, people have used fire as a means of encouraging the growth of grass and discouraging 'scrub'. But for probably 70 million years, fires have been started naturally by another agent – *lightning*. There are now species and whole communities that cannot persist unless fires occur with a certain minimum frequency. Such **fire climax** systems (Book 3) are typical of areas with a Mediterranean climate (hot dry summers and warm, wet winters) but in virtually all regions with a definite dry season, fire has become an essential part of cyclic change in ecosystems. It is necessary to understand this role of fire in order to manage ecosystems both for conservation and for timber production.

The ecological impact of fire depends on three interacting variables:

1 The timing, frequency and severity of fires (Section 3.9.1).
2 The changes caused by fire to local conditions of soil and climate (Section 3.9.1).
3 The extent to which plants and animals are killed or damaged by fire and the relative ease of regeneration or re-invasion after fire (Section 3.9.2).

3.9.1 The severity of fire and effects on soil and local climate

The *severity* of a fire depends mainly on the temperatures reached and the speed with which the fire spreads. These in turn are a function of climate, vegetation, topography and soil. *Dry conditions* are the first prerequisite for a severe fire – which is why rainforests virtually never burn naturally. In dry weather, the *greater the mass of combustible material* and the drier the material and soil (especially the litter layer), the hotter the fire. In addition, species such as pines with a high wax or oil content burn at especially high temperatures. The rate of fire spread and to some extent its temperature are influenced by *wind speed* and fires also move fastest *up slopes*, the rate of spread approximately doubling for every 10° increase in slope. Where the *soil has a high content of organic matter* (e.g. peat or landfill sites), a severe fire will cause the soil to smoulder, sometimes for weeks, greatly delaying recolonization and stabilization of the site.

Crown fires, which move along the top of vegetation and spread down to soil level, are the hottest and most devastating: temperatures at the soil surface may reach 800 °C and all litter is consumed, leaving bare earth. However, soil is a good insulator and temperatures rarely exceed 100 °C at a depth of 5 cm unless the soil is peaty, in which case the peat may ignite and burn down to 30 cm or more. **Surface fires** at ground level are much less severe and are the type commonly used for land management (as on heather moors); soil surface temperatures rarely exceed 200 °C and litter is scorched but not always consumed. An important point is that human suppression of fires has often caused a decrease in fire frequency but the resulting increase in litter and plant biomass has led to increased severity of fires when they *do* happen. Some of the after-effects of fire on local conditions of soil and climate are listed below.

- *Temperature*: removal of soil and litter exposes soil to full sunlight and blackened soil absorbs heat readily. At 10 cm depth, temperatures may increase by up to 10 °C and increases in maximum surface temperatures of up to 26 °C have been recorded.

- *Erosion*: soil is more susceptible to erosion after removal of vegetation and litter, especially on steep slopes.

- *Soil nutrients*: when plants and litter are burnt, inorganic nutrients are released and the ash increases nutrient levels in the upper layers of the soil. This is the basis of 'slash and burn' agriculture, still used by local people in some tropical rainforests. The exception is *nitrogen*, of which about 70% may be lost in smoke as gaseous compounds.

❑   From Figure 3.62, identify the three factors that influence nitrogen loss.

■   (i) Temperature – the hotter the fire, the greater the loss of N.
    (ii) Water content – the drier the soil and litter, the greater the loss.
    (iii) Location – higher temperatures are required to cause N
    volatilization from litter than from soil organic matter.

•   *Soil pH and microbial activity:* fire typically increases soil pH and this
    (plus warmer conditions) causes an increase in microbial and especially
    bacterial activity. Residual organic matter is rapidly decomposed and
    nitrification (conversion of organic matter to ammonium and then to
    nitrate) increases. The net result may be a rise in the soil nitrogen
    available for plant growth.

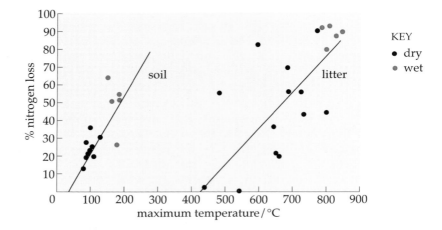

Figure 3.62   Volatilization of nitrogen from soil and litter in chaparral systems in
relation to maximum fire temperature.

In general, post-fire conditions are very favourable for plant growth:
conditions of light and temperature are improved and the supply of
nutrient resources increased. What matters is ability to survive the fire or
recolonize quickly. We discuss this below.

However, there is one further effect of fire which is receiving increasing
attention and that is its *'polluting' effect on the atmosphere.* Satellite images
have made it possible to estimate the extent of biomass burning (Figure
3.63) and then, from the nature of the vegetation burnt, estimate the
amounts of various trace gases released to the atmosphere. The extent of
fires is astonishing: in Africa alone, about 10 million square kilometres of
savannah grassland are burnt annually (Figure 3.63 – Cahoon *et al.*, 1992).
Biomass burning releases 50–100% as much CO_2 as does burning fossil
fuels; it is the largest source of atmospheric carbon monoxide (CO) and
releases significant amounts of methane (CH_4), ammonia (NH_3), nitric
oxides (NO and NO_2), ozone and methyl cyanide (CH_3CN) (Brasseur and
Chatfield, 1991). All of these trace gases appear to be increasing; some (e.g.
CO_2 and methane) are greenhouse gases that contribute to global warming;
others, e.g. CO and methyl cyanide, influence atmospheric chemistry and
the rates of removal of other trace gases. The point is that burning
vegetation by natural or human agents is now seen to occur on a large-
enough scale to have global effects on the atmosphere.

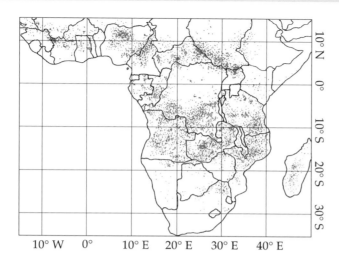

Figure 3.63 Map constructed from satellite imagery showing occurrence of fires in African savannah during 1987.

3.9.2 Effects on plants and animals

Species can be classified into three broad groups with respect to fire:

(i) **resistant species** can survive as adults all but the most severe fires;

(ii) **rapid recolonizers** are usually killed by fire but either have propagules (seeds or eggs) that survive or some means of rapid recolonization from outside the burnt area (e.g. wind-blown seeds or flying animals);

(iii) **sensitive species** are killed by fire and have no special mechanisms for rapid recolonization.

An obvious but important point is that the more frequently a community experiences fire, the higher the proportion of types (i) and (ii) species and the more specialized they tend to be. Communities that have been regularly burnt over long periods of time are often described as a **fire climax** and an interesting suggestion is that the structure and high flammability of plants in these communities actually *increases* the frequency of fire.

Resistant species

Animals that can retreat underground or move rapidly out of the way can be classified here but the great majority of truly resistant species are plants. Thick insulating *bark* can protect the living tissues of trees, e.g. cork oaks *Quercus suber* in chaparral, and Harmon (1984) found a linear relationship between bark thickness and survival of low-intensity surface fires for dry oak–pine forest in Tennessee, USA. *Buds* may be protected beneath bark, by dense tussocks at ground level, by moist, densely packed leaf bases in aerial organs (as in some palms), or on underground organs such as tubers and rhizomes – soil being an effective insulator against heat. Regrowth from these protected buds is the key mechanism by which resistant plants recover. Trees such as cork oak and some Australian eucalypts sprout rapidly after fire because dormant buds protected by bark are stimulated to grow by loss of foliage; this is called **epicormic growth**. Many shrubs sprout from buds on roots or at the stem base and some have large woody swellings called **lignotubers** which have numerous buds and are regarded as a fire-survival mechanism. Monocotyledons which have basal leaf meristems protected in dense tussocks can simply regrow new leaves and are often the first plants to show new greenery after grassland fires.

All these protective mechanisms vary with plant age so that very young plants are usually poorly protected and resprouting ability often declines with age. This is illustrated for heather by the data in Table 3.14: heather older than about 15 years gradually loses the ability to produce new shoots after burning, so burning rejuvenates only relatively young plants. Harmon's (1984) study of oak–pine forest highlighted another effect of this age-related variation in fire resistance: changes in fire frequency can have dramatic effects on species composition. In his study area, fires occurred about every 10 years before 1940 and the community was dominated by fast-growing, thick-barked trees such as *Pinus rigida*. After 1940, active fire suppression (this was a National Park area) had allowed slow-growing, thin-barked species such as *Acer rubrum* and *Quercus coccinea* to reach an age and size that was resistant to fire. So fire could no longer return this community to its pine-dominated state.

Table 3.14 Effect of age on the ability of heather (*Calluna vulgaris*) to regenerate after burning.

Age of *Calluna* stand/yr		12–13	16–18	23–25	32–39
% of stem bases sprouting	mean	57.7	14.3	14.3	2
	range	28–80	9–29	0–20	0–5

Among fire-resistant monocots, including grasses, lilies and orchids, fire often has the effect of *stimulating flowering*. The result is higher seed production (with fewer seed predators around to eat them) and more seedlings, with a fine open seed bed in which to grow. This is a nice example of how an abiotic factor (fire) can affect the population dynamics of plants (Book 2).

Rapid recolonizers

The simplest examples here are species such as birch (e.g. *Betula pendula*), spruce (*Picea glauca*) and rosebay willowherb or fireweed (*Chamaenerion angustifolium*) which produce huge numbers of light, wind-dispersed seeds. Their shade-intolerant seedlings grow rapidly on fire-sites, which they can recolonize very effectively *provided* seeds reach the site at an appropriate time of year, which depends on chance. Seeds rapidly lose viability and so do not form a seed bank in the soil. Other species in fire-prone habitats produce seeds with thick, hard seed coats which protect them from fire. These seeds may form a seed bank in the soil and are stimulated to germinate either by heat from the fire (e.g. many shrubs), by removal of competitors and possibly chemical inhibitors (many annuals), or by removal of litter (some conifers). In some South African plants, seeds germinate only after exposure to *smoke*.

Finally, and most specialized of this group, there are species that retain all or a substantial proportion of seeds on the mature plant. Seeds are held in long-lived woody fruits or cones which are often attached to trunks or large branches and open to release seeds only after fire. This phenomenon is called **serotiny** ('late appearance'). The jack pine *Pinus banksiana* in North America has serotinous cones that can withstand temperatures of up to 360 °C for one minute before igniting; heating to above 140 °C melts the resin that keeps the cones closed and liberates the protected seeds inside after the fire has passed. Mediterranean pines such as *P. halepensis*, Australian eucalypts and many members of the family Proteaceae such as *Banksia* and *Hakea* spp. (from South Africa and Australia) show serotiny. Interestingly, the degree of serotiny (i.e. the extent to which cones or fruits remain closed) may vary within a species depending on fire frequency: the lower the frequency, the less serotiny. And in *Pinus banksiana* and *P. contorta*, serotiny decreases with age as bark becomes thicker – an apparent case of changing strategy from being a recolonizer to a resister.

Animals

You can see from the above that plants may benefit from fire or, indeed, depend on it for their persistence in a community. For animals it was thought that the effects of fire were largely negative but it now appears that this may not always be true. First, any herbivores able to move back rapidly into burnt areas where plants are regrowing have a feeding bonanza. Stein *et al.* (1992), working in the mountains of northern Arizona, USA, burnt willow shrubs (*Salix lasiolepis*) and then studied grazing by elk (after a May burn) and grasshoppers (after a July burn) on the new sprouts compared with grazing on unburnt control shrubs throughout the season. The change in herbivore behaviour was dramatic (Figure 3.64).

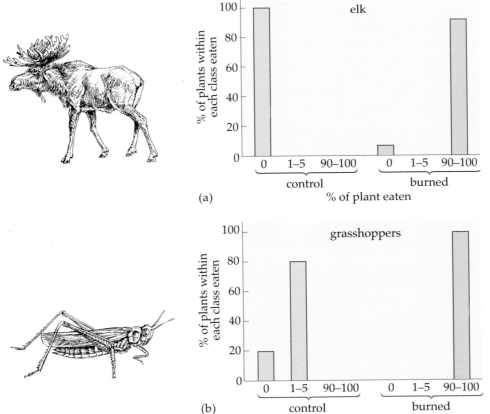

Figure 3.64 Grazing by (a) elk (×0.012) and (b) grasshoppers (×2) on control willow shrubs over a whole season and on the regrowth shoots of burnt willows.

❑ Describe these changes.

■ On the control plants over the whole season (i.e. including young spring growth), there was very little grazing by grasshoppers and no grazing by elk. The sprouts from burned shrubs were virtually all grazed intensively by both grazers.

Regrowth after burning is clearly very palatable, probably highly nutritious and causes a marked shift in herbivore diet. Any insect herbivore able to survive a fire would probably be at a considerable advantage compared with those that recolonized from outside the burnt area

3.9.3 Ecological significance of fire frequency

You should now have some idea of what the ecological effects of fire are, and we emphasized at the beginning of Section 3.9 how important this is for conservation and land management. In this Section, we give an overview of the significance of fire by considering its influence on vegetation over a short time-scale (decades to hundreds of years) and a longer time-scale (hundreds to thousands of years).

Short-term effects: north-east American temperate forests

North American temperate forests are both more extensive and, for reasons discussed in Book 3, more diverse than European forests. They are also well studied and it has become increasingly clear that the essential nature of many forests – the species present, their size, age and density – depends on the frequency of disturbance, with fire being the main agent of disturbance before European settlement in the 19th century. The triangular area shown in Figure 3.65 is the tall grass prairie region, maintained in pre-settlement times through burning every one to ten years by native Americans. These fires affected forests bordering the prairie and forest pockets within it, all of which were oak-dominated. Native oaks (of which there are 15 common species, compared with two in the UK) have good resistance to surface fires and good recolonizing ability after severe crown fires. Intolerance of seedlings to shade (Section 3.8.2) is the other relevant characteristic: oaks cannot regenerate under a dense, closed canopy. Since European settlement, however, fire frequency has decreased markedly and the nature of the forests has changed and is still changing. The presettlement prairie border forests of Missouri, for example, were dominated by white oak (*Q. alba*).

❑ What sort(s) of species, with what characteristics, do you think are likely to be replacing white oak?

■ Species with lower resistance to fire, good dispersal ability and probably casting a dense shade (thus preventing oak regeneration) but having Type II shade tolerance.

In fact, white oak and sugar maple (*Acer saccharum*) are now co-dominants in these forests – and the understorey is almost entirely of sugar maple, which does show most of the characteristics above. Thus, within a hundred years, reduced fire frequency is causing white oak forest to change into sugar maple forest.

Oak frequency has *increased* in parts of the northern hardwood zone (Figure 3.65) where fires and/or logging have increased since settlement;

Figure 3.65 Eastern USA showing the original tall grass prairie region (darker-green triangular area) and the northern hardwood zone (lighter-green area).

fires occurred at a natural, low rate (e.g. every 800–1000 years) in most parts of this zone prior to settlement. Clearly, the nature of these North American forests is influenced strongly by fire; oak is favoured by an intermediate fire frequency (150–200 years), pines or spruce by high frequency (50 years or less) and maples and beech by moderately low frequency (every 300–350 years). This poses something of a problem for conservation: *what* exactly should be conserved? The presettlement type of forest which may (northern hardwood zone) or may not (prairie border forest) be 'natural'; the early post-settlement forest which has become familiar to people; or the next stage, likely to be maple- or beech-dominated?

Long-term effects: climate–fire interactions

Palaeoecological studies have produced a wealth of evidence that the frequency of natural fires increases when the climate becomes drier and (usually) warmer. This occurred, for example, in southern Wisconsin (Figure 3.65) and caused a shift from species-rich, mixed forests to oak savannah over the period 6500–3500 years ago. Subsequently, an increasingly cool wet climate with fewer fires led to the development of dense oak forests.

Tree cover in the **forest–tundra zone** in northern Quebec, Canada – a zone comprising scattered areas of trees or krummholz (Section 3.4.6) extending for 100–400 km from closed conifer forest to the treeline bordering treeless tundra – has been linked similarly with climate–fire interactions. The dominant species here is black spruce *Picea mariana*, which is serotinous; it has a minimum germination temperature of 15 °C and seeds shed from serotinous cones are viable in the soil for only 5–8 years.

❑ So what do you think might happen during episodes of climatic cooling accompanied by occasional fires?

■ Fires will result in seed shedding but germination will be reduced in the cooler climate, so little regeneration will occur. With repeated fires, black spruce would disappear completely.

Studies of charcoal and wind-blown sand horizons in the soil over the past 5000 years (Filion, 1984; Payette and Gagnon, 1985) indicate that this had, indeed, happened. Subsequent warming episodes allowed trees to re-establish from wind-blown seed in bare or thinned areas south of the treeline. Within the forest–tundra zone, there is thus a constantly shifting mosaic of tree cover with cycles of deforestation–reforestation determined by fire and climate (temperature). It has been suggested that climate change caused by global warming could alter fire frequencies and, if this happened, it would clearly have wide ecological effects.

Summary of Section 3.9

- Fire affects both resources (increased soil nutrients) and conditions (warmer but more erodable soil) and **fire climax** communities depend on certain minimum frequencies of fire. It is a natural agent of disturbance although often caused by human activities; currently, biomass burning is a major cause of global atmospheric pollution and contributes significantly to the rise in greenhouse gases.

- Post-fire conditions are very favourable for plant growth and any animal herbivores that can survive a fire (e.g. underground) or recolonize rapidly have a strong advantage over those which cannot.

- Fire-tolerant plant species can be classified as:
  - **resistant** – where adults survive because of protective mechanisms such as insulating bark and protected meristems that can re-sprout;
  - **rapid recolonizers** – where seeds are widely wind-dispersed or are present in a soil seed bank or on serotinous fruits. Fire commonly stimulates germination or release from fruits of these last two types.

- Repeated fires increase the proportion of resistant and rapidly recolonizing species. Changes in fire frequency have caused changes in the dominant species in north-east American forests over the past 150 years and fire–climate interactions in the forest–tundra zone have altered tree cover over the past 5000 years.

Question 3.19   (Objectives 3.1, 3.5 & 3.8)

In parts of the south Californian chaparral (a type of Mediterranean scrub), fires occur every 2–10 years. Which of the types of plants (a) to (e) are likely to be common in these areas and which are not? Explain your answer.

(a) Pine trees with serotinous cones.

(b) Shrubs with numerous underground buds at the crown of the root system.

(c) Plants that produce seeds which germinate only after fire.

(d) Plants with a capacity for epicormic growth.

(e) Herbaceous plants with underground bulbs.

Question 3.20   *(Objective 3.8)*

In the Californian chaparral described in Question 3.19, most of the shrubs that reproduce exclusively by seed and lack the ability to sprout after fire have an average lifespan of 20 years. Most of the shrubs with a capacity for sprouting will continue to grow for 50–100 years if left undisturbed. How might the composition of this community change (a) in the short term (20–40 years) and (b) in the long term if fire frequency decreased to once every 40 years? When answering (b), you should bear in mind variation between species in seed longevity and in sprouting ability with increasing age.

Question 3.21   *(Objectives 3.8 & 3.10)*

On heather (*Calluna vulgaris*)-dominated areas of the North Yorkshire moors, UK, fire is used as a management tool. Following a light burn, there is almost 100% heather cover present next year; after a more severe fire, cover is reduced for 3–4 years but increases progressively; after a very severe fire (usually accidental), the ground may remain bare of vascular plants for up to 10 years.

(a) Suggest an explanation for the relationship between fire severity and subsequent cover of heather.

Legg *et al.* (1992) suggested two hypotheses that might explain the effects of very severe fire: slow restoration of heather cover could result mainly from:

A – absence of heather seeds; or

B – conditions or lack of resources that inhibited the establishment of heather seedlings.

Observations and experiments (i)–(iii) were carried out to test these hypotheses.

(i) Most of the seed bank of heather is in the surface soil and litter with few seeds present below 6 cm depth.

(ii) The initial dispersal of heather seeds is immediately adjacent to the parent plant, with few dispersed beyond 10 m. However, winds blowing over bare peat were observed on a few occasions to effect secondary dispersal and carry litter and seeds 100 m or more.

(iii) After severe fires, 50% or more of the peat surface becomes covered by a crust of lichens and algae. Figure 3.66 shows the results of an experiment in which germination of heather seeds was measured on blocks of peat with crust present (control) and crust removed by scraping.

(b) For each of (i)–(iii), state whether it supports, refutes or is irrelevant to hypotheses A and B and give your opinion as to the most likely explanation for the long delay in restoring cover after a very severe fire.

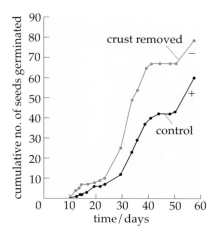

Figure 3.66 For use with Question 3.21. Germination of heather seeds on peat blocks with a crust of lichens and algae present (control) and with the crust scraped off.

3.10 Effects of climate and climate change: an overview

Most of this Chapter has been concerned with the impact on organisms of particular climatic factors (and we include fire here). In this last Section, we step back and review the overall effects of climate on living organisms (the *biota*); consider the reverse – how the biota affects climate; and discuss briefly current work on climate change and its possible consequences.

3.10.1 Effects of climate on organisms

There are many examples in this Chapter which show that climatic factors, acting singly or together, influence the distribution of organisms, i.e. where they live, grow, reproduce and, therefore, persist in the long term.

Recall some examples.

Obvious ones are the treeline (Section 3.4.6), biomes (Figure 3.15) and fire-resistant and fire-sensitive species. Strongly seasonal climates, hot versus cold or dry versus wet, also clearly impose restrictions on organisms: they must be able to survive the 'unfavourable' season, and the 'favourable' season must be long enough to allow growth and reproduction. These examples identify temperature, precipitation, fire frequency and length of growing season (which involves light, temperature and precipitation) as factors of major importance in determining distribution. Apart from fire, such factors rarely act on adult survival but are more likely to influence ability to reproduce or the survival of immature stages.

Often, however, although one is forced to conclude that the geographical limits of a species must be influenced by climate (frost-sensitive tropical species, for example, could not survive winter cold at higher latitudes), actual evidence is hard to find and identifying which climatic factor(s) is acting on which stage of the life cycle and in what ways is even harder. One reason for these difficulties is that the distribution boundary of a species and its climatic limits do not correspond. For example:

1   A species may occur regularly outside its climatic limits because it frequently disperses there (by chance or through regular migrations) but does not reproduce or survive in the long term. This applies to butterflies such as the clouded yellow *Colias crocea* which migrates to the UK in most years but cannot survive the winter there. Equally, a species may be absent from climatically suitable areas because it has limited powers of dispersal and cannot reach them. The success in the UK of many Australasian and American species introduced by humans testifies to this, as does the pest status of introduced rabbits and prickly pear cactus in Australia.

2   Long-lived species (such as trees) may persist in areas which are now outside their tolerance limits for reproduction (flowering, seed set or germination and seedling survival) but were within limits at some time in the past, as described for some species at the treeline.

3   If a species declines in some aspect of its performance as it approaches climatic limits, it may be eliminated by competition, predation or some other biotic interaction before reaching those limits. The vertical distribution of barnacle species (*Semibalanus* and *Chthamalus* spp.) on rocky shores in the UK (Section 4.1) illustrates this on a small scale.

Example 1 above demonstrate the importance of *dispersal* in determining distribution and the numbers of individuals in an area; this is discussed in Book 2, Chapter 2; example 2 illustrates the importance of longevity and past history; and example 3 demonstrates the way in which abiotic and biotic factors interact to determine distribution, which was emphasized in Section 3.1.

Two other general points about the impact of climate on organisms are summarized below:

- A species may show wide climatic tolerance either because of phenotypic plasticity plus an ability to acclimate (e.g. the eurythermal organisms (Section 3.4.3), herb robert and shade tolerance (Section 3.8.2)); or because there are populations that are genetically adapted to different climatic conditions (climatic ecotypes, Sections 3.4.3, 3.8.3).

- Occasional extreme climatic events (severe winters, prolonged drought, hurricane winds or fires) influence greatly the numbers of individuals and species locally. Numbers usually recover provided that dispersal capacity is adequate or there is a reservoir of resistant stages (e.g. a seed bank). The critical factor is the *frequency* of such disturbances and, as described for wind and fire (Sections 3.4.1, 3.9.3), this is a major factor determining which species are present in a community.

3.10.2   Effects of organisms on climate

The impact of organisms on climate has not been emphasized in this Chapter but is considerable and merits discussion here. On the scale of micro- and local climates, the effects are very obvious: if you walk into a forest on a sunny day, it is cooler, more humid, less windy and light levels are lower. Thus, vegetation profoundly affects local climatic conditions. Less obvious are larger scale regional or global effects and three examples are described below.

1   It has long been suspected that large-scale clearance of tropical rainforest affects regional climate and a number of modelling studies support this view. For example, Lean and Warrilow (1989) suggest that replacement of Amazonian rainforest by pasture will result in a 20% reduction in regional rainfall (because about half of the rainfall in this area derives from evapotranspiration, which is greater from forest than from grassland) and a rise in surface temperature of about 2 °C.

2   The structure of vegetation also profoundly affects the amount of rain which enters soil (and therefore contributes to water supply, Figure 3.35) as opposed to running off the surface and often causing soil erosion. Table 3.15 illustrates this point for a tropical area in Tanzania with a gradient of 3.5° (Sundberg, 1983).

Table 3.15   Influence of vegetation type on the percentage of rainfall which runs off the soil surface and on the amount of erosion (sediment carried away in runoff). Data relate to an area in Tanzania (slope 3.5°) where the natural vegetation (A, ungrazed scrub, a mix of woody and herbaceous plants) was replaced by different types of cultivation (B–D).

Vegetation	Annual runoff/ % of total rainfall	Annual erosion/ tonnes sediment ha^{-1}
A ungrazed scrub (natural)	0.4	0
B grass	1.9	0
C arable (maize crop)	26	78
D fallow arable (soil left bare)	50	146

❑ From Table 3.15, what four factors appear to influence most strongly the amount of runoff and erosion in this tropical area?

■ (i) The complexity of the vegetation above ground, i.e. the variability in height and number of layers of leaves (LAI, Section 3.8.1): this affects the force with which rain hits the soil and will decrease from A to D. (ii) Complexity below ground in root type and depth: this affects ease of soil erosion and decreases from A to D. (iii) Amount of bare ground between plants. (iv) Amount of disturbance by cultivation: the contrast here is between A/B and C/D, where arable cultivation involves considerable disturbance at least once annually.

3 Another effect of plants on climate is less direct. Terrestrial vegetation is a reservoir for carbon which is reduced when forests are cut down. The carbon returns to the atmosphere as CO_2 through either burning or natural decay and contributes significantly to the global rise in atmospheric CO_2 which, it is predicted, may cause global warming. Equally, the net uptake of CO_2 in photosynthesis by terrestrial plants and, even more important, marine unicellular algae (phytoplankton) is crucial in regulating atmospheric CO_2 levels (discussed further in Book 4).

3.10.3 Climate change

The possibility and likely effects of climate change in the near future is such a burning issue (not least to ecologists) that we end this Chapter with a brief review of the subject. One point to bear in mind is that climate change is *nothing new*: there were glaciations in the recent geological past and they will probably occur in the future. The Earth's climate has undergone major shifts over geological time and the only way in which the current predicted change differs is that it is caused by human activities and may be faster than past, natural changes.

In this Chapter, several types of possible climate change, arising from two basic causes, have been mentioned.

Recall the two main causes of climate change and then list as many types of possible change as you can.

The two causes are (a) increase in atmospheric greenhouse gases (Section 3.4.7) and (b) decrease in thickness or local destruction of the ozone layer (Section 3.8.3). Possible consequences of (a) are:

(i) global warming (Section 3.4.7) and a consequent rise in sea-level;

(ii) increased frequency of fires (Section 3.9.2);

and not discussed (but you may know from general knowledge):

(iii) altered patterns of precipitation (especially, lower rainfall in northern mid-continental areas and higher rainfall in arid tropical areas such as North Africa);

(iv) increased frequency of strong winds (because of larger temperature differentials between land and oceans).

Resulting from cause (b) is:

(v) increased levels of UV-B radiation, mainly at higher latitudes.

Only (v) has actually been detected. Although greenhouse gases have indeed risen (see Figure 3.67a, b), the *consequences* of this rise are still a matter for debate and none of (i) to (iv) has been detected and linked unequivocally to cause (a). In Figure 3.67c, you can see that there appears to be a recent warming trend in global mean temperature. But Figure 3.67d shows that, in Fennoscandia, summer temperatures have been fluctuating in a similar way for the past 500 years (and, in fact, these tree-ring data (Briffa *et al.*, 1990) extend back for 1400 years). So we still cannot be sure that recent warming is caused by the rise in greenhouse gases rather than being a 'normal' fluctuation. Briffa *et al.* (1990) suggest that it will be 2020 at the earliest before greenhouse warming could be detected with 95% confidence.

Such uncertainties do not mean that we should ignore possible ecological consequences of global climate change. But with so many possible changes occurring simultaneously and no means, at present, of predicting what would happen in a small area (such as the UK), the nature of ecological change is *very* difficult to forecast. We describe below two approaches to this problem, which can be described broadly as 'learning from the past' and 'experimental'.

'Learning from the past'

The essence of this approach is that changes of distribution accompanying past changes in climate (which are determined by a range of techniques outside the scope of this Chapter) are used as a basis for predicting future change.

❑   Recall from Section 3.4 an example of this approach.

■   Shifts in treeline during and after glaciation (Section 3.4.6).

This use of palaeoclimatic data necessarily provides fairly coarse information: only abundant organisms that left recognizable remains (e.g. woody tissues, pollen, insect cuticles) can be studied and the majority of smaller animals or rare plants cannot. Another limitation is that palaeoclimatic change did not include elements such as increased UV-B that are expected in the future. Nevertheless, this is perhaps the best technique available for predicting broad changes in vegetation pattern.

More recently, long-term records of weather extending back to the 19th century have been correlated with long-term monitoring, mostly of vegetation, birds and plankton (Leigh and Johnston, 1994). With such fine-grained data, weather fluctuations can obscure climatic trends but these kinds of data could be of great value in detecting the earliest ecological consequences of climate change and discovering also the speed with which species can adapt genetically to altered climatic conditions. The UK is fortunate that a few high-quality monitoring studies have been carried out over the past 30 to 150 years. One example is the study of vegetation on road verge plots at Bibury, Gloucestershire, carried out since 1958 by Professor A.J. Willis. The original aim was to examine the long-term effects of herbicide application but changes in the untreated control plots give some indications of responses to weather. For example, there were significant correlations between 'good' summers (mild and dry), growth retardation in fast-growing species (the competitive strategy, discussed further in Book 2, Chapter 5) and growth promotion in weedy (ruderal) species and slow-growing, stress-tolerant species (Grime *et al.*, 1994). 'Good' winters (mild and dry) promoted growth of the fast-growing competitors but had no effects on the other two types.

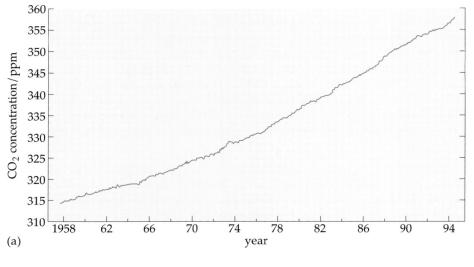

(a)

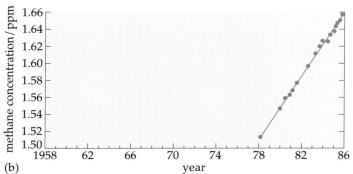

(b)

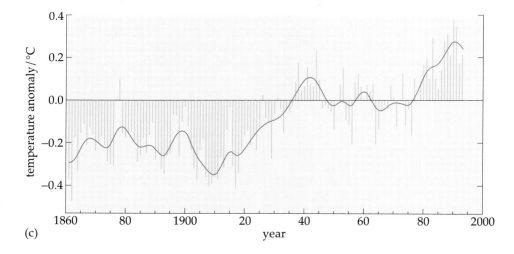

(c)

Figure 3.67 Recent increases in (a) atmospheric CO_2 and (b) methane in the Northern Hemisphere. (c) Recent warming trend shown as the deviation of mean annual global surface temperature from the average value for 1861–1993. (d) Longer-term temperature trends in northern Europe (Fennoscandia) reconstructed from tree-ring width data; the deviation of the 10-year mean summer temperature (April to August) from the 1951–1970 mean is shown.

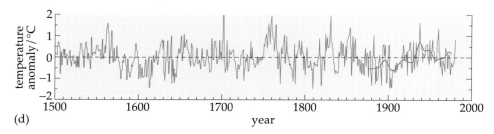

(d)

A problem with analysing data from the recent past, as at Bibury, is that vegetation responses to climate cannot be distinguished with certainty from responses to other changes. Atmospheric inputs of nitrogen, for example, have risen as a result of air pollution and atmospheric CO_2 levels have increased (Figure 3.66a). To resolve this question and also questions about the relative effects of temperature and moisture supply, plants from Bibury are being grown in large pots where they can be exposed to simulated climates – in other words, the experimental approach is being used to supplement the monitoring approach.

Experimental approaches

The method here is to grow organisms (mostly plants) in controlled conditions where specific climatic factors are altered and note the effects. At one end of the spectrum are experiments where plants in monoculture are raised in controlled environment cabinets. These can provide relevant data about crop plants but of a very limited kind.

❑   Suggest some limitations of this approach.

■   Conditions in growth cabinets are far from natural and altered interactions of crops with pests and weed species will be missed.

At the other end of the spectrum are experiments where climatic conditions are altered for small samples of natural communities in the field. Elaborate systems have been developed to mimic climate (and also CO_2) changes in as natural a way as possible; growth and reproduction can be measured and altered interactions between species studied. Most of these experiments started in the 1990s and only preliminary results are available for the majority.

In the UK, there are indications that warmer conditions and higher CO_2 levels stimulate the growth of bracken (*Pteridium aquilinum*) but not heather (*Calluna vulgaris*). So could invasion of heather moorlands by bracken increase? Possibly, but there are so many other interacting factors – increased risk of summer drought, higher wind speeds, raised UV-B, changes in pathogens or competition with other species, for example – that this conclusion must be very tentative. It is not even possible to make broad generalizations about types of plants such as C3 and C4 species (Section 3.6.2).

❑   (a) If conditions became hotter and drier, how might this affect the growth of C3 and C4 plants in open habitats? (b) Would you expect rising CO_2 levels to influence the growth of C3 or C4 plants?

■   (a) Given their high water use efficiency and lack of photorespiration at high temperatures, C4 plants would probably grow as well or better but C3 plants probably less well. (b) Growth of C3 plants would probably be stimulated with little or no effect on C4 plants.

To assess the *ecological* impact of such climatic changes, however, looking at growth alone is not enough. For example, when Morse and Bazaz (1994) studied the effects of elevated temperature and CO_2 on a C3 and a C4 annual grown from seed in tubs, the C3 species had fewer but much larger individuals (increased intraspecific competition caused higher mortality (self-thinning)) whereas the C4 species had more individuals (increased survival) but no increase in their size. Such shifts in relative size would probably affect seed output and, in the wild, competitive relationships. To understand and predict the effects of climate change, therefore, one needs to consider population properties (Book 2), community interactions (Book 3) and ecosystem processes such as nutrient cycling (Book 4).

But why bother to try to predict the effects on organisms of climate change? On the one hand are pragmatic reasons linked to human food and health. How will crop growth be affected? Will it be possible for UK farmers to grow sub-tropical crops with longer, warmer growing seasons? And how will weeds, pests and pathogens be affected? On the other hand, ecologists involved with long-term conservation need to know the most appropriate actions to take now in anticipation of future climate change. Cool, high altitude habitats might be crucial, for example, as refuges for organisms with low optimum temperatures, and conserving these habitats would then be given high priority. Equally, expensive measures to conserve coastal wetlands may be inappropriate if these areas are likely to be inundated by the rise in sea-level that would accompany global warming. Regional conservation priorities might well change, given reliable data about the ecological effects of climate change.

Question 3.22   (*Objective 3.6*)

(a) In what ways may dispersal ability influence the current geographical distribution of a species and its probability of survival in the face of future, rapid climate change?

(b) Why are phenotypic plasticity and the capacity to adapt genetically to altered abiotic conditions also likely to influence a species' chance of surviving climate change?

Question 3.23   (*Objective 3.6*)

Is large-scale forest clearance likely to have the same impact on climate for all geographical regions?

Question 3.24   (*Objectives 3.6 & 3.10*)

From 1736 to 1947, annual observations were made on the Marsham family estate, Norfolk, of the first 'signs of spring', e.g. dates of first flowering, leafing of trees, arrival of migratory birds such as swallows. Climatic data for the period were obtained from mean monthly records of temperature for Central England and from annual average rainfall records for England.

- Sparks and Carey (1995) carried out regression analysis to determine which climatic data were the best predictors for each first appearance date. They used monthly temperature from May of the preceding year to April of the recorded year and annual rainfall data of the recorded and preceding year and obtained significant relationships. For example, 73% of the variation in leafing date for oak trees could be explained by March to April temperatures with some influence from rainfall and October temperature of the preceding year.

- The relationships between climate and first appearance dates were extrapolated in a linear way to predict the effect of future climate change in Norfolk, i.e. a 3.5 °C rise in temperature from December to February, a 3 °C rise from June to August and a 10% increase in annual precipitation. One prediction using this method was that oak trees would come into leaf 22 days earlier.

(a) In the context of Section 3.10.3, what kind of study is this?

(b) In what ways are the climatic data used in the study less than ideal? What would have been better?

(c) In what ways might the dates of first appearance be unreliable as indicators of biological events?

(d) The dates at which migratory birds such as swallow and cuckoo first appeared correlated less strongly with climatic data than did dates of flowering and leafing. Suggest a possible reason for this poor correlation.

(e) If oak trees came into leaf three weeks earlier as predicted, how might this affect (i) oak trees and (ii) insects such as winter moth whose larvae feed on young oak leaves (Chapters 1 and 2)?

(f) If, based on the Marsham study, you saw a statement 'oak trees will come into leaf three weeks early in Norfolk as a result of global warming', what criticisms would you express about the statement's validity?

Objectives for Chapter 3

After completing this Chapter, you should be able to:

3.1   Define and use correctly all the terms printed in bold. (*Questions 3.1, 3.4, 3.7, 3.12, 3.15, 3.18 & 3.19*)

3.2   Describe the main types of organisms and their zonation on rocky shores in the temperate zone. (*Questions 3.1 & 3.2*)

3.3   List the main abiotic and biotic factors that influence the zonation and abundance of organisms on rocky shores and, given relevant information about these factors, assess their relative importance on particular shores. (*Questions 3.1–3.3*)

3.4   Given relevant information about an abiotic factor, decide whether it is acting as a resource or a condition. (*Question 3.6*)

3.5   Describe and give examples of the various ways in which temperature, wind, water supply, light and fire may singly or in combination affect: (a) survival and/or reproduction and (b) potential distribution of named plants or animals. (*Questions 3.5, 3.6, 3.8–3.10, 3.18 & 3.19*)

3.6   Discuss and give examples of: (a) the ways in which global climate change might affect the distribution of plants and animals; (b) the ways in which the biota influences local climate; (c) studies that aim to predict the effects of climate change. (*Questions 3.18, 3.22–3.24*)

3.7   Describe for specified animals and plants, the factors that influence: temperature tolerance, water supply to or loss from, shade tolerance and tolerance of fire. (*Questions 3.7, 3.9, 3.16 & 3.17*)

3.8   Predict or explain the effects of fire on communities or individuals given relevant information about their characteristics. (*Questions 3.19–3.21*)

3.9   Describe the main features of the environment in hot deserts and the various mechanisms which allow particular organisms to survive in this environment. (*Questions 3.11–3.13*)

3.10 Interpret experimental data in ways that are consistent with the principles described in this Chapter. (*Questions 3.3, 3.6, 3.10, 3.14, 3.16, 3.17, 3.21 & 3.24*)

References for Chapter 3

Ballantine, W. J. (1961) A biologically-defined exposure scale for the comparative description of rocky shores, *Field Studies* **1**(3), 1–19.

Baskin, C. C., Chesson, P. L. and Baskin, J. M. (1993) Annual seed dormancy cycles in two desert winter annuals, *Journal of Ecology*, **81**, 551–56.

Berthold, P., Mohr, G. and Querner, U. (1990) Control and evolutionary potential of obligate partial migration – results of a 2-way selective breeding experiment with the blackcap (*Sylvia atricapilla*), *Journal of Ornithology*, **131**, 33–45.

Boorstein, S. M. and Ewald, P. W. (1987) Cost and benefits of behavioural fever in *Melanoplus sanguinipes* infected by *Nosema acridophagus*, *Physiological Zoology*, **60,** 586–95.

Brasseur, G. P. and Chatfield, R. B. (1991) The fate of biogenic trace gases in the atmosphere, in: *Trace Gas Emissions by Plants* (eds T. D. Sharkey, E. A. Holland and H. A. Mooney), pp. 1–27, Academic Press.

Briffa, K. R., Bartholin, T. S., Eckstein, D., Jones, P. D., Karlen, W., Schweingruber, F. H. and Zetterberg, P. (1990) A 1,400-year tree-ring record of summer temperatures in Fennoscandia, *Nature*, **346**, 434–39.

Cahoon, D. R., Jr., Stocks, B. J., Levine, J. S., Cofer III, W. R. and O'Neill, K. P. (1992) Seasonal distribution of African savanna fires, *Nature*, **359**, 812–15.

Canham, C. D. and Loucks, O. R. (1984) Catastrophic windthrow in the presettlement forests of Wisconsin, *Ecology*, **65**, 803–9.

Chandler, C., Cheney, P., Thomas, P., Trabaud, L. and Williams, D. (1983) *Fire in Forestry, Vol. 1: Forest Fire Behaviour and Effects*, Wiley.

Christian, K., Tracy, R. and Porter, W. P. (1983) Seasonal shifts in body temperature and use of microhabitats by Galapagos land iguanas (*Conolophus pallidus*), *Ecology*, **67**, 169–89.

Clark, J. B. and Lister, G. R. (1975) Photosynthetic action spectra of trees. II. The relationship of cuticle structure to the visible and ultraviolet spectral properties of needles from four coniferous species, *Plant Physiology*, **55**, 407–13.

Cloudsley-Thompson, J. L. (1977) *Man and the Biology of Arid Zones*, Arnold.

Coe, M. J. (1967) Microclimate and animal life in the equatorial mountains, *Zoolog. Africana*, **4**, 102–28.

Connell, J. H. (1961a) Effects of competition, predation by *Thais lapillus* and other factors on natural populations of the barnacle *Balanus balanoides*, *Ecological Monographs*, **31**, 61–104.

Connell, J. H. (1961b) The influence of interspecific competition and other factors on the distribution of the barnacle *Chthamalus stellatus*, *Ecology*, **42**, 710–23.

Dale, M. P. and Causton, D. R. (1992) The ecophysiology of *Veronica chamaedrys*, *V. montana* and *V. officinalis*. I: Light quality and light quantity, *Journal of Ecology*, **80**, 483–92.

Farrell, T. M. (1991) Models and mechanisms of succession: an example from a rocky intertidal community, *Ecological Monographs*, **61**, 93–113.

Filion, L. (1984) A relationship between dunes, fire and climate recorded in the Holocene deposits of Quebec, *Nature*, **309**, 543–46.

Galen, C. and Stanton, M. L. (1993) Short-term responses of alpine buttercups to experimental manipulations of growing season length, *Ecology*, **74**, 1052–58.

Grime, J. P. (1979) *Plant Strategies and Vegetation*, Wiley.

Grime, J. P., Willis, A. J., Hunt, R. and Dunnett, N. P. (1994) Climate–vegetation relationships in the Bilbury road verge experiments, in: *Long-Term Experiments in Agricultural and Ecological Sciences* (eds R. A. Leigh and A. E. Johnson), pp. 271–85, CAB International, Wallingford, Oxon, UK.

Harmon, M. E. (1984) Survival of trees after low-intensity surface fires in Great Smoky Mountains National Park, *Ecology*, **65**, 796–802.

Hernd, G. J., Muller-Niklas, G. and Frick, J. (1993) Major role of ultraviolet-B in controlling bacterioplankton growth in the surface of the ocean, *Nature*, **361**, 717–19.

Hinton, H. E. (1960) Cryptobiosis in the larva of *Polypedilum vanderplankae* Hint (Chironomidae), *Journal of Insect Physiology*, **5**, 286–300.

Jordon, P. W. and Nobel, P. S. (1981) Seedling establishment of *Ferocactus acanthodes* in relation to drought, *Ecology*, **62**, 901–6.

Lean, J. and Warrilow, D. A. (1989) Simulation of the regional climatic impact of Amazon deforestation, *Nature*, **342**, 411–13.

Lee, D. W. and Lowry, J. B. (1975) Physical basis and ecological significance of iridescence in blue plants, *Nature*, **254**, 50–1.

Legg, C. J., Maltby, E. and Proctor, M. C. F. (1992) The ecology of severe moorland fire on the North York Moors: seed distribution and seedling establishment of *Calluna vulgaris*, *Journal of Ecology*, **80**, 737–52.

Leigh, R. A. and Johnston, A. E. (1994) *Long-Term Experiments in Agricultural and Ecological Sciences*, CAB International, Wallingford, Oxon, UK.

Louw, G. N. (1993) *Physiological Animal Ecology*, Longman Scientific & Technical.

MacDonald, G. M., Edwards, T. W. D., Moser, K. A., Pienitz, R. and Smol, J. P. (1993) Rapid response of treeline vegetation and lakes to past climate warming, *Nature*, **361**, 243–46.

Morse, S. R. and Bazaz, F. A. (1994) Elevated CO_2 and temperature alter recruitment and size hierarchies in C3 and C4 annuals, *Ecology*, **75**, 966–75.

Moyse, J. and Nelson-Smith, A. (1963) Zonation of animals and plants on rocky shores around Dale, Pembrokeshire, *Field Studies*, **1**(5), 1–31.

Mulroy, T. W. and Rundel, P. W. (1977) Annual plants: adaptations to desert environments, *BioScience*, **27**, 109–14.

Negbi, M. and Evenari, M. (1961) The means of survival of some desert summer annuals, in: *Plant-water relationships in arid and semi-arid conditions*, Proc. Madrid Symp., Arid Zone Research, **16**, 249–59, UNESCO, Paris.

Payette, S. and Gagnon, R. (1985) Late Holocene deforestation and tree regeneration in the forest-tundra of Quebec, *Nature*, **313**, 570–72.

Payette, S., Fillon, L., Delwaide, A. and Bégin, C. (1989) Reconstruction of tree-line vegetation response to long-term climate change, *Nature*, **341**, 429–32.

Pigott, C. D. and Pigott, S. (1993) Water as a determinant of the distribution of trees at the boundary of the Mediterranean zone, *Journal of Ecology*, **81**, 367–72.

Retuerto, R. and Woodward, F. I. (1992) Effects of windspeed on the growth and biomass allocation of white mustard *Sinapis alba* L., *Oecologia*, **92**, 113–23.

Roughgarden, J., Gaines, S. D. and Pacala, S. W. (1987) Supply side ecology: the role of physical transport processes, in: *Organization of Communities Past and Present* (eds J. H. R. Gee and P. S. Giller), pp. 494–518, Blackwell.

Rugg, D. A. and Norton, T. A. (1987) *Pelvetia canaliculata*, a high-shore seaweed that shuns the sea, in: *Plant Life in Aquatic and Amphibian Habitats* (ed. R. M. M. Crawford), pp. 347–58, Blackwell.

Schonbeck, M. and Norton, T. A. (1980) Factors controlling the lower limits of fucoid algae on the shore, *J. Exp. Mar. Biol. Ecol.*, **43**, 131–50.

Sparks, T. H. and Carey, P. D. (1995) The responses of species to climate over two centuries: an analysis of the Marsham phenological record, 1736–1947, *Journal of Ecology*, **83**, 321–9.

Stein *et al.* (1992) The effect of fire on stimulating willow regrowth, *Oikos*, **65**, 190–96.

Sundberg, A. (1983) *Nature and Resources*, UNESCO, **19**(2), 10.

Sutton, S. (1972) *Woodlice*, Ginn & Co.

Wardle, P. (1974) Alpine timberlines, in: *Arctic and Alpine Environments* (eds J. D. Ives and R. G. Barry) pp. 371–402, Methuen.

Yeaton, R. I. (1978) A cyclical relationship between *Larrea tridentata* and *Opuntia leptocaulis* in the northern Chihuahuan Desert, *Journal of Ecology*, **66**, 651–56.

THE CHEMICAL ENVIRONMENT   CHAPTER 4

Prepared for the Course Team by Irene Ridge

4.1   Introduction

Chapter 3 was concerned mainly with physical factors, particularly aspects of climate, that affect living conditions and the supply of resources for organisms. The emphasis in this Chapter is on chemical resources (e.g. mineral nutrients) and chemical conditions (e.g. salinity).

❑   Before looking at Table 4.1, try to list the main chemical resources that (a) plants and (b) animals generally require.

■   Check your list with that in Table 4.1.

Table 4.1   Chemical resources required by plants and animals. √, required; – not required.

Resource	Plants	Animals
essential elements	√ taken up as inorganic ions	√ obtained from food (sometimes supplemented by external, inorganic sources)
vitamins	(√ required by some algae)	√
water	√	√
O_2 (for respiration)	(√ sometimes required for non-green organs and/or in the dark)	√
CO_2 (for photosynthesis)	√	–

The chemical and physical nature of the soil determines in large part the chemical environment of terrestrial rooted plants and soil organisms, and Part I of this Chapter is about soils. Also important for terrestrial organisms is the chemical make-up of the atmosphere, especially levels of CO_2 (a resource for plants) and pollutants such as ozone, O_3 (a condition); this is not considered here but is touched on in Books 4 and 5. Part II looks more broadly at the factors that influence plant distribution and survival in saltmarshes, a *semi-terrestrial* habitat which is intermittently exposed to air or submerged under water and where fluctuating salinity is a major feature of the chemical environment. Fluctuating physical and chemical conditions are also a feature of estuaries, a semi-aquatic habitat that we consider in Part II. The habitat approach is continued in Part III, which looks at the general nature of aquatic habitats and organisms and the ways in which chemical, physical and biotic factors interact to influence plants and animals.

You can see from Table 4.1 that animals obtain most of their chemical resources from living or dead organisms – food – and it follows that food availability may have a strong influence on animal distribution and abundance. *May* is the operative word, however. Book 2 describes animal populations that fluctuate widely with no variation in food supply and others where food supply and population size are closely matched. In a given situation, the basic question is whether factors operating from the 'bottom up', i.e. food supply, or from the 'top down', e.g. predation, are more important. We touch on this problem in Part III when discussing aquatic animals.

A final point to bear in mind as you read this Chapter is the large and increasing effect of human activities on soils, water bodies and even the atmosphere. Drainage and fertilizer addition to soils, nutrient and pollution inputs to water and large-scale engineering changes to river channels are just a few examples of the ways in which resources and conditions are altered. The subject is considered further in Book 5 but many examples are described in this Chapter.

TERRESTRIAL HABITATS: SOILS   PART I

4.2 The nature of soils

The soil in a particular place is not a fixture: it changes constantly as climate and organisms modify and reorganize raw materials. So to understand why soils vary in ways that can affect the distribution of plants and animals, you need to know something about this process of *soil development* as well as about the nature of the raw materials. What follows is a very simplified account and you will find more details (especially about how to name and describe soils – as may be necessary when carrying out ecological projects) in the video 'Soils' and the audiocassette programme 'Understanding soils'.

4.2.1 Non-living soil components

Soil is a mixture of inorganic or mineral particles that derive from the breakdown or weathering of rocks and organic material that derives from the decomposition of dead organisms and their waste products.

Inorganic material

Table 4.2 lists some properties of clay and quartz, the main components of the **inorganic soil fraction**. Not surprisingly, 'sandy soils', with a high proportion of coarse, free-draining quartz, dry out rapidly and often cause problems of water availability to plants.

Table 4.2   Major properties and components of the inorganic soil fraction.

Property	Clay	Quartz
chemical nature	mixed silicates of aluminium and other metals: very variable	silica, SiO_2
particle size	very small, < 0.002 mm	medium (0.002–0.02 mm) = **silt** large (> 0.02 mm) = **sand**
chemical activity	high – binds cations and organic matter	inert – no binding capacity
water percolation	slow – poor drainage	fast – free-draining

❑   From Table 4.2, why do you think that soils with a high proportion of clay are potentially more fertile than sandy soils?

■   Because of the high chemical activity, particularly the ability of clay to bind cations (positively charged ions such as K^+ and Ca^{2+}) and organic matter. Bound cations are retained in the soil rather than being washed away by percolating water and, provided that they can be released for uptake by plants, the soil is potentially fertile.

Cation binding occurs because of negative charges associated with clays, and the 'tightness' of binding varies for different cations and also with soil conditions such as pH. However, organic matter and iron-rich complexes in clay soils may have positive charges on their surfaces and consequently bind anions, again with varying degrees of 'tightness'. Nitrate and chloride ions, for example, are bound only weakly and hence are readily lost from soil by percolating rain – a process called **leaching**. Phosphate ions bind tightly and are much less easily leached and, in addition, phosphates form sparingly soluble salts (such as calcium phosphates) whereas nitrates are freely soluble.

Weathering can be physical (the erosive action of wind, water and fluctuating temperatures), chemical (when acid solutions dissolve certain rock components) or even biological: in the Namib desert, for example, snails rasping at rocks in order to consume lichens growing just below the surface have been estimated to produce about 1 tonne of rock dust per hectare per year, about the same as produced by wind erosion. As well as clay or quartz particles, weathering liberates various ions (depending on the nature of the **parent rock**) and sometimes large quantities of iron and aluminium oxides, the former being responsible for the red colour of some soils. The nature of the parent rock is, therefore, an important factor influencing the composition of the inorganic soil fraction. A complication here is that soil can move and so does not always overlie its *original* parent rock. On steep slopes, for example, soil may creep downwards whilst wind, water and, especially, glaciers transport enormous quantities of soil. Over large areas of Britain, rocks are covered by a thick layer of **boulder clay**, which is a mixture of clay and small rocks deposited by glaciers after the last glaciation and on which new soils have developed.

Organic material

When plants and animals die, they are broken down by the action of micro-organisms and soil animals until an organic material is produced that is relatively resistant to further decay. Just as weathering is the main process by which the inorganic soil fraction is produced, so *decomposition* produces the organic fraction (described in detail in Book 4, Chapter 2). The decay-resistant material or **humus** forms intimate complexes with clay particles and plays a major role in 'improving' soil because the clay–humus complexes aggregate to give a more open, free-draining texture and also bind cations effectively. On very sandy soils, humus is the only natural agent by which water-holding and ion retention can be improved (although gardeners may also add clay to good effect).

Now or after reading the next Section you should view the video 'Soils', and listen to the AC programme 'Understanding soils'. These describe the appearance of different soil types (so that you can recognize them in the field) and explain the processes that modify soils and lead to the formation of distinct strata or **horizons**. The next Section gives a brief overview of these processes.

4.2.2 Soil structure and processes

Because soil organic matter derives mainly from dead roots or material deposited on the surface (and termed **litter**), the upper layer of soil has an even mixture of organic and mineral components. This is the **A horizon** and it is here that most plant roots are found and much ion uptake occurs.

The **B horizon** – the sub-soil of gardeners – lies below the A and is composed largely of inorganic (mineral) material; it overlies the **parent material**, which may be the original parent rock or some transported substrate such as boulder clay or blown sand. On many soils, **organic horizons** of undecomposed and partially decomposed litter lie above the A horizon and, particularly in forests, roots and soil animals may be abundant here also. (There are diagrams of these horizons and of the various layers into which they may become subdivided, in the AC Notes in the *Companion for Book 1*.)

The factors that influence soil development most strongly are listed below.

1   The *rate of decomposition*, which is affected by the chemical nature of litter, by the activity of soil animals and by climate (Book 4). It is high in warm, humid conditions and low in cold, dry or waterlogged conditions. If the rate of decomposition is slower than the rate of litter deposition, incompletely decomposed material builds up above the A horizon as **peat**.

2   The *activity of soil animals* not only affects (1) but also plays a key role in mixing together mineral and organic material to produce the A horizon. Moles, for example, can move soil from the B horizon onto the surface; dung and carrion beetles bury dead organic matter; and earthworms, the supreme 'soil improver', not only drag leaves into their burrows but also deposit on the surface *casts* which are an intimate mixture of inorganic material, fresh organic matter, gums and lime passed through the worm gut. In Charles Darwin's classic book *The Formation of Vegetable Mould through the Action of Earthworms* (1881), based on 40 years' observations of earthworms, he demonstrates that a layer of casts around 25 cm deep may be deposited on a pasture during a 50-year period.

3   The *balance between rainfall and evapo-transpiration* (water evaporating from soil and vegetation) profoundly affects the extent to which materials can be leached from upper to lower soil horizons. In Britain, especially the northern and western parts, rainfall greatly exceeds evapo-transpiration, so easily soluble ions (especially nitrate) tend to be leached and lost in drainage water whilst clay particles, humus and iron compounds move down the soil profile and are often re-deposited in the B horizon. The more acidic the percolating water, the greater the extent of leaching.

4   *Drainage* – the rate at which water can move downwards – will obviously affect the extent of leaching but it also affects soil chemistry because poorly drained and waterlogged soils tend to be anaerobic. Oxygen for roots and soil heterotrophs is then in short supply and certain elements, notably nitrogen, iron and manganese, are present in their reduced forms – ammonium (NH_4^+), iron(II) (Fe^{2+}) and manganese(II) (Mn^{2+}) respectively. All these reduced ions can be toxic to plants and soil organisms.

❏   Think of at least two factors that will affect drainage.

■   Soil composition (the relative amounts of clay and sand), slope, and height above the water table are three obvious ones. Roots which die, worm burrows and even mole tunnels all create channels in the soil that improve drainage. If you have viewed the video 'Soils' you should also recall that the impermeable iron pan that may form in podzols greatly impedes drainage.

All these processes are affected to some extent by climate (especially temperature and precipitation), and other factors such as slope, aspect, vegetation and, of course, cultivation modify soil processes. Figure 4.1 summarizes these interactions. The result is huge variation in soil type and properties, which are discussed in the video 'Soils' and the AC programme 'Understanding soils'. If you have not already done so, view and listen to these tapes now.

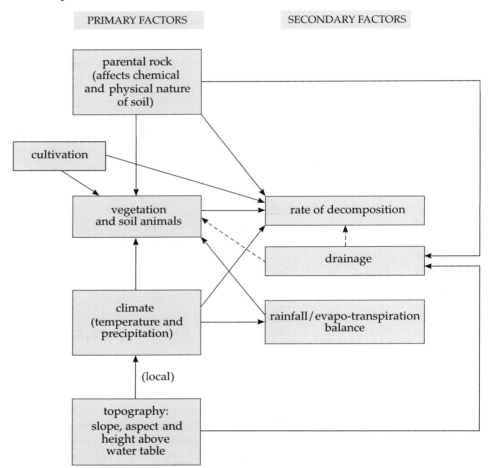

Figure 4.1   Interactions between factors that influence soil development.

Summary of Section 4.2

- Soil is a mixture of inorganic material derived from the weathering of rock and organic material produced by the decomposition of living organisms and their wastes.

- The two main types of inorganic material in soil are clay (with small particles that readily bind ions) and quartz (with medium (silt) to large (sand) particles that are chemically inert). The main organic component is humus, which is the decay-resistant end-product of decomposition and binds both to clay minerals and to soluble ions.

- Distinct horizontal layers (horizons) develop in undisturbed soils. The main types are the organic horizons at the surface, the A horizon of mixed mineral and organic material and the B horizon of mineral material, which overlies the parent material.

- The main processes that influence soil development, structure and properties are: rate of decomposition, numbers and activity of soil animals, the balance between rainfall and evapo-transpiration and drainage. These processes are influenced by climate, slope, aspect and vegetation.

A fuller description of soil structure and processes is given in the video 'Soils' and the AC programme 'Understanding soils'; questions to test your understanding are in their accompanying notes in the *Companion for Book 1*.

4.3 Soils and nutrient supply

For rooted land plants, soil properties determine to a large extent the supply of mineral nutrients, i.e. chemical resources. Six elements (plus carbon, oxygen and hydrogen which are obtained from the air and/or water) are required in relatively large amounts and it is the supply of these **major** or **macro-nutrients** (Table 4.3), especially nitrogen, phosphorus and potassium, that has the greatest influence on plant growth and the species composition of vegetation (discussed in Section 4.3.1). Only rarely does shortage of **trace** or **micro-nutrients** (Table 4.3) appear to influence plant distribution but *over*-supply can. These elements become toxic above the very low levels that are required for plant growth and then represent a soil *condition* (rather than a resource); we discuss this in Section 4.4.4.

Table 4.3   Essential nutrient elements for higher plants.

| Major or macro-nutrients | | | | Trace or micro-nutrients | | | |
Cationic		Anionic		Cationic		Anionic	
potassium	K	nitrogen	N*	iron	Fe	boron	B
magnesium	Mg	phosphorus	P	manganese	Mn	molybdenum	Mo
calcium	Ca	sulphur	S	copper	Cu	chlorine	Cl
				cobalt	Co		
				zinc	Zn		

* N is anionic when present as nitrate (NO_3^-) but may also be present in cationic form as ammonium (NH_4^+).

4.3.1 Plants and their response to major nutrients

Over the last 20–30 years, one of the most striking changes in the British countryside has been the loss of species-rich grassland: chalk downs, grassy heaths and flower-rich meadows, for example. The most important causes of this change have been the **application of fertilizers** containing the major nutrients N, P and K, usually accompanied by ploughing and re-seeding with 'productive' grasses or arable crops. With rising agricultural surpluses and the introduction of 'set-aside' policies in Europe, some land is being taken out of intensive agriculture but this does not result in a rapid reversion to flower-rich grassland: the effects of fertilizers may persist for many years and, in particular, phosphate may persist for long periods (Section 4.2.1) and does not leach out of the soil (as nitrates and, to a lesser extent, potassium do). Even if steps are taken to prevent invasion by scrub, abandoned agricultural land is likely to remain dominated by grasses and a few vigorous weedy dicots for a long time. We discuss below the reasons for high soil fertility having such a dramatic effect on vegetation and in Section 4.3.2 the ways in which plants obtain nutrients on very infertile soils.

The massive crop yields obtained with intensive agriculture depend to a large extent on the application of fertilizers and modern crop varieties represent an extreme example of plants that grow, survive and reproduce best on nutrient-rich soils. But, as anybody who grows alpine plants knows, some plants grow, survive and reproduce best on relatively nutrient-poor soils: even in the absence of competition, these plants perform less well on rich soils. In broad terms (and with many exceptions) plants can be divided into two groups with respect to their response to and requirements for major nutrients, illustrated by species X and Y in Figure 4.2.

❑   From Figure 4.2, describe two major differences between species X and Y.

■   (1) The maximum growth rate of species X is much lower than that of species Y. (2) Species X attains its maximum growth rate with much lower supplies of major nutrients than does species Y and is *insensitive* to or inhibited by high nutrient levels.

Nutrient-insensitive plants (species X) have intrinsically low rates of growth and are characteristic of nutrient-poor soils; **nutrient-responsive plants** (species Y) are commonly dominant on nutrient-rich soils and are more plastic – they grow slowly on poor soils but very fast indeed on rich soils. The way in which these differences in response to major nutrients correlate with differences in plant distribution is illustrated by the data in Figure 4.3, which relate to a Derbyshire soil where, because of mining activity, nutrients were very patchily distributed.

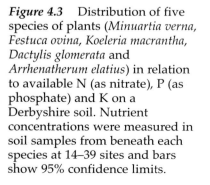

Figure 4.2   The relationship between average growth rate and the supply of major nutrients for two hypothetical species of plants.

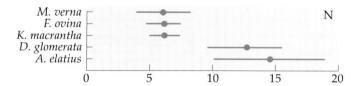

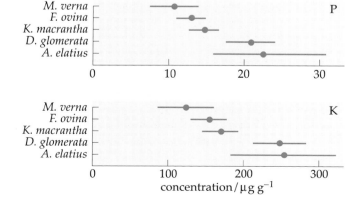

Figure 4.3   Distribution of five species of plants (*Minuartia verna, Festuca ovina, Koeleria macrantha, Dactylis glomerata* and *Arrhenatherum elatius*) in relation to available N (as nitrate), P (as phosphate) and K on a Derbyshire soil. Nutrient concentrations were measured in soil samples from beneath each species at 14–39 sites and bars show 95% confidence limits.

❑   In Figure 4.3, identify the nutrient-insensitive and nutrient-responsive
    species.

■   Three species occur on soils with low nutrient levels (*Minuartia verna* (a
    low-growing dicot), *Festuca ovina* (a grass) and *Koeleria macrantha*
    (crested hair-grass)): these are relatively nutrient-insensitive types and
    are excluded by competition from the richer soils. *Dactylis glomerata*
    and *Arrhenatherum elatius* (false oat-grass) are both vigorous grasses,
    and are nutrient-responsive types that grow on much richer soils.

Minuartia verna
vernal sandwort (× 0.35)

Plants that show the smallest response to nutrients are typically herbaceous
dicots; many are low-growing (such as wild thyme *Thymus praecox*) and
easily shaded out by taller-growing competitors, so they are often
associated with sites that are both nutrient-poor and grazed. Recall from
Chapter 1 the effects of rabbit grazing. Nearly all grasses show some
response to nutrients and, although *Festuca* and *Koeleria* above were
classified as unresponsive, this is only by comparison with *Dactylis* and
Arrhenatherum; sheep's-fescue (*Festuca ovina*), in particular, shows a clear
response to increases in major nutrients.

One problem for conservation, therefore, is that fertilizer application
favours nutrient-responsive species which compete with and displace less
responsive species. Reducing soil nutrient levels can be achieved only by
repeatedly cutting or grazing the vegetation and *removing* the crop, which
can take many years. Soil phosphate is a particular problem because it may
be released continuously from insoluble complexes for hundreds of years:
Roman middens can still be detected by their higher soil phosphate and
still support plants that require high phosphate!

Festuca ovina
sheep's-fescue (× 0.35)

The supply of major nutrients, however, is strongly influenced by soil
conditions, in particular *soil pH*. We discuss this in Section 4.4.3 but, as a
crude rule of thumb, nitrogen is likely to limit plant growth on upland and
acid soils whereas phosphate is more likely to be limiting on lowland and
neutral to alkaline soils. The latter arises mainly because phosphate is
locked up as insoluble calcium phosphate on alkaline soils rich in calcium
carbonate. The situation is more complex for nitrogen because it occurs as
two inorganic ions – ammonium and nitrate – with NO_3^- the dominant ion
on neutral or alkaline soils and NH_4^+ on acid soils. Many plants do not
readily take up and use NH_4^+ and it may indeed be toxic at high levels.
High levels of nitrogen in either form sometimes inhibit the growth of
nutrient-unresponsive plants and this is thought to explain the loss of
Sphagnum moss from some Pennine blanket bogs. These upland areas have
received increasing inputs of nitrogen from atmospheric deposition in rain:
NO_3^- derives from nitrogen oxide gases (e.g. released from car exhausts)
and NH_4^+ derives mainly from ammonia gas, NH_3, which volatilizes from
animal dung spread as a fertilizer.

Dactylis glomerata
cock's-foot (× 0.1)

Nutrient supply, both shortage and excess, can thus influence plant
distribution, the ability of plants to grow, survive and reproduce and the
composition of vegetation. As an example of how plant–nutrient
interactions are studied by ecologists, we describe below a classic
investigation by Pigott and Taylor (1964) into the distribution of common
stinging nettle *Urtica dioica*, a nutrient-responsive species characteristic of
rich soils.

Nettle distribution

Pigott and Taylor carried out their investigation at a number of open woodland sites where nettles and dog's mercury *Mercurialis perennis* occurred in adjacent and more or less pure stands. The question asked was whether major soil nutrients could explain this distribution pattern. They found:

1   On a unit area basis, vegetation dominated by nettles contained 3–5 times more N and P in the aerial shoots than vegetation dominated by *Mercurialis*. This is characteristic of nutrient-responsive plants.

2   When nettles were grown from seed in pots of soil taken from nettle- and *Mercurialis*-dominated sites and with and without fertilizer addition, the results shown in Table 4.4 were obtained.

Table 4.4   Growth of nettles from seed on three woodland soils. Plants were grown in pots in a greenhouse with and without addition of fertilizer: P, 10–20 mg per pot as calcium phosphate; N, 10–20 mg per pot as ammonium nitrate; N+P, both fertilizers added. Results are means of nine replicates ± SEM.

Natural dominant at site of soil origin	Seedling age at harvest /days	Dry weight per nettle plant/mg			
		no addition	+P	+N	+(N+P)
Mercurialis perennis (mull soil*)	33	6 ± 2	92 ± 12	5 ± 1	125 ± 21
M. perennis (protorendzina soil)	50	1[†]	248 ± 26	9[†]	282 ± 31
Urtica dioica[‡] (mull soil)	30	41 ± 5	30 ± 9	—	—

* Mull soils are characterized by having rapidly decomposing litter (mull) and are commonly brown earths.

† Many dead.

‡ Nettles were regenerating from seed on this site.

3   On soil from nettle-dominated sites where mature plants were stunted and *not* regenerating naturally from seed, nettle seedlings grew poorly (and responded to phosphate addition).

4   Seedlings of *Mercurialis* grew well on soil from *Mercurialis*-dominated sites and there was no response to added P.

These data confirm the nutrient-responsive character of nettles (items 1 and 2) and indicate that *shortage of phosphate* probably excludes nettles from *Mercurialis*-dominated sites (item 2). *Mercurialis* has a lower requirement for P than do nettles (item 4).

❑   What is the significance of item 3?

■   It suggests that mature nettles (which are long-lived perennials) may persist on soils that contain too little phosphate to support the growth of nettle seedlings. This implies that P supply is most critical at the seedling stage – which is not surprising, given their small root system.

Because plants in greenhouse pots may behave quite differently from plants in the field, Pigott and Taylor carried out a further experiment.

5   In April, nettle seeds were sown on two woodland sites where
    Mercurialis was dominant; one site had been recently coppiced and had
    high light intensity and the other site was deeply shaded. On each site,
    three matched pairs of 1 m square quadrats were marked out and one
    of each pair was watered with a solution of calcium phosphate.
    Seedling dry weight was measured after five months (Table 4.5) and
    the appearance of nettles, with and without added P, on the site with
    high light intensity is shown in Figure 4.4.

Table 4.5   Response of *Urtica dioica* to addition of phosphate: seeds were sown on
two woodland sites in April and harvested in September. Results are means ± SEM.

| Site | Dry weight per plant/mg | |
	no P added	P added
high light intensity	15.2 ± 2.5	177.7 ± 24.0
low light intensity	2.0 ± 0.2	10.5 ± 1.1

❑   Using all the information provided, suggest an explanation for the non-
overlapping distribution of nettles and dog's mercury in these
woodlands.

◼   Nettles require *two* factors in order to grow and compete strongly: high
soil P (items 2 and 5) and high light intensity (item 5). Wherever these
resources are plentiful in these woodlands, nettles are likely to be
dominant. Dog's mercury tolerates lower soil P and lower light
intensity and cannot compete with vigorous nettles, so it is dominant at
sites where either P or light intensity or both are low.

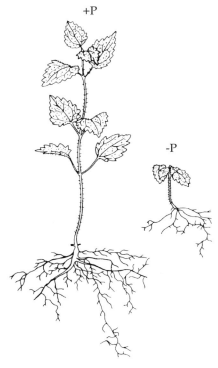

Figure 4.4   Response of *Urtica dioica* to addition of phosphate; seeds were sown on a woodland site with high light intensity and where *Mercurialis* had been dominant. Results are means ± SEM. (× 0.6)

The *interaction* of two factors to determine plant distribution is clearly
shown in this example. Other studies have shown that nettles also require
high soil N but, in these woodlands, soil nitrogen levels were uniformly
high so that soil P alone affected distribution in open sites. The
nettle/dog's mercury example also suggests two kinds of questions that
can be asked if field observations show that a plant is confined to a
particular kind of soil:

1   Does the plant have some special *requirement* (e.g. high or low levels of
    a nutrient) which can be met only on that soil?
2   Does the plant *tolerate* conditions on that soil (e.g. high or low supply
    of nutrients) but would it actually grow better on other types of soil
    from which it is normally excluded by factors such as climate or
    competition?

Nutrient-poor soils

In sharp contrast to nettles, nutrient-insensitive plants grow on soils where
phosphate is barely detectable. Some of these plants actually require low-
nutrient supply (e.g. *Sphagnum* mosses and some alpine plants). Many,
however, grow and reproduce well on rich soils but cannot compete with
nutrient-responsive species. The question we ask here is how plants
manage to obtain sufficient nutrients to grow at all on very nutrient-poor
soils. Three sorts of mechanisms facilitate growth:

1   *Highly effective nutrient-uptake systems.* Key elements are a large root
    system (relative to shoot size) and the capacity to absorb nutrients
    present in the soil at very low concentrations. This last depends on ion

uptake systems in cells at the root surface. On alkaline soils, phosphate availability and uptake can be increased by maintaining an acid zone around individual roots. Calcium phosphate is solubilized in this acid zone, which is achieved either by secreting hydrogen ions (H^+) or, if enough respiratory CO_2 is retained close to roots, by the formation of carbonic acid ($CO_2 + H_2O \rightleftharpoons H_2CO_3 \rightleftharpoons HCO_3^- + H^+$). On soil with much organic matter (including acid peats), plants may have *phosphatase enzymes* on the root surface which allow them to release phosphate from organic complexes.

2   Good **nutrient conservation**. This means, firstly, that once nutrients have been taken up, few are lost when shoots die down or leaves fall off. Instead, nutrients are withdrawn from the dying organ, stored in some perennial part and re-used when new organs grow. Such **internal cycling** is well developed in trees such as birch *Betula* spp., which can grow on poor soils, but is weakly developed in sycamore *Acer pseudoplatanus*, characteristic of richer soils. A second aspect of nutrient conservation concerns the rate at which organs are shed: perennial plants may conserve nutrients by having long-lived evergreen leaves (e.g. many rainforest trees and also heather and heaths, *Erica* spp.).

3   **Nutrient supplementation**, i.e. some mechanism for acquiring nutrients other than by uptake via roots. Symbiotic association with nitrogen-fixing bacteria (Chapter 1) is the commonest way of supplementing the supply of nitrogen; and phosphate supply is most commonly supplemented by root associations with mycorrhizal fungi (Chapter 1). An unusual system of nitrogen supplementation was demonstrated for the epiphyte *Dischidia major* (Treseder *et al.*, 1995); this grows in the Malaysian 'kerangas' forests which occupy very nutrient-poor soils. *Dischidia* is an ant–plant mutualist (Chapter 1) and produces sac-like 'ant-leaves' (Figure 4.5).

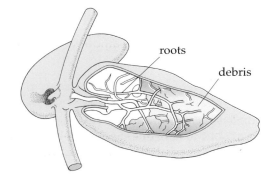

Figure 4.5   Two sac-like 'ant-leaves' of the epiphyte *Dischidia major* ($\times$ 1). One leaf is cut away to show roots and ant debris in the cavity.

❑   Study Figure 4.5 and work out how the presence of ant partners supplements nitrogen supply for *Dischidia*.

■   Roots penetrating the ant debris are the clue: they absorb the nitrogen released when the debris decomposes.

On average, *Dischidia* obtains 29% of its nitrogen from ant debris – a very substantial supplement. On acid, boggy soils, another mechanism is **insectivory**. Insects are trapped in a variety of ways ranging from sticky hairs (e.g. sundews *Drosera* spp., see Figure 4.6), to the elaborate hinged structures of Venus fly traps *Dionea* spp. or the slippery-sided vessels of pitcher plants *Nepenthes* spp. Trapped insects are digested by extracellular enzymes and small, organic molecules containing N and P are absorbed into the plant – one of the rare instances where higher plants utilize organic rather than inorganic sources of nitrogen.

Survival on nutrient-poor soils may thus involve considerable specialization (e.g. insectivory) and/or considerable expenditure of energy in acquiring nutrients: ectomycorrhizal trees such as birch, for example, may supply a quarter or more of their fixed carbon to mycorrhizal partners. In addition, there is often a high expenditure of energy on *anti-herbivore mechanisms*, such as the production of very tough leaves with a high proportion of woody sclerenchyma fibres (Choong *et al.*, 1992) and the synthesis of products such as tannins or terpenoids. Both of these mechanisms reduce palatability and deter generalist chewing herbivores so that, by influencing the 'food quality' of vegetation, nutrient-poor soils indirectly affect the survival and distribution of animals. Such anti-herbivore mechanisms are thought to have evolved because slow-growing plants on nutrient-poor soils cannot easily replace lost leaves.

4.3.2 Animal nutrition and soils

By influencing the palatability of herbage as described above, soil nutrient levels may affect indirectly the distribution of grazers. The nutrient *content* of vegetation, which depends to some degree on soil nutrient levels, can also affect the behaviour and distribution of herbivores. Grazing insects occur at very low densities in the 'Fynbos' of South Africa, for example, because nutrient supplies to this Mediterranean scrub-type vegetation are exceptionally low. Even very local increases in nutrient supply to plants can affect insect abundance: Port and Thomson (1981) showed that there are much higher densities of grazing insects close to roads because of the nitrogen input from car exhausts. Similar effects occur among vertebrate grazers.

It used to be quite common to see farm animals licking blocks of salt (mostly sodium chloride) in pastures. Farmers supplied salt because the vegetation contained insufficient amounts of sodium (an essential element for animals, but not plants, and now supplied mainly in feed concentrates). On acid pastures extra copper is still commonly provided. Wild, grazing mammals may also fail to obtain adequate micro-nutrients from their plant diet, particularly on the African savannah where vegetation is notoriously nutrient-poor. They can obtain their own supplements by eating soil or licking rocks at special **salt-licks** where there are high levels of the minerals in short supply. Because salt licks are often widely scattered, their availability can have a considerable influence on herbivore distribution. This situation contrasts with that of plants, where distribution is very rarely influenced by the supply of soil micro-nutrients.

A study by McNaughton (1988) suggests that a major factor explaining the remarkably patchy distribution of large African herbivores over (apparently) uniform grasslands is the mineral content of food. Using records over 19 years from the Serengeti National Park, Tanzania, he identified areas which were grazed throughout the year and had consistently high (H) or low (L) numbers of grazers, and he measured mineral nutrient concentrations in both young vegetation and soil in these areas. There were significant differences in the vegetation of H and L areas for 10 out of the 19 elements tested, and from Table 4.6 you can see that three elements judged to be in critically short supply for herbivores in L areas were indeed more abundant in H vegetation.

Figure 4.6   *Drosera rotundifolia,* round-leaved sundew, an insectivorous plant of acid, boggy places. Insects are trapped by the sticky hairs on leaf surfaces. (× 1)

Table 4.6   Concentrations (in p.p.m.) of three elements in young herbage and in soil at Serengeti sites where there were high (H) or low (L) numbers of large herbivores. Significant difference between H and L sites: * $P < 0.05$; ** $P < 0.01$.

Element	Herbage		Soil	
	L sites	H sites	L sites	H sites
sodium, Na	894	3097*	180	162
magnesium, Mg	1602	1960*	529	377
phosphorus, P	2722	4162**	449	481

❑   Can the differences in nutrient content of herbage between H and L areas be explained in terms of differences in soil nutrient concentrations?

■   No: from Table 4.6 there are no significant differences in the soil concentrations of these (or any other) nutrients tested for H and L areas.

This lack of correlation between soil and plant nutrient concentrations is a (still unexplained) puzzle; it may relate, for example, to differences in nutrient cycling from dung to plants or to differences in mycorrhizal partners of plants in H and L areas, i.e. underlying differences in availability and/or uptake systems of nutrients. The main point, however, is that the mineral content of plants can influence choice of grazing sites by herbivores.

Summary of Section 4.3

- Soil is the main source of supply for inorganic nutrient ions required by plants in large (**major** or **macro-nutrients**) or small (**trace** or **micro-nutrients**) amounts.

- Plants can be classified broadly as **nutrient-responsive** or **nutrient-insensitive**. For nutrient-responsive plants, growth rate and size increase with nutrient supply. For nutrient-insensitive plants, growth rate is intrinsically low and does not increase with nutrient supply.

- Nutrient-responsive plants are commonly dominant on rich, fertile soil and include many grasses and fast-growing, 'weedy' species such as nettles. Mature plants (but not always seedlings) may survive on nutrient-poor soils but they grow and compete poorly. The nettle/dog's mercury example illustrates how two resources – soil phosphate and light – interact to determine the distribution of nettle patches in a woodland.

- Nutrient-insensitive plants commonly occur on nutrient-poor soils but whereas some (e.g. alpines, *Sphagnum* moss) require these conditions, and survive and reproduce better than on rich soils, others are excluded from fertile soils only by their inability to compete with nutrient-responsive species.

- The supply of nutrients is strongly influenced by soil conditions, especially **soil pH**. Nitrogen is the most common limiting nutrient on acid soils and phosphate on neutral or alkaline soils.

- The long-term persistence of plants in very nutrient-poor conditions involves three sorts of mechanism: (i) highly effective nutrient uptake; (ii) good **nutrient conservation**; and (iii) **nutrient supplementation** (e.g. by symbiotic partners or **insectivory**). These mechanisms involve considerable specialization and/or expenditure of energy.

- Slow-growing plants on nutrient-poor soils often have well-developed anti-herbivore mechanisms which deter generalist grazers. The nutrient content of herbage may also affect the behaviour and distribution of grazing insects and of large herbivores: the latter may seek salt-licks if sodium or micro-nutrient supply is inadequate, and the Serengeti study showed that they select vegetation with the highest nutrient content.

Question 4.1   (Objectives 4.1–4.3)

As coastal sand dunes stabilize and age, they often pass through a stage where the surface is stabilized by lichens and the soil is then colonized by low-growing perennials such as wild thyme (*Thymus drucei*) and some very small annual species.

(a) What characteristics of the soil and the plants are likely to explain the presence of these colonizers?

(b) The low-growing colonists may persist for many years if the dunes are heavily grazed by rabbits but otherwise they tend to disappear. Suggest an explanation for this observation.

(c) What changes in the vegetation would you predict in the short and long term if a heavy application of general fertilizer was given to dunes shortly after colonization by low-growing plants?

Question 4.2   (Objectives 4.5 & 4.15)

Table 4.7 shows the results of experiments in which various plants were grown, in pots in the open, on soil taken from a wood where dog's mercury *Mercurialis perennis* was the dominant species of the ground flora. Additional nitrogen and/or phosphorus was added to some pots as solutions of NH_4NO_3 and $Ca(H_2PO_4)_2$.

(a) For each species, decide whether N and/or P was in limiting supply during the experiment and classify it as nutrient-responsive or nutrient-insensitive.

(b) From Table 4.7 and information in Section 4.3, which species would you expect to find growing naturally on nutrient-rich soils, which on nutrient-poor soils and which, if any, on both soil types? For which, if any, species can you predict the habitat where it is likely to grow abundantly and show local dominance?

Table 4.7   For use with Question 4.2. Response to addition of N and P of various species grown from seed or transplanted seedlings in pots of soil collected from a site dominated by *Mercurialis perennis*. Values are means of shoot dry weight per pot (mg) with nine replicate pots per treatment. Figures followed by a different superscript letter are significantly different from others in that row ($P < 0.05$).

Species	Duration of experiment /days	No addition	+N	+P	+(N+P)
1 *Urtica dioica* (nettle)	33	6	5	92[a]	125[a]
2 *Chamaenerion angustifolium* (rose-bay willow-herb)	43	2	3	20[a]	33[a]
3 *Alliaria petiolata* (hedge garlic)	40	23	41	257[a]	281[a]
4 *Silene dioica* (red campion)	30	63	115[a]	208[b]	235[b]
5 *Galium aparine* (goosegrass, cleavers)	55	257	330[a]	264	587[b]
6 *Campanula trachelium* (nettle-leaved bellflower)	130	1189	1059	2175	1890
7 *Brachypodium sylvaticum* (slender false-brome)	30	123	–	141	153
8 *Deschampsia cespitosa* (tufted hair-grass)	47	49	37	43	38

Question 4.3   *(Objective 4.4)*

A nutrient-insensitive shrub became established on very poor soil in a recently abandoned quarry in England. It was a member of the pea and bean family (Fabaceae (Leguminosae)) and had deciduous, smooth, flat leaves. Which of the characteristics (i)–(v) would you expect to find in this species and why would you expect to find them?

(i)   insectivory;

(ii) high levels of tannins in mature leaves;

(iii) nitrogen-fixing bacteria in root nodules;

(iv) ectomycorrhizal fungi associated with the roots;

(v) a root to shoot ratio of about 1 : 10.

4.4 Soil conditions

Conditions in soil may affect soil organisms and plant roots in two distinct ways. First, there may be a *direct* effect: high levels of metal ions, for example, represent a potentially toxic chemical environment. Secondly, soil conditions may have an *indirect* effect by influencing the supply of essential resources such as oxygen and mineral nutrients; recall that soil pH strongly affects the availability to plants of major nutrients (Section 4.3.1).

❑   From Section 4.2.2, recall a soil condition that has both direct and indirect effects.

■   Waterlogging: it can be associated with toxic chemical conditions (direct effect) and it reduces the availability of oxygen (indirect effect via resource supply).

In this Section, we consider the impacts on plant growth and distribution of four soil conditions: physical structure and depth; aeration; soil pH; and the presence of toxic ions. Soil moisture was discussed in Chapter 3 (Section 3.5) and is not considered further here but do bear in mind that it strongly affects plants and soil animals and depends not only on rainfall but also on soil properties such as depth, ease of drainage, and organic matter and clay content.

4.4.1 Physical structure and depth of soil

Soils with a high clay content or subject to heavy trampling often have a dense, stiff texture (or high **impedance**) which means that drainage is poor, gas diffusion is slow and it is difficult for roots to grow or soil animals to move through. A study of Australian *Banksia*, a genus of shrubs characteristic of very nutrient-poor, sandy heaths, illustrates the importance of soil impedance. Enright and Lamont (1992) were trying to find out why some species of *Banksia* failed to grow on old mining sites where they had grown before mining operations and where the soil apparently resembled closely the *Banksia*'s native soil. The mining company was required by law to rehabilitate the site by restoring natural vegetation.

Closer examination of the soil from a site where *Banksia* grew naturally (N) and at the rehabilitation site (R) showed that there were no significant differences in nutrient levels, except for iron which was much more abundant at R, especially in the sub-soil (B horizon) (Table 4.8). However, the R site had a much higher impedance in the sub-soil and slightly less coarse sand and more fine sand (Table 4.8). Table 4.9 gives data for survival, growth and xylem pressure potential* of *B. attenuata* seedlings.

Table 4.8   Soil chemical and physical properties at a site in western Australia where banksias grow naturally (N) and at a rehabilitation site (R) where soil had been reconstituted after mining operations ceased and where banksias did not grow well. A and B refer to average values for the A and B horizons (top-soil and sub-soil respectively). Units are p.p.m. for iron, % for soil mineral fractions and Megapascals, MPa (pressure units) for soil impedance. Impedance values with a different superscript letter in the same column are significantly different at $P < 0.001$

Site	Iron		Coarse sand		Fine sand		Silt/clay		Impedance top of A	Impedance top of B
	A	B	A	B	A	B	A	B		
N	53	55	91	88	6.4	9.1	1.7	1.5	8.6[a]	14.6[a]
R	210	381	80	84	13	11	1.8	1.6	12.6[b]	52.0[b]

more difficult to obtain water from R soil.

Table 4.9   Data for *Banksia attenuata* on natural (N) and rehabilitation (R) sites. Survival (%) and relative growth rate (RGR, g g^{-1} week^{-1}) relate to a growing season; mean root : shoot ratios (g g^{-1}) and tap-root length (m) are for seedlings in May; xylem pressure potentials (pre-dawn values, in MPa) are for midsummer. Column values with a different superscript are significantly different ($P < 0.05$).

Site	Survival	RGR	Root : shoot ratio	Tap-root length	Xylem pressure potential
N	67[a]	0.047[a]	0.39[a]	1.74[a]	−0.24[a]
R	35[b]	0.025[b]	0.29[b]	0.11[b]	−1.01[b]

❏   What conclusions do you draw from the data in Table 4.9?

■   *B. attenuata* certainly survived and grew less well at the rehabilitation site and had a small root system with a very short tap-root. The data show that seedlings on the R site also had a much lower (more negative) xylem pressure potential, i.e. more water stress.

Water availability (measured as soil water potential) was much the same in the A horizon (which was about 10 cm deep) at the two sites and there was actually more water in the B horizon at the rehabilitation site.

❏   So why do you think *B. attenuata* did so poorly on the R site?

■   Seedlings were not short of mineral nutrients but did appear to be short of water (Table 4.9): their short tap-root meant that they could not obtain water from the B horizon. Something inhibited root extension into the sub-soil – possibly iron toxicity or, more likely, the high impedance of the soil (Table 4.8).

* The pressure in water-conducting xylem cells – xylem pressure potential (in units of pressure) – is usually *below* atmospheric pressure: you can think of it as a pressure deficit, tension or suction force. Transpiration from leaves sucks up water from the soil via the roots. The more negative the value, the greater the tension in xylem, the more difficult it is for water to reach the leaves and the greater the degree of water stress of the plant.

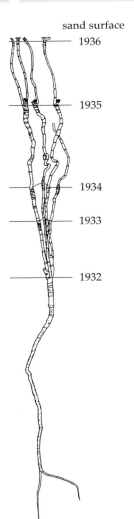

sand surface
1936
1935
1934
1933
1932

Figure 4.7 Sea holly *Eryngium maritimum* (× 0.08) showing flotation in relation to four changes in dune level.

Figure 4.8 Marram grass *Ammophila arenaria* (× 0.07), a common plant of young sand dunes which requires frequent burial by sand.

High impedance was the hypothesis proposed by Enright and Lamont and they suggested that this was caused by the high iron levels, which acted to cement soil particles together. Their recommendation was that the reconstructed soil at rehabilitation sites should have a deep, loose sandy layer between top-soil and sub-soil.

The above example illustrates the importance not only of soil impedance for plant survival but also of soil depth. On very freely-draining soils that dry out rapidly, such as sand dunes, long-lived plants survive only if they can reach water deep underground (unless they have the special water-conserving features described for desert plants, Chapter 3). Another problem for sand-dune plants is *soil stability*. Loose, unconsolidated material readily blows away, so seedling establishment is difficult and established plants are in constant danger of being buried. The problem is particularly acute on young dunes and perennials here often have a remarkable ability to grow up through smothering layers of sand and produce new shoots. Figure 4.7 illustrates this property of *flotation* in the sea holly *Eryngium maritimum*: note the long roots. Marram grass *Ammophila arenaria* (Figure 4.8), which dominates and helps to stabilize young dunes, actually seems to *require* frequent burial by sand and does not grow well on fully stabilized dunes.

4.4.2 Soil aeration

Most soil organisms (including roots) require oxygen for respiration during at least the periods of active growth. The supply of oxygen depends on the amount of air in the soil – its **aeration** – and two main factors influence this property. One is *particle size* because there is little space for gases between very small particles, which is why a clay soil with low organic content and high impedance is described as 'heavy'. The other is *soil water content*, since water displaces air, and this depends on several conditions which include drainage and rainfall. At one extreme are soils that are more or less permanently airless, either because they are constantly waterlogged or because they comprise almost pure clay. And at the other extreme are loose, sandy soils or well-cultivated loams which are nearly always well-aerated to a considerable depth. The question of interest here is how does soil aeration influence the survival, growth and distribution of plants and soil organisms?

Brief periods (hours or days) of anaerobiosis are tolerated by most soil organisms, using the simple device of switching to anaerobic respiration. The problem then is that end-products of anaerobic respiration (e.g. ethanol or lactate) may accumulate to toxic levels. The shorter the period of anaerobiosis tolerated, the more likely it is that organisms are restricted to well-aerated soils such as freely-drained alpine screes or sand dunes. Roots of the least tolerant plants die and rot if anaerobiosis is prolonged – usually the result of waterlogging – but it is unclear if death is caused by lack of oxygen *per se* or by other, associated chemical changes. These last include increases in toxic gases such as hydrogen sulphide (H_2S), produced by anaerobic microbial metabolism, as well as increases in reduced ions of, for example, iron and manganese (mentioned in Section 4.2.2 and discussed further in Section 4.4.3).

On the majority of soils, aeration is quite good in the surface layers (litter and A horizon) and rather poor in the sub-soil (B horizon). Most plant roots and soil animals occur in the surface layers and it is partly as a result of root growth and the burrowing of soil animals such as earthworms that aeration *is* good here. However, the common observation that deep digging or ploughing which aerates the deeper soil layers increases root growth and plant size or yield in crop plants, suggests that plant growth can be limited by root penetration determined by soil aeration.

❑   Try to think of two reasons why shallow root penetration may cause a reduction in shoot growth.

■   First, there will be greater competition for mineral nutrients in the surface layers and, secondly, growth is more likely to be limited by water supply if roots cannot penetrate into the sub-soil (recall the *Banksia* example in Section 4.4.1).

Where soil aeration is chronically poor, plants have evolved two mechanisms that permit survival. Some, such as the cross-leaved heath, *Erica tetralix*, are just very tolerant of anaerobic root respiration and of the chemical environment in waterlogged soil (discussed in Section 4.4.4). Others, including all truly aquatic plants and many of those growing in fens and bogs, have a greatly modified structure characterized by extensive aerenchyma tissue (Chapter 3). Aerenchyma can serve as as an internal ventilation system which pipes oxygen (produced during photosynthesis) from shoots to roots. Oxygen may then diffuse out of the roots into the soil, creating an aerobic micro-environment where toxic reduced ions are oxidized: roots of marsh plants are commonly surrounded – and protected – by a rusty-coloured sheath of oxidized iron (iron(III) hydroxide). So these plants avoid rather than tolerate the effects of poor soil aeration, although at some stages in their life cycle – for example, during seedling germination or when new leaves emerge in spring – they must also be able to survive periods without any oxygen at all (anoxia).

Areas prone to flooding, such as valley bottoms and dune slacks (the 'valleys' between sand dunes), often have distinctive wetland communities and there may also be sharp boundaries between dominant species *within* wetlands. In nearly all cases studied, these boundaries relate to varying tolerance (or avoidance) of low soil oxygen. Tolerance of intermittent or permanent oxygen shortage around roots allows wetland plants to colonize some of the most productive habitats, such as river deltas and tropical swamps, where there are abundant water and nutrients. But the converse is that these plants rarely, if ever, grow in closed communities on soils that never flood, even when there are abundant supplies of water and nutrients.

❑   What is the most likely explanation for this exclusion of wetland plants from non-wetland communities?

■   Competition. Wetland plants are presumably at a competitive disadvantage on non-wetland sites.

The fact that, for example, wild iris *Iris pseudacorus* grows perfectly well in many British gardens (without competition) but occurs naturally only in wetlands tends to support this view, but the exact nature of the disadvantages operating on non-wetland sites is still uncertain. Crawford (1992) discusses the subject further in a wide-ranging review of the effects of oxygen shortage on plant distribution.

4.4.3 Soil pH and the calcicole–calcifuge problem

Soil pH is a measure of hydrogen ion concentration, $[H^+]$, in the soil solution, being defined as $\log_{10}(1/[H^+])$ which is the same as $-\log_{10}[H^+]$.

❑   Will a soil of low pH have a high or a low $[H^+]$?

■   High: because of the reciprocal relationship between pH and $[H^+]$, the lower the pH the higher the $[H^+]$ (acidity). A soil of high pH has a low $[H^+]$ and is said to be alkaline.

pH is a soil condition that can affect directly soil micro-organisms and some soil animals. For example, fungi tend to be the dominant decomposers in acid soils while bacteria dominate in alkaline soils; most nitrifying bacteria, which convert ammonium to nitrate ions, have a pH optimum around 7 (neutral). Many bryophytes that grow on soil are similarly affected and so are earthworms (see Figure 4.9): earthworms are much scarcer on acid soils, especially at deeper levels. However, apart from extreme and rarely encountered pH values (above about pH 9 and below pH 3), there is little evidence that soil pH has direct effects on plant roots.

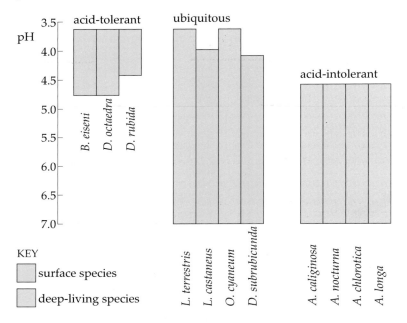

Figure 4.9   Distribution of earthworm species in soils of differing pH. (*B.* = *Bimastos*; *D.* = *Dendrobaena*; *L.* = *Lumbricus*; *O.* = *Octolasium*; *A.* = *Allobophora*.)

The main way in which pH affects rooted plants is through its effect on the *availability of ions*, because ion solubility is strongly affected by pH. Study Figure 4.10 but note that 'availability' is not synonymous with 'level': an element may be present in large quantities, but be insoluble or in some form which plants cannot absorb.

❑   At which soil pH values is nitrogen likely to be in short supply?

■   Acid soils below about pH 4.5–5, as mentioned in Section 4.3.1.

It is because nitrifying bacteria (see above) commonly have low activity on acid soils that nitrogen is available here chiefly as the ammonium ion (NH_4^+), which some plants are unable to take up and use (Section 4.3.1). On strongly acid soils – below pH 4.5 – all nutrients except iron and manganese are in relatively short supply.

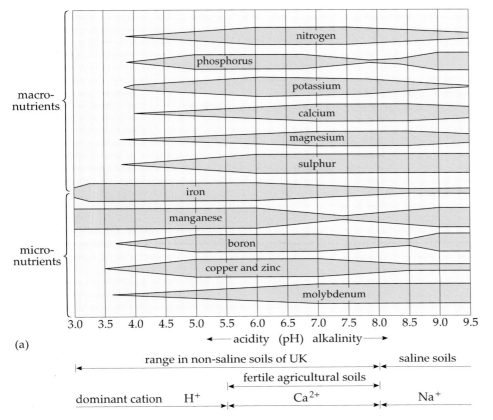

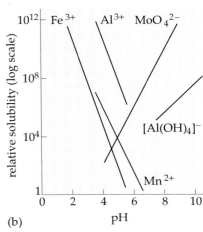

(a)

range in non-saline soils of UK   saline soils

fertile agricultural soils

dominant cation   H⁺   Ca²⁺   Na⁺

Figure 4.10 (a) Availability of mineral nutrients (indicated by thickness of grey bars) in relation to soil pH. (b) Relationship between pH and solubility for some ions.

❑ From Figure 4.10b, why are iron(III) (Fe^{3+}), manganese(II) (Mn^{2+}) and the non-nutrient aluminium ion (Al^{3+}) not usually in short supply in acid soils?

■ Because the solubility of these ions *increases* with acidity as they are released from insoluble compounds or complexes. On very acid soils, these ions may reach levels that are actually toxic to plant roots.

In contrast, the solubility of the molybdate ion (MoO_4^{2-}, the form in which Mo is taken up by plants) *decreases* with acidity (Figure 4.10b), so this nutrient is more available on neutral or alkaline soils. The macro-nutrient phosphorus and the micro-nutrients manganese and boron all have minimum availability on soils of pH 7–8 because they form insoluble complexes (Figure 4.10a). On very acid soils, apart from P, which may be locked up as insoluble iron or aluminium phosphates, the low availability of most nutrients is because they are not there: low pH tends to mobilize ions bound to soil particles and they are then readily leached. Soil pH, therefore, affects both the *availability of essential resources* (nutrient ions) and the *conditions* for root growth (e.g. toxic levels of some ions on acid soils).

It is worth considering briefly here the factors that influence soil pH, which is not a fixed property but depends on several variables, of which two are discussed below.

1   The main factor is the level of basic cations (e.g. Ca^{2+}, Na^+) and this is usually influenced most strongly by the amount of calcium carbonate present ($CaCO_3$, limestone or chalk), commonly referred to as the lime content. $CaCO_3$ is insoluble in pure water but dissolves in rainwater (a dilute solution of carbonic acid, which is weakly acidic), according to the system shown in Equation 4.1. (*Note*: hydrogen carbonate is more commonly known as bicarbonate.)

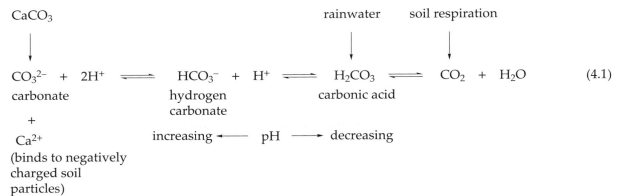

$$CaCO_3$$

$$CO_3^{2-} + 2H^+ \rightleftharpoons HCO_3^- + H^+ \rightleftharpoons H_2CO_3 \rightleftharpoons CO_2 + H_2O \qquad (4.1)$$

carbonate   hydrogen   carbonic acid
                  carbonate

$+$

Ca^{2+} increasing $\longleftarrow$ pH $\longrightarrow$ decreasing
(binds to negatively
charged soil
particles)

rainwater   soil respiration

Addition of carbonate ions causes the equilibrium to be pushed to the right of the carbon dioxide–hydrogen carbonate system, and pH tends to increase. Note that if calcium is added to soil in the absence of high carbonate levels (e.g. as calcium sulphate), this does not affect soil pH. In arid areas where the soil was once saline and has low calcium but high sodium levels, *alkaline soils* may develop. Na_2CO_3 replaces $CaCO_3$ in the equilibrium system and may dissolve to give a soil pH of 9 or more.

2   A second factor that may influence soil pH is CO_2 *level*. Respiration by roots and micro-organisms often causes this to rise, at least locally, to values of 1% or more and the higher the CO_2 concentration, the lower the pH (following Equation 4.1). Neutral or alkaline soils with relatively low [CO_2] are especially sensitive to this effect.

❑   From Figure 4.11, how will pH change in a clay–$CaCO_3$ system if [CO_2] increases from atmospheric level to 1%?

■   It will fall from about pH 8.6 to around pH 7.7.

Localized acidification around plant roots, caused not only by root respiration but also by active secretion of hydrogen ions, was mentioned in Section 4.3.1 as an important mechanism for increasing the availability of phosphate by solubilizing insoluble calcium phosphate. A third factor that

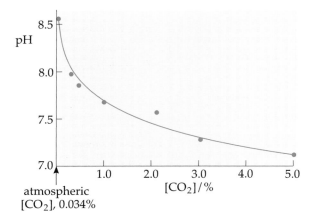

Figure 4.11 Increasing CO_2 concentration in a clay–$CaCO_3$ system causes pH to fall.

may influence pH in certain soils is the presence of readily oxidized ions or organic substrates that release H^+ ions when oxidized. For example, when a waterlogged soil containing much Fe^{2+}, Mn^{2+} and NH_4^+ dries out, these ions become oxidized and contribute to a fall in soil pH.

Because soil conditions and nutrient availability differ so markedly in lime-rich (calcareous) and lime-poor (acid) soils, it is hardly surprising that the vegetation differs: compare acid moorland covered by heather (*Calluna vulgaris*) and chalk grassland with a great diversity of species. Plants confined to lime-rich or lime-poor soils are called, respectively, strict **calcicoles** or strict **calcifuges** and they can be useful indicators of soil type. For more than 70 years, ecologists have studied the relationship between these plants and the soils they grow on to try to explain it.

Calcicoles and calcifuges

These plants grow in a range of plant communities, including woodland and marshes, but the most characteristic communities are grassland or heath. Common features of soils here are listed in Table 4.10 and they differ in many respects. It is quite possible, therefore, that different species are confined to one or other soil type for different reasons. Three general kinds of reasons could help to explain calcicole–calcifuge distribution:

1 *Competition*: calcicole and calcifuge species may be able to *grow* on both types of soil but grow much better on one type and are excluded from the other because of competition.

2 *Lack of resources*: calcifuges might be excluded from calcareous soils by shortage of iron, manganese, phosphate or nitrogen (if unable to use nitrate) while calcicoles might be excluded from acid soils by shortage of calcium or nitrogen (if unable to use ammonium).

3 *Soil conditions*: soil pH or physical nature could be involved whilst, for calcicoles, high levels of Fe, Mn and Al could be toxic on acid soils and high calcium could be toxic to calcifuges on calcareous soils.

The implication underlying factors 1–3 is that calcicoles and calcifuges differ in their ability to obtain particular scarce resources and/or their ability to tolerate certain conditions. Early experiments and observations showed that the physical nature of soil, calcium levels and, for most species, soil pH *per se* were not primary factors influencing distribution.

Table 4.10 The characteristics of calcareous and acid soils which support calcicole and calcifuge floras respectively.

Soil property	Calcareous soil (+ calcicoles)	Acid soil (+ calcifuges)
pH (at least in the top 3 cm)	above 6.5	below 6.5
calcium content	high, present as $CaCO_3$	low as $CaCO_3$ but may be quite high as other salts, e.g. $CaSO_4$
physical nature	usually light, well-drained and aerated; humus well mixed with mineral material	variable, but often heavy with poor drainage and aeration; humus tending to form a distinct superficial layer
availability of elements:		
Fe	low	high
Mn	often low	often high
P	moderate to very low	usually low to very low
N	low or high (most present as NO_3^-)	often low (may be much NH_4^+)
Al	low	high

❑ (a) What information in Table 4.10 indicates that calcium is not a critical factor? (b) Suggest experiments to investigate this further.

■ (a) Acid soils may have high levels of calcium sulphate (which, as mentioned earlier, does not cause high soil pH because $CaSO_4$ is not part of the CO_3^{2-}–HCO_3^-–CO_2 system). So *some* calcifuges must tolerate high [Ca] and calcicoles are still excluded from such $CaSO_4$ soils. (b) In experiments, you could add to soils different levels of calcium in forms that did not affect pH (e.g. $CaSO_4$ or $Ca(NO_3)_2$) and in forms that did (e.g. $CaCO_3$ or $Ca(OH)_2$) and observe how this affected the growth and survival of calcicole and calcifuge species. Plants could also be grown in water culture with different levels and forms of calcium.

Rorison (1960a, b) carried out such water culture experiments which confirmed that, for the species he studied, calcium is not the primary factor limiting the growth of calcicoles or restricting the growth of calcifuges. It may, however, be a contributing factor because high levels of Ca are required for good germination and growth of most calcicoles but are not required by calcifuges and may even be *toxic* for some.

Other experiments indicated that competition cannot be the *sole* factor explaining calcicole–calcifuge distribution.

❑ How do the data in Figure 4.12 support this conclusion?

■ They show that seedlings of a calcicole on an acid soil and of a calcifuge on a calcareous soil did not survive long *irrespective* of competition: there would be no interspecific competition when seeds were sown on bare soil but there would when sown into turf.

Scabiosa columbaria
(small scabious)

Galium saxatile (heath bedstraw)

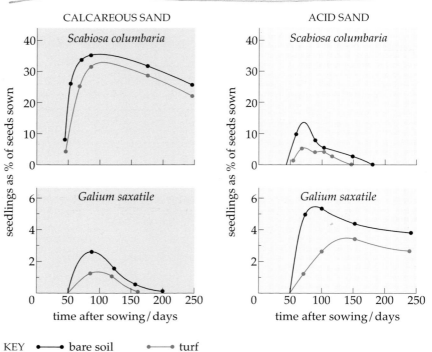

Figure 4.12 Seedling survival of a calcicole *Scabiosa columbaria* (small scabious) (× 0.2) and a calcifuge *Galium saxatile* (heath bedstraw) (× 0.5) on calcareous-sand and acid-sand (i.e. physically identical soils). Seeds were sown on bare ground or in turf.

The only other factors that might explain distribution patterns relate to the availability or toxicity of mineral ions (lower part of Table 4.10). Considering calcicoles first, three factors have been identified as being of major importance:

1 Rorison (1960a and b) noted symptoms of *phosphate deficiency* in seedlings of calcicoles grown on acid soil (leaves turned red). Later work generally confirmed that strict calcicoles cannot take up sufficient P on soils below about pH 5.

2 Symptoms of *aluminium toxicity* (stunted roots) were also observed in Rorison's early experiments. It was subsequently shown that the calcicolous sedge *Carex lepidocarpa* tolerates only 1 p.p.m. (mg l^{-1}) Al^{3+} in solution culture whereas the closely related calcifuge sedge *Carix demissa* tolerates 30 p.p.m. So high levels of Al on acid soils are a second major factor that may contribute to the exclusion of calcicoles and also to the poor growth of crop plants on many acid soils throughout the world.

3 A third factor is *ammonium toxicity*.

❑ What conclusions do you draw from the data in Figure 4.13?

■ When grown in water solution (an important proviso), the calcicole *Bromus erectus* is much more sensitive to Al if nitrogen is supplied as nitrate. Ammonium inhibits growth so drastically that the effects of Al are largely masked.

Since nitrogen is often available mainly as NH_4^+ on acid soils (Table 4.10), this is likely to be of considerable significance for calcicoles. Calcifuges typically grow equally well with ammonium-N or nitrate-N and they are insensitive to Al *unless* high levels of nitrate are present. We must conclude that the form in which nitrogen is available affects sensitivity to aluminium.

What about calcifuges? We mentioned earlier that calcium toxicity may be a problem for some species on calcareous soils, but of much wider significance are shortages of one or both of two nutrients: iron and phosphorus. Calcifuges commonly show a pronounced yellowing (chlorosis) between leaf veins when grown on calcareous soil and this **lime chlorosis** is a classic symptom of *iron deficiency* – an inability to take up or translocate to shoots adequate amounts of iron. *Phosphate deficiency* can occur on calcareous soils even when P is present in quite large amounts as insoluble complexes because, unlike calcicoles, calcifuges lack mechanisms that solubilize phosphate and make it available for uptake. The data in Table 4.11 indicate that P rather than Fe deficiency is of greater significance for heathers (family Ericaceae) on calcareous soil.

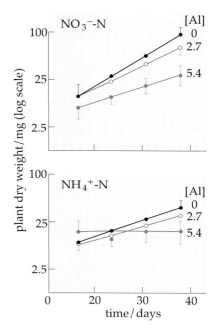

Figure 4.13 Effect of aluminium on the growth of upright brome *Bromus erectus* with different sources of nitrogen. Plants were grown for six weeks in nutrient solution containing either nitrate-N or ammonium-N and different concentrations of aluminium: [Al] values are in mg l^{-1}. Only curves that differ significantly from each other ($P < 0.05$) are included.

Table 4.11 Concentrations (mg g^{-1}) of iron and phosphorus in the shoots of five species of Ericaceae grown in pots of soil of different pH for 12 months. Relative growth over this period is given as dry wt per plant on acid soil / dry wt per plant on calcareous soil.

	Fe	P	Relative growth (acid/calcareous)
Acid gley soil			
Erica cinerea	0.78	1.36	35.65
E. erigena	0.46	0.95	0.55
E. herbacea	0.21	1.20	0.76
E. tetralix	0.37	1.56	11.14
Calluna vulgaris	0.26	1.38	8.38
Calcareous soil			
Erica cinerea	2.06	< 0.41	–
E. erigena	0.42	0.97	–
E. herbacea	0.31	0.91	–
E. tetralix	0.58	< 0.17	–
Calluna vulgaris	0.99	0.17	–
LSD*			
acid gley soil	0.45	1.02	
calcareous soil	0.26	0.54	

* Least significant difference (95% confidence level).

❑   From Table 4.11, (a) identify the calcicole and calcifuge species. Then (b) explain how the data support P as the more important factor limiting growth of these calcifuges.

■   (a) Relative growth on the two kinds of soil is the best guide. *Erica cinerea*, *E. tetralix* and *Calluna vulgaris* grew much better on the acid soil and are calcifuges; *E. erigena* and *E. herbacea* grew slightly better on the calcareous soil and are calcicoles.

(b) The three calcifuges actually contain more iron in their shoots when grown on calcareous soil than on the acid soil, which argues against shortage of Fe being a major problem. However, Marrs and Bannister (1978) did observe symptoms of iron chlorosis in shoot tips on the calcareous soil so possibly Fe was not being translocated to shoot tips. Much more striking are the minuscule amounts of P in the calcifuge plants on calcareous soil; they were almost certainly phosphate-deficient. Note that P levels in calcicoles were much the same on both types of soil.

Now look at the data of Tyler (1994) in Table 4.12, which again concern the relative importance of Fe and P in limiting calcifuge growth.

Table 4.12 Shoot dry weight of four calcifuge species grown for 60–80 days in a calcareous soil with (+) and without (–) additions of iron or phosphate. Values are means of 10 plants. Means in each row carrying different superscript letters are significantly different ($P < 0.01$).

Species		–P –Fe	(+P)–Fe	–P +Fe
			Dry weight/mg	
Deschampsia flexuosa (wavy hair-grass)	P	122[a]	244[b]	106[a]
Galium saxatile (heath bedstraw)	Fe	23[a]	< 1[b]	80[c]
Holcus mollis (creeping soft-grass)	P	80[a]	294[b]	76[a]
Nardus stricta (mat-grass)	P+Fe	139[a]	214[b]	194[c]

❑   For each species in Table 4.12, decide whether growth on a calcareous soil is limited more by P or by Fe or by both.

■   For *Deschampsia flexuosa* and *Holcus mollis*, P is more significant: growth is increased by P but not by Fe addition. For *Galium saxatile*, Fe is more important and P addition actually decreases growth. For *Nardus stricta*, P and Fe are of equal importance.

Clearly, different calcifuge species vary with respect to the factor(s) that exclude them from calcareous soils. To summarize the whole issue, the chief reason for the different distribution of strict calcicoles and calcifuges appears to be soil pH acting on the supply of mineral ions in soils with rather low nutrient availability. On acid soils, calcicoles may be affected by one or more of three factors: Al toxicity, NH_4^+ toxicity, or P deficiency – to all of which calcifuges are tolerant. On calcareous soils, calcifuges may suffer from lime chlorosis (shortage of iron) and/or an inability to absorb sufficient P (and sometimes K, potassium) in conditions where calcicoles can obtain adequate supplies. Other ions have been implicated, e.g. manganese toxicity and calcium deficiency on acid soils and calcium toxicity on calcareous soils; the only thing that can be said with certainty is that no single mechanism explains calcicole–calcifuge distribution.

4.4.4 Toxic soil conditions

When chemical conditions in soil are extreme in some way – usually because of very high levels of certain ions – the majority of plants and soil animals will be excuded from that site and only organisms with special characteristics will be able to survive there. Like other 'extreme' conditions, soil toxicity is therefore a relative concept: what is lethal for the majority is quite acceptable to the specialized few, and examples already discussed (tolerance of very acid soils by strict calcifuges and of waterlogged, anaerobic soils by marsh plants) illustrate the point. In this Section we examine further the impact of toxic soil conditions caused by waterlogging and also those caused by high levels of 'heavy' metals.

Heavy metal toxicity and soil ecotypes

Metallic elements such as copper (Cu), zinc (Zn), lead (Pb), cadmium (Cd), mercury (Hg) and several others are toxic to both plants and animals at very low levels – even though some are essential micro-nutrients and *required* in trace amounts (Figure 4.10). Some soils contain naturally high (i.e. toxic) levels of one or more *heavy metals* (as they are commonly termed); for example, the copper soils of Zambia and Australia and the *serpentine soils*, which are discussed further below, and are characterized by high nickel and chromium. Even more common are metal-polluted soils contaminated because of human activities; mine spoil heaps, for example, or industrial waste dumps. Nearly all these metal-rich soils are vegetated to some extent and the questions of interest here are:

1   What kinds of plants can grow on these soils and how do they manage to do so?

2   Are these plants confined to metal-rich soils or can they grow elsewhere?

Such questions are of more than academic interest because of the recurring need to revegetate toxic waste dumps.

If you look at the plants growing on the spoil heap of (say) an old lead mine, the chances are that the most abundant species will be bent-grasses (*Agrostis* spp.) or fine-leaved fescue grasses (*Festuca* spp.). Either or both of these are likely to grow on nearby 'normal' soils but if seeds or young plants of *Agrostis* from normal soil are placed on the metal-rich soil, only a tiny proportion will survive. The explanation is that plants on the toxic soil are a genetically distinct variety or **ecotype** (Chapter 3, Section 3.4.3) that carries genes conferring tolerance of the heavy metal; all plants on the toxic soil carry these genes but very few individuals on the normal soil do so. The degree of tolerance varies (within and between species) depending on the metal concentration and, furthermore, different genes and tolerance mechanisms are involved for each metal. Heavy metal tolerance was one of the first characteristics used to define an ecotype, and the term is still used to describe varieties that grow in habitats with extreme physical or chemical conditions.

Usually, metal-tolerant populations of a species look the same as those on normal soil but, under identical (non-toxic) soil conditions, the former show physiological differences such as slower growth or lower seed production. Herein lies the clue as to why genes conferring metal tolerance are so rare in populations on normal soil: individuals carrying these genes are at a competitive disadvantage and most are removed by natural selection. On the toxic soil, by contrast, natural selection acts in the opposite direction because only individuals carrying genes for metal tolerance survive. In a few species, metal-tolerant ecotypes actually look different and, for example, zinc-tolerant ecotypes of the mountain pansy *Viola lutea* and alpine penny-cress *Thlaspi alpestre* (Figure 4.14) are sometimes classified as separate species – *V. calaminare* and *T. calaminare*. They are indicators in continental Europe of zinc ore (calamine) sites and grow virtually nowhere else.

The plant community on a metal-rich soil is often highly distinctive because genes for metal tolerance occur in only a limited number of species. In addition, naturally metal-rich soils commonly have other distinctive chemical features, notably a shortage of one or more macro-nutrients, so that plants growing there must tolerate both adverse chemical conditions and shortages of chemical resources. A good example is provided by the **serpentine vegetation** which grows on infertile soils that derive from serpentine rock – essentially a magnesium iron silicate with appreciable amounts of chromium and nickel. Serpentine soils occur all over the world from the wet tropics almost to polar latitudes and support a

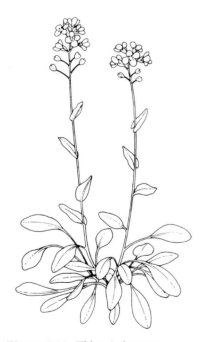

Figure 4.14 *Thlaspi alpestre* (alpine penny-cress), a species with an ecotype that is largely confined to zinc-rich soils. (× 0.3)

sparse vegetation consisting mainly of serpentine ecotypes of widely distributed species and many rare species which often occur on no other type of soil. The unique flora of the Lizard peninsula, Cornwall, is a famous example of serpentine heath and is the only natural site in the UK for the Cornish heath *Erica vagans*. Table 4.13 shows the concentrations of certain ions in the soil solution of various serpentine soils (Proctor *et al.*, 1981).

Table 4.13 The concentrations of ions (mg l^{-1}) in the soil solution of some serpentine soils and the usual range in fertile, non-agricultural soils. (Values for serpentine soils are means ± 95% confidence limits.)

Site	Ni^{2+}	Ca^{2+}	Mg^{2+}	K$^+$	NO$_3^-$	PO$_4^{3-}$	pH	Mg : Ca ratio
Scotland 1	< 0.1	0.45 ± 0.21	9.4 ± 1.8	0.36 ± 0.13	50 ± 12	0.95 ± 0.92	6.6 ± 0.15	23 ± 8.1
Scotland 2	0.13 ± 0.04	5.4 ± 2.3	28 ± 15	5.0 ± 1.6	29 ± 14	0.97 ± 0.10	5.8 ± 0.3	5.0 ± 0.7
Scotland 3	0.67 ± 0.07	11 ± 1.2	180 ± 33	5.4 ± 1.2	830 ± 69	13 ± 5.0	ND	16 (av.)
Zimbabwe 1	0.58 ± 0.48	19 ± 10	28 ± 21	20 ± 10	ND	ND	ND	1.5 ± 0.36
Zimbabwe 2	0.67 ± 0.29	23 ± 36	38 ± 56	19 ± 32	ND	ND	ND	1.8 ± 1.6
'ordinary soil'	0	50–100	25–200	11–200	50–150	0.03–0.3		

ND = not determined.

❑   From Table 4.13, what distinctive chemical features occur (a) on all and (b) on some of the serpentine soils examined?

■   (a) High levels of nickel is a universal feature (although Ni is patchily distributed on the Scotland 1 site) and so is low calcium. (b) Low nitrate and/or potassium occurs on some serpentine soils.

Early studies emphasized low Ca and the high ratio of Mg to Ca (because magnesium influences absorption of calcium) – but you can see from Table 4.13 that this is not a viable explanation for the Zimbabwean serpentine soils. What, therefore, are the main causes of distinctive serpentine vegetation? The current view is that nickel toxicity plays a central role and serpentine plants nearly always contain high levels of Ni and show Ni tolerance when grown in water culture. The other major factor is shortage of one or more major nutrients such that plants inevitably grow very slowly; as a result, Ni accumulates in their tissues, even if Ni levels in the soil are not exceptionaly high, and only plants tolerant of Ni can survive. An interaction between resource shortage (inhibiting growth) and nickel accumulation seems, therefore, to be the most plausible explanation of serpentine vegetation. Low calcium due to a high Mg : Ca ratio may be a further significant factor on some soils.

Waterlogged soils

Some of the toxic soil conditions that arise when soil is waterlogged were described in Section 4.4.2 and some are illustrated in Figure 4.15, which shows that various changes occur in sequence after waterlogging. As oxygen falls to zero, certain micro-organisms begin to use other substances (such as nitrate ions) instead of O$_2$ as terminal electron acceptors for respiration. These substances are thus *reduced* (acquire electrons) and the overall tendency of the soil to accept electrons, i.e. the oxidation–reduction or **redox potential**, E_h, decreases, meaning that the soil system has a greater tendency to donate electrons and is a *more reducing environment*. The sequence of chemical changes that occur in soil after waterlogging reflects the decline in E_h.

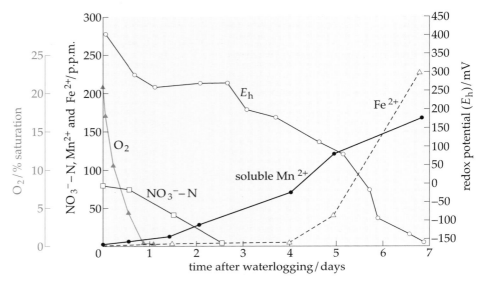

Figure 4.15 Effects of waterlogging. Changes in oxygen, nitrate, soluble manganese and iron (Mn^{2+} and Fe^{2+}) and redox potential (E_h at pH 7) of a silty clay soil during seven days following waterlogging.

❑   From Figure 4.15, identify in sequence three changes that occur as E_h falls.

■   The first change is a decline in nitrate (which is reduced to nitrogen gas by the action of denitrifying bacteria): nitrogen is available only as ammonium ions, NH_4^+, in waterlogged soil. The second is a rise in soluble manganese(II) (Mn^{2+}) as the relatively insoluble manganese(III) (Mn^{3+}) is reduced, followed by a rise in soluble iron(II) (Fe^{2+}) as iron(III) (Fe^{3+}) is reduced.

Further changes which occur after the formation of iron(II) include solubilization of insoluble phosphate and accumulation of the end-products of anaerobic microbial metabolism (e.g. methane, ethylene, hydrogen sulphide and other sulphides, and organic acids). Most of these end-products are toxic to plants and so are high levels of Fe^{2+}, Mn^{2+} and NH_4^+ so that, in general, the longer a soil remains waterlogged the more hostile it becomes as an environment for roots. It follows that the sensitivity of plants to waterlogging (i.e. how long before they die) depends on which factors they tolerate: the most sensitive species will be intolerant of anoxia *per se* while others may tolerate anoxia, NH_4^+ and high Mn^{2+} – but perhaps not high levels of Fe^{2+}.

Plants that grow on permanently waterlogged soil must tolerate all these hostile conditions although there is the compensating luxury of abundant water. However, wetland soils vary considerably with respect to (for example) the concentrations of iron(II), soluble phosphate and manganese(II). So we can now frame some questions about the distribution of wetland plants:

1   Do wetland plants grow on waterlogged soils because they tolerate chemical conditions here and: (a) require large, constant supplies of water so that they grow and compete poorly on drained soils unless these are very well-watered; or (b) simply grow and compete poorly on drained soils irrespective of water supply?

2   Does the occurrence of particular wetland plants on certain sites and
their absence on others reflect chemical conditions; or is it determined
by other factors such as climate, dispersal capacity or past history?

Figure 4.16, showing the distribution of three common moorland species, is
relevant to the first question. Figure 4.16a shows that cross-leaved heath
Erica tetralix is usually confined to the wettest parts; common heath or bell-
heather *Erica cinerea* to the driest parts; and heather or ling *Calluna vulgaris*
grows mainly in intermediate conditions of wetness.

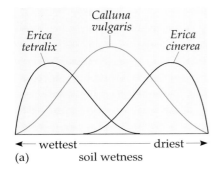
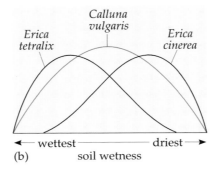

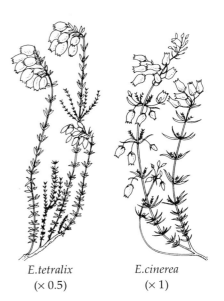

Figure 4.16 The distribution of
Erica tetralix, E. cinerea and
Calluna vulgaris on acid
moorland in relation to the
average water content of the soil.
(a) Actual distribution; (b)
potential distribution when
grown on bare plots.

❑ By comparing Figure 4.16a and b, decide which of options (a) and (b)
in the first question best explains the actual distribution of *E. tetralix*.

■ Option (b) appears to offer the best explanation. *E. tetralix* does not
grow on the driest soils even in the absence of competition, so it must
have a lower tolerance of drought. Note, however, that it can grow on
much drier soils in the absence of competition, so it does not *require* a
constantly high water supply.

Regarding the second question, we mentioned in Section 4.4.2 that
variation in soil oxygen influences strongly the distribution of species
within a particular wetland. But this begs the question asked here, which is
why on different wetland sites with similar oxygen levels the dominant
plants vary. Earlier studies indicated that variation in iron(II) could be
involved and Snowdon and Wheeler (1993) carried out a study of iron
tolerance in 44 wetland species to test this idea. They grew plants in water
culture with different levels of iron(II) present and obtained an index of
iron tolerance based on growth rates relative to controls without added
iron (Figure 4.17). All species were placed in one of four groups on the
basis of their iron tolerance and iron tolerance indices were plotted against
the maximum iron concentrations in soil at sites where a species actually
grew (Figure 4.18).

E.tetralix   *E.cinerea*
(× 0.5)   (× 1)

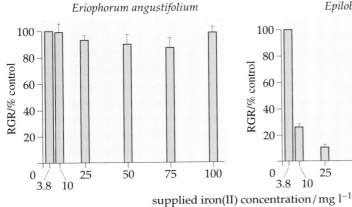
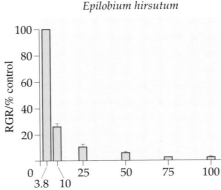

supplied iron(II) concentration/mg l^{-1}

Figure 4.17 Iron tolerance of two
fen species, *Eriophorum
angustifolium* (cotton-grass) and
Epilobium hirsutum (great hairy
willow-herb), as shown by their
relative growth rates (RGR) in
solution culture (as a percentage
of control treatments) at different
levels of iron(II).

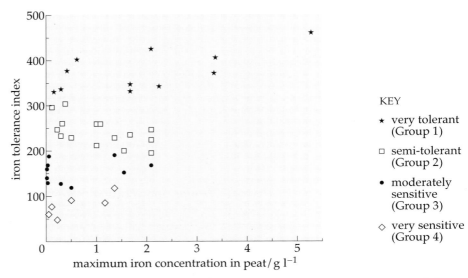

Figure 4.18 Relationship between the iron tolerance index of a species and the maximum iron concentration of soil at sites where the species occurred in the field. Symbols indicate the four iron tolerance groups.

❑ From Figure 4.17, which species has the highest index of iron tolerance?

■ Cotton-grass, which grew as well as controls even at the highest concentration of iron tested.

❑ From Figure 4.18, is the occurrence of different wetland species at different sites explicable solely in terms of their iron tolerance, i.e. can the second question be answered in terms of iron toxicity?

■ No. Species in Group 1 (very high iron tolerance) are the only species to occur on soils with high iron levels *but* these species also occur on soils with low iron levels. Thus you cannot predict where a Group 1 species will occur solely on the basis of its iron tolerance index. The 'match' is good only for the Group 4 species with the lowest iron tolerance.

What we can say, therefore, is that increasing iron levels in soil *exclude* an increasing number of wetland species so that iron(II) certainly influences distribution and the species composition of fen vegetation. Very iron-tolerant species (Group 1) also tend to have low growth rates and Snowdon and Wheeler's study showed that the sites where they grow have low levels of available phosphate and low 'fertility' (assessed by measuring the growth of a test plant). This illustrates the difficulty of ascribing to any one factor a dominant role in determining distribution: perhaps, for example, low fertility is more important than high iron levels for Group 1 species. Low fertility is also characteristic of very acid or alkaline soils and of serpentine soils, and species that grow here also have low growth rates. So there is a correlation between the ability to tolerate 'difficult' soil conditions, low growth rates and low soil fertility.

Summary of Section 4.4

- Soil conditions may affect organisms directly or indirectly (via effects on the supply of essential resources).

- Physical structure (e.g. **impedance** and stability) and depth of soil directly affect plant growth. The *Banksia* example illustrates how poor root growth because of high impedance can lead to shortage of water – an indirect effect.

- Soil **aeration,** which depends on clay content, compaction and water content, influences oxygen supply and also the likelihood of toxic chemical conditions.

- **Soil pH** has direct effects on soil heterotrophs and rootless plants but not usually on plant roots. Its main effect on rooted plants is through the supply of essential mineral nutrients (Figure 4.10) and it is usually influenced most strongly by the calcium carbonate (lime) content of soil.

- The distribution of strict **calcicoles** and **calcifuges** illustrates the indirect effects of pH. For calcicoles, inability to grow on acid soils may relate to shortage of phosphate, aluminium toxicity or ammonium toxicity. For calcifuges, inability to grow on calcareous soils may be related to phosphate or iron deficiency (lime chlorosis). The relevant factors vary for different species; factors interact (e.g. the nitrogen source affects sensitivity to Al); and other factors (e.g. Ca toxicity or deficiency) may be involved.

- Two extreme types of toxic soil conditions were discussed:

    (a) Soils *rich in heavy metals*: plants survive here mainly because they carry genes that confer tolerance to metals. There may be genetically distinct populations of widely distributed species (**ecotypes**) or species largely confined to metal-rich soils. The distinctive vegetation of serpentine soils appears to result mainly because of high soil nickel and low levels of some macro-nutrients; plants grow slowly and accumulate (and tolerate) high Ni levels.

    (b) On waterlogged soils: lack of oxygen results in reducing conditions (low redox potential), potentially toxic levels of ammonium, reduced iron (Fe^{2+}) and manganese (Mn^{2+}) and high levels of toxic microbial metabolites. Wetland plants tolerate these conditions to varying degrees and are absent from drained soils usually because they grow and compete poorly there. Differences in iron(II) and macro-nutrients on different wetland sites influence the distribution of wetland species between sites.

Question 4.4   *(Objective 4.1)*

Identify the terms defined or described in (a)–(c).

(a) Genetically distinct populations within a species that occur in a specific habitat and grow or compete less well in other habitats.

(b) A property of soil that influences drainage and the ability of plant roots and soil animals to grow or move through the soil.

(c) A property of soil that is influenced by lime and CO_2 content and has a strong effect on the availability of nutrients for uptake by plant roots.

Question 4.5 *(Objectives 4.2, 4.3 & 4.5)*

Explain (a)–(c).

(a) If calcium is added to soil of pH 7 or less as $Ca(NO_3)_2$ or $CaSO4$, there is usually no change in soil pH; if $CaCO_3$ is added, soil pH usually rises.

(b) The rusty-coloured iron-rich deposit which is often found on the roots of plants growing in waterlogged mud probably helps these plants to survive in this habitat.

(c) On soils with medium to high levels of macro-nutrients, the vegetation of localized areas rich in zinc, such as adjacent to a galvanized zinc fence, is usually indistinguishable from the surrounding area; on soils with very low nutrient levels, the vegetation of zinc-rich spoil heaps often includes species not found in the surrounding area.

Question 4.6 *(Objectives 4.5 & 4.15)*

Banksia robur grows on dry, acid and very nutrient-poor heaths in western Australia. Seedlings were raised to maturity in pots of coarse sand irrigated with nutrient solution and levels of N (as nitrates), P and Ca were varied for different sets of replicate pots (Figure 4.19; Grundon, 1972). For each treatment one nutrient at a time was varied, levels ranging from a maximum (level 5), which was the same as in the standard nutrient solution, through a series of fivefold dilutions to a minimum (level 1). For Ca, for example, concentrations for levels 1 to 5 were, respectively, 4, 20, 100, 500 and 2500 μmol l^{-1}, all other treatments having level 5 concentrations. Total dry weight (shoots and roots) were measured at the end of the experiment and results in Figure 4.19 are expressed in terms of the percentage of the maximum dry weight for each treatment. From Figure 4.19, what evidence is there that *B. robur* is (a) a calcifuge and (b) a nutrient-insensitive species?

Figure 4.19 For use with Question 4.6. The dry matter response of *Banksia robur* to five levels of nitrogen, phosphorus and calcium. The response is shown as the % of the maximum yield within the particular nutrient series. Green bars represent mean plant dry weights for these maximum yields. Statistical analysis: treatments sharing the same letter (within each diagram) are not significantly different, $P > 0.05$.

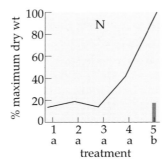 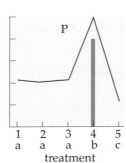 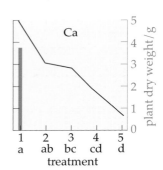

Question 4.7 *(Objectives 4.2 & 4.15)*

Many fens in the lower Rhine area, Germany, have experienced a decrease in the groundwater table. A study was carried out (Kazda, 1995) at one fen in order to determine the effect of lowered water tables on soil properties and whether changes in the soil could explain observed changes in the vegetation. Originally, the fens studied were fed by calcareous water and had a peat soil with pH between 5 and 6 and high levels of organic N and P in the peat. Alder *Alnus glutinosa* was the dominant tree. Table 4.14a relates height of the water table (distance from the soil surface) to the area of fen with a particular soil pH.

Table 4.14 For use with Question 4.7. The relationship for a German fen between (a) groundwater levels and the area with a particular soil pH and (b) levels of soluble (readily exchangeable) aluminium ions and the area with a particular soil pH.

(a)

Ground water level/cm	≤ 3.5	3.5–4	4–4.5	pH 4.5–5	5–5.5	> 5.5
				Area/m^2		
0–50	0	653	3184	2971	1725	286
50–100	3802	3766	1572	365	91	6
100–150	7608	4442	1018	85	0	0
>150	17 723	14 678	1778	0	0	0

(b)

Exchangeable aluminium /mg g^{-1}	< 3.5	3.5–4	4–4.5	pH 4.5–5	5–5.5	> 5.5
				Area/m^2		
< 0.2	0	790	3895	2798	1765	292
0.2–0.4	1564	8970	2929	616	51	0
0.4–0.6	7412	9323	697	7	0	0
0.6–0.8	10 132	4211	31	0	0	0
0.8–1.0	7924	245	0	0	0	0
> 1.0	2101	0	0	0	0	0

(a) From Table 4.14a, describe the relationship between groundwater level and soil pH and suggest an explanation for this relationship.

Table 4.14b shows the relationship between levels of soluble aluminium and the area with particular soil pH.

(b) Describe this relationship and suggest what processes may have caused changes in Al levels.

Other information about the fen is:

(i) In drained parts of the fen where pH was 4 or higher, dense stands of nettles *Urtica dioica* grew up to 2 m tall.

(ii) In the driest areas where pH was below 4, soil particles or aggregates were often covered by rusty deposits of iron.

(iii) In drained areas, there was a correlation between decrease in pH and decrease in the amount of calcium and organic nitrogen in the peat. Note that when organic nitrogen compounds are broken down (decomposed), they yield ammonium ions which may be oxidized to nitrate.

(c) Suggest brief explanations for (i)–(iii).

(d) Use all the information provided to work out a sequence of changes which occurred in the fen peat as the water table fell.

PART II

SALINITY, ESTUARIES AND SALTMARSHES

4.5 Salinity as a condition

Salinity is a condition relating to the concentration of ions or 'salts' in soil or water, which profoundly affects the movement of water and ions into tissues. You can see from Table 4.15 the huge difference in salt concentration between marine and freshwater environments and the range of differences between the body fluids (blood) of different animals. This last reflects fundamental differences in physiology between marine and freshwater or terrestrial organisms: life evolved in the sea, followed by invasion of the land, and freshwaters were then colonized by terrestrial organisms. Usually, the main 'salt' in saline environments is sodium chloride (as in the sea) but in the African soda lakes, for example, and in many saline desert soils, sodium carbonate or sulphate predominates.

Table 4.15 Concentration of major ions in sea- and freshwaters expressed as: (a) g or mg per kg (= litre) (note the different units for sea- and freshwater); (b) mmol per litre (= mol m^{-3}) and including the blood of various animals.

(a)

Ion	'Average' seawater /g kg^{-1}	Lake Windermere water/mg kg^{-1}
Na$^+$	10.56	3.8
K$^+$	0.38	0.6
Ca^{2+}	0.40	6.2*
Mg^{2+}	1.27	0.7
Cl$^-$	18.98	6.7
SO$_4^{2-}$	2.65	7.6
HCO$_3^-$	0.14	11.0*
total	c. 35	c. 40*
salinity as ‰ (parts per thousand or mille)	35	c. 0.04

* These values are much higher in hard, alkaline waters.

(b)

	Na$^+$	K$^+$	Ca^{2+}	Mg^{2+}	Cl$^-$	SO$_4^{2-}$
'average' seawater	470	10	10	54	548	28
range of 'average' freshwaters	0.2–0.5	0.06–1.5	0.07–4.0	0.04–2.0	0.2–2.5	0.05–4.0
Blood of:						
Arenicola, lugworm (marine polychaete or segmented worm)	460	10	10	52	537	24
Anodonta, freshwater mussel (lamellibranch mollusc)	16	0.5	8	0.2	12	0.8
Lophius, angler fish (marine teleost or bony fish)	185	5	6	5	155	–
Canis, dog (terrestrial mammal)	150	4	5	2	106	–

Organisms that exchange water and ions freely with the surrounding medium (most aquatic and many soil organisms and all rooted plants) can live in saline or non-saline conditions – but rarely both. If you place a marine organism such as *Arenicola* (Table 4.15b) in freshwater, the inward movement of water by osmosis and the outward diffusion of salts from the body rapidly cause death. Most marine organisms, therefore, tolerate only very small changes in external salinity and are described as **stenohaline**. Similarly, most freshwater and terrestrial organisms cannot tolerate saline environments (nor survive by drinking saltwater). However, there are some habitats where salinity fluctuates widely, including saltmarshes, estuaries and inland saline areas where salts are concentrated as water evaporates. Only organisms tolerant of widely fluctuating salinity (described as **euryhaline**) can survive in these habitats and their tolerance range is one factor determining where they occur.

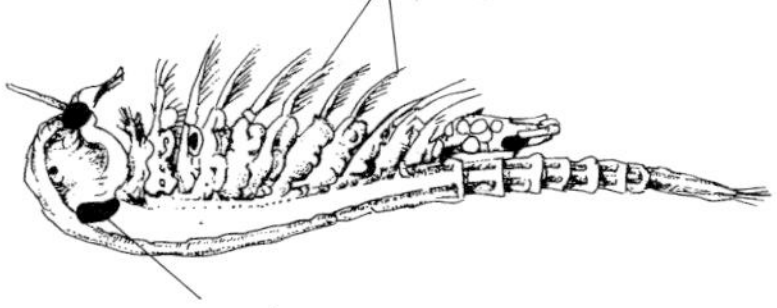

The **brine shrimp *Artemia salina*** (Figure 4.20) is legendary in its tolerance of fluctuating salinity, occurring in warm and often temporary pools of 3–300 parts per thousand, which is written as ‰ and is equivalent to grams per litre (see Table 4.15a). The solute concentration of the blood remains constant over this range, a feat achieved in relatively dilute media by having a body surface with low permeability to water and ions. In concentrated media, however, *Artemia* avoids desiccation by employing the same mechanism as marine teleost fishes: it swallows saltwater, absorbs water, Na^+ and Cl^- ions from the gut and then secretes these ions against the concentration gradient through the first ten body appendages (which also function as gills).

Such ion secretion requires an expenditure of energy and illustrates the general rule that tolerance of extreme chemical conditions – whether of salinity, heavy metals or organic pollutants – is energetically expensive and often accompanied by slow growth rates. This applies to the relatively small number of salt-tolerant angiosperm plants (termed **halophytes**), which are discussed briefly below. On the other hand, organisms able to tolerate extremely salty conditions (as in the hypersaline Dead Sea or temporary saline pools) have few competitors and a few species may occur each with large numbers of individuals. This pattern of low species diversity/high numbers of individuals applies also to invertebrates in the higher reaches of estuaries.

4.5.1 Estuaries

A number of invertebrate species are confined to estuaries, which are a food-rich habitat because of the abundant detritus carried down by rivers. The two estuarine specialists most common in Britain are the snail *Hydrobia ulva* (operculate gastropod mollusc) and the shrimp-like *Corophium volutator* (amphipod crustacean) – see Figure 4.21.

Up to 42 000 m^{-2} *Hydrobia* and 28 000 m^{-2} *Corophium* have been recorded in Scottish estuaries and they provide a rich source of food for enormous numbers of birds, especially in winter. Both of these species tolerate widely fluctuating salinity, 3–94 and 2–50‰ for *Hydrobia* and *Corophium* respectively, and can survive on mudflats when the tide is out. *Hydrobia* crawls over the mud, feeding on the deposits, or floats in the water and collects particles in a mucus bag. *Corophium* is a filter-feeder that constructs U-shaped burrows and prefers anaerobic mud of a certain particle size. Together with the marine mussel *Mytilus*, these euryhaline estuarine

Figure 4.20 A female brine shrimp *Artemia salina* ($\times$ 5).

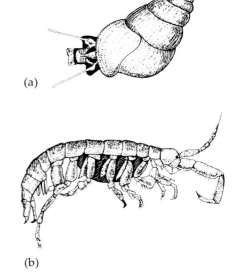

Figure 4.21 Two common British estuarine invertebrates: (a) *Hydrobia ulva* ($\times$ 6); (b) *Corophium volutator* ($\times$ 5).

invertebrates are of great importance as processors of detritus and as a food source for birds. It is for this reason that conservationists often object so strongly to the creation of estuarine barrages which destroy much of the habitat by drastically altering conditions.

4.5.2 Halophytes

The majority of angiosperms cannot tolerate saline environments so that, like metal-tolerant ecotypes (Section 4.4.4), halophytes are a specialized group. They have a quite different mechanism for surviving high salinity from that of the brine shrimp: internal solute concentrations (which determine the osmotic pressure) are maintained *above* those of the exterior (so that water can always enter by osmosis) and excess, potentially toxic Na^+ and Cl^- ions are got rid of by, for example, salt glands on the leaf or secretion into leaf cell vacuoles, the leaves being subsequently shed.

Some of the plants from saline habitats are ecotypes of widely distributed species such as red fescue grass *Festuca rubra*. Others occur only in saline habitats although they nearly always grow as well or better in cultivated, non-saline soil.

❑   So why is their distribution usually so restricted?

■   Largely because of competition – there is good evidence of an inability to compete with non-halophytes on non-saline soils.

At higher latitudes, including Britain, most halophytes are coastal, growing either on sea cliffs exposed to spray or on saltmarshes. However salting of roads in winter can expose roadside verges to 50 times more salt than exposed cliffs, resulting in intense selective pressure for salt tolerance and the spread inland of several coastal species. At lower latitudes, halophytes occur also on saline desert soils which are often very alkaline (Figure 4.10), and increasingly, in arid areas, they occur on abandoned farm land that has become too saline to cultivate because of **salinization**. This process results from faulty irrigation practice: watering from above washes salts down the soil profile but water and salts are drawn up by capillary action as evaporation occurs and salts are deposited near the surface. Unless a slight excess of water is used at each irrigation cycle, surface salts gradually accumulate until the soil is essentially saline.

You can see from this description of halophyte habitats that salinity often accompanies other 'difficult' conditions and resource shortages: mineral nutrients and water are in short supply in alkaline deserts and temperatures are high; strong winds and a risk of desiccation are a hazard on exposed coasts; and saltmarsh plants must tolerate submersion, waterlogging and the effects of sediment deposition and water currents. So although salt tolerance is a necessary property for survival in these places, rarely if ever does it explain exactly which plants occur where. Even in the case of salted road verges, type of seed dispersal seems to have been the main factor determining which coastal species spread inland; sea plantain *Plantago maritima*, whose seeds are spread by vehicle tyres, has been one of the most successful. To illustrate the way in which salinity interacts with other factors – biotic and abiotic – to influence vegetation, we consider next the ecology of saltmarshes and the reasons for variation between and within marshes.

4.6 Saltmarshes

4.6.1 General characteristics

Saltmarshes are saline grasslands that develop on sediments deposited during periods of tidal submergence. To the untutored eye they can look like flat, muddy-green expanses dissected by creeks and with scattered bare *salt pans*, where evaporation from temporary pools leads to very high salt levels. In fact, saltmarshes are of considerable ecological importance and interest, as feeding grounds for birds, as protective buffer zones between land and sea and as processors of nutrients which pass to the sea (see Book 4); in summer, some are vividly coloured by carpets of flowering dicots, such as thrift *Armeria maritima* or sea lavenders *Limonium* spp. They occur world-wide (although in the tropics, mangroves replace herbaceous vegetation at the seaward edge) and are highly dynamic communities – periods of sediment accumulation, when the marsh increases in height and area, alternating with loss periods, when shifting currents or catastrophic storms lead to net erosion.

Two general features of marsh vegetation are that it has *low species diversity* (number of species per square metre) and shows **zonation**: as on rocky shores (Chapter 3, Section 3.1) there are zones dominated by different species and running parallel to the shore. Figure 4.22 shows the positions of zones at Morecambe Bay, Lancashire.

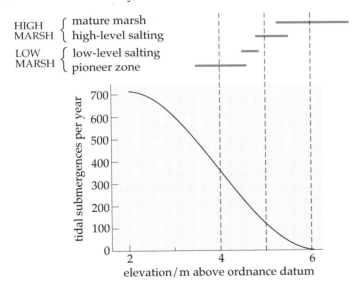

Figure 4.22 Elevation ranges of the main saltmarsh zones at Morecambe Bay, Lancashire.

☐ In what general way do saltmarshes and rocky shores differ on the lower shore?

■ Rocky shores show zonation to the lowest levels whereas saltmarshes do not extend so far down the shore. In general, rocky shores are marine communities that extend towards land and saltmarshes are terrestrial communities that extend towards the sea.

Conditions obviously vary with elevation on the marsh and we consider later how this affects zonation. First, there is the question of why saltmarshes in different geographical regions show variation in species composition and zonation pattern.

4.6.2 Variation between saltmarshes

Variation between saltmarshes in Britain is illustrated in Figure 4.23, which shows a broad division between marshes on the south and east coasts (**Type A**) and those on the west (**Type B**) (Adam, 1990). Why does this *regional* variation occur?

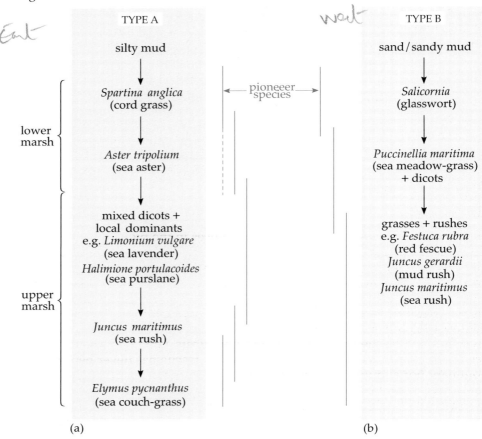

Figure 4.23 General, simplified versions of plant zonation for British saltmarshes (a) on south and east coasts and (b) on western coasts. Species named are those dominant for that zone. Vertical lines indicate the approximate extent of zones.

❑ From Figure 4.23, what is one possible reason for the difference in pioneer species on Type A and B marshes?

■ Substrate (soil) differences: Type A marshes usually develop where there is silty mud whereas Type B are more usual where there is a high proportion of sand in the mud.

Spartina anglica (Figure 4.24) is indeed somewhat intolerant of sandy substrates although the converse is not true: *Salicornia* (Figure 4.25) can colonize a wide range of substrates and used to be the commonest pioneer on the firm mud of south-eastern marshes in Norfolk, with *Spartina* confined mainly to the sloppy mud of the southern coast. The story of the spread of *Spartina* is an interesting one.

Figure 4.24 *Spartina anglica* (cord grass) (× 0.12), a pioneer species on silty mud.

Spartina anglica is a 'new' species whose origins go back to 1870 when a now rare British species of *Spartina* interbred with an introduced American species. The resulting sterile hybrid (*Spartina × townsendii*) spread rapidly by vegetative means and was a strong competitor. Eventually, through a doubling of the chromosome number, the fertile tetraploid species *S. anglica* was produced. This species and its hybrid parent can tolerate up to 6 hours of submergence per day and trap silt extremely effectively so that the height of the lower marsh and its seaward spread tend to increase. *Spartina* covers and stabilizes mudflats with astonishing speed, provided only that the site is rather sheltered and is accreting coarse silt; it has been widely planted in the temperate zone as a means of land reclamation. Frequently, *Spartina* displaces *Salicornia* and *Puccinellia* (Figure 4.23b) and spreads onto lower mudflats which these species cannot colonize. The result has been a considerable reduction in the area of open mudflats in north-west Europe and as these are major feeding grounds for wildfowl and wading birds, this has caused much concern. Saltmarshes are often described as being 'infested' with *Spartina* but this could be merely a temporary state: at long-established sites such as Poole Harbour, Dorset, *Spartina* cover has decreased dramatically over the last 70 years because of an unexplained 'die-back'.

Figure 4.25 *Salicornia europaea* (glasswort) (× 0.5), an annual pioneer species on saltmarshes.

Substrate, therefore, seems to explain differences in the pioneer zone between south-east and west coast marshes. Less easily explained are differences in the upper marsh (Figure 4.23) but interactions between grazing and climate are two of the factors involved. Type B marshes have a long history of grazing and this appears to have favoured growth of red fescue grass *Festuca rubra* at the expense of dicots such as sea lavender (Figure 4.26) and sea purslane. Even if grazing pressure is relaxed, rank growth of red fescue prevents reinvasion by dicots. Grazing pressure has been more variable in Type A marshes but red fescue has never become so dominant nor eliminated completely dicots such as sea lavender. The most likely explanation is that hotter, drier summers in the south-east, perhaps in combination with the fine, silty substrate, dry out the upper marsh considerably and cause very high salt levels (*hypersalinity*): *Festuca rubra* is much less salt-tolerant than *Halimione* or *Limonium*.

On the west coast saltmarshes of Scotland, high rainfall and freshwater seepage from inland produce low-salt conditions with very low incidence of hypersalinity. Species with low salt-tolerance (essentially non-halophytes) occur in the upper marsh and red fescue is abundant. However, species such as *Halimione portulacoides*, *Limonium vulgare* and *Spartina anglica* are absent for a different reason: they reach the northern limit of their distribution along a line which runs north-east from the Solway Firth. So high rainfall and other climatic factors (low temperature or a short growing season) have a major influence on the vegetation of these Scottish marshes. Overall, therefore, regional variation in Britain can be explained in terms of interactions between substrate, climate and grazing.

Figure 4.26 *Limonium vulgare* common sea lavender (× 0.2).

4.6.3 Zonation within marshes

There remains the question of what determines *zonation within marshes*. Below are listed six factors which might be expected to influence zonation:

1   *Frequency and duration of submergence* ranges from about 6 hours per day in the pioneer zone to a few hours per year at the top of the high marsh

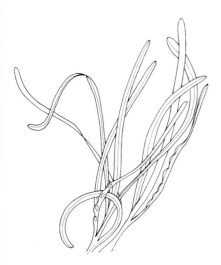

Figure 4.27 *Zostera marina* (eel-grass) (× 0.25), a marine angiosperm that grows on mudflats below the pioneer zone of saltmarshes.

(Figure 4.22). Below the pioneer zone there is also a marine community comprising mats of algae on the mud and, sometimes, eel-grass *Zostera marina* (a marine angiosperm, see Figure 4.27) which stabilizes sediments and prepares the way for saltmarsh development.

2   *Soil aeration* is typically poor on saltmarshes, with a gradation from permanently waterlogged, anaerobic mud at low elevations to drier soils that are only rarely waterlogged at higher elevations. Drainage and aeration are much improved when the sediments contain a fair proportion of sand.

3   *Sediment deposition* occurs more or less constantly in the lower marsh and can smother seedlings or low-growing plants. Even in the higher marsh, however, severe storms may cause considerable deposition of coarse, sandy sediments.

4   Closely linked with factor 3 is the problem of *substrate instability* and *erosion* caused by water currents, which affect particularly any seedlings in the pioneer zone and plants on the edge of creeks.

5   *Salinity* in the lowest part of the marsh is equal to that of seawater and fairly constant. But higher up it may vary considerably where the soil dries out, concentrating salt at the surface, or when rainfall is low and salt is not leached out. If rainfall is high, as on most west European coasts, salinity of the surface soil is low in the upper marsh (as mentioned earlier for Scottish marshes).

6   A biotic factor which affects some but not all saltmarshes is *grazing* by domestic stock and/or wildfowl. Herbage in the middle and upper marsh is highly nutritious and close-cropped.

❑   Which British saltmarshes show the clearest effect of grazing?

■   The Type B marshes (Figure 4.23b) on the west coast.

There is certainly a gradient of conditions (and possibly of the supply of some resources) from the seaward to the landward end of saltmarshes. Snow and Vince (1984) suggest three models to explain zonation along this environmental gradient:

1   Species are *physiologically restricted to* (tolerate) different parts of the gradient.

2   Each species *grows and survives best* in its usual zone – but tolerates conditions in other zones from which it is excluded because of inadequate dispersal, competition, grazing or some combination of these.

3   Most species *grow and survive best in the same part* of the gradient but are displaced along the gradient according to their breadth of tolerance, dispersal, competitive abilities and susceptibility to grazing.

There is no evidence to support model 1. Even annual pioneer species such as *Salicornia* occur in the upper marsh, especially in disturbed or bare areas such as creek sides. The available evidence indicates that model 2 fits some species and model 3 others; below are some examples.

•   *Aster tripolium* sea aster (Figures 4.23a and 4.28) requires five days free of submergence for seedling establishment although, once established, this long-lived perennial tolerates greater submergence and can spread lower down the marsh.

Figure 4.28 *Aster tripolium* sea aster (× 0.5).

- *Armeria maritima* thrift (Figure 4.29) is less tolerant of submergence and also of anaerobic soil: it requires good drainage and aeration of the surface sediments, which is more likely to occur in sandy areas.

- In a study of an Atlantic saltmarsh in south-west Spain, Castellanos *et al.* (1994) found that the relative abundance of perennial glasswort *Arthrocnemum perenne* (Figure 4.30) and a species of *Spartina*, *S. maritima*, in the pioneer zone also depended on drainage and surface aeration of sediments. Both species are equally tolerant of prolonged submergence but whereas *S. maritima* tolerates and can establish on very anaerobic sediments (strongly reducing with very low redox potential), *Arthrocnemum* cannot. On this marsh, *S. maritima* was the first to colonize bare mud and produced scattered tussocks within which sediment rapidly accumulated. Shoots in the centre of large tussocks commonly suffered from an unexplained 'die-back' and *provided* that drainage occurred freely, *Arthrocnemum* invaded this central, raised area and, with its dense, sprawling growth, gradually ousted *Spartina*. In areas where construction of a dyke had impeded marsh drainage, there was no invasion by *Arthrocnemum*.

Figure 4.29 *Armeria maritima* thrift (× 0.2).

Table 4.16 Average redox potentials (in millivolts) of surface sediments under *Spartina* or *Arthrocnemum* or on bare, uncolonized mud for well-drained and poorly drained areas of the pioneer zone in a saltmarsh in south-west Spain.

| Site | Redox potential /mV | | |
	Spartina	*Arthrocnemum*	**Bare mud**
poorly drained	−275	− *	−444
well drained	−148	+133	−368

* No *Arthrocnemum* present.

❑ From the information above, including Table 4.16, what factor determined the ability of *Arthrocnemum* to invade *Spartina* tussocks?

■ The redox potential (i.e. aeration) of surface sediments. This depended to some extent on the raised elevation within tussocks but, primarily, on the overall drainage. You can see from Table 4.16 that redox potential was much lower (more reducing) beneath *Spartina* on the poorly drained soils where *Arthrocnemum* did not grow.

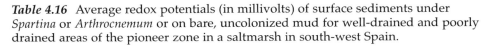

In this example you can see that prior colonization by *Spartina* alters conditions so that invasion by *Arthrocnemum* can occur; this *facilitation* is one mechanism of *succession* – the replacement of one community or species by another – and as saltmarshes extend seawards as silt accumulates and the level rises, pioneer communities are indeed replaced by those characteristic of the higher marsh. Succession is discussed further in Book 3, Chapter 2, and is of great ecological importance. In general, the upward spread of saltmarsh species and their restriction to particular zones seems to be limited mainly by competition; the downward spread more often depends on abiotic conditions. There is no evidence that inadequate dispersal or moderate grazing plays a major role in zonation, although heavy grazing in the upper marsh can be important. Recall the contrasting effects of grazing on the western and south-eastern marshes of England and Wales.

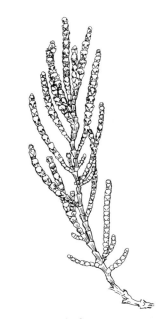

Figure 4.30 *Arthrocnemum perenne* perennial glasswort (× 0.3).

Summary of Sections 4.5 and 4.6

- Salinity is a chemical condition relating to the concentration of ions in soil or water. Marine organisms, which may be **stenohaline** or **euryhaline,** have a physiology adapted to saline conditions. For terrestrial and freshwater organisms, salinity inhibits growth.

- Euryhaline animals are most abundant in estuaries, where there is abundant food. They are an important food source for birds.

- **Halophytes** are terrestrial angiosperms with mechanisms that increase their tolerance of salinity. They occur on saltmarshes and some inland saline soils but tend to compete poorly in non-saline habitats.

- Saltmarshes develop and extend seawards as sediments accumulate. The vegetation has low species diversity, predominantly halophytes, and shows **zonation,** the pioneer zone being at the seaward edge.

- In the UK, regional variation between saltmarshes relates mainly to differences in substrate (affecting particularly the pioneer zone) interacting with climate and grazing.

- Zonation within a marsh reflects the gradient of conditions and resources from the seaward to landward edges. The gradient arises because of differences in duration of submergence, soil aeration, sediment deposition and stability, salinity and grazing pressure. A species may occupy a particular zone either because it grows best there and competes less well in other zones; or because it competes well there and tolerates conditions, although without competition it grows better in other zones. Critical factors may be tolerance of anaerobic soil and the period free of submergence required for seedling establishment.

Question 4.8   *(Objectives 4.1 & 4.5)*

Suggest explanations for (a)–(c).

(a) If a stenohaline invertebrate is placed in seawater diluted to 10%, its body swells up and within a short time the organism dies; if the same is done to a euryhaline invertebrate, it shows little change in size and it survives.

(b) In salt deserts, there is often a distinct zonation of vegetation around shallow depressions or in valley bottoms but the causes of this zonation are not the same as on a saltmarsh.

(c) Mosses are virtually absent from most saltmarshes in southern and eastern England. In northern and western Britain, mosses are often abundant; they occur mainly in the upper marsh and especially where grazing is heavy and there is much disturbance (e.g. along paths).

Question 4.9   *(Objective 4.7)*

It is proposed to plant *Spartina anglica* on a saltmarsh to hasten stabilization and you are asked to advise on the suitability of this action and its probability of success. What sorts of observations and experiments would you make before giving a recommendation?

Question 4.10   *(Objectives 4.3, 4.6 & 4.15)*

In the upper parts of saltmarshes, bare patches often appear as a result of disturbance, e.g. from grazing animals or deposition of floating plant debris which smothers and kills the underlying vegetation before it rots away. Bare patches are first colonized by plants (patch pioneers) that are either rare or absent in the rest of the marsh, but plants dominant in the zones where patches occur then invade gradually. Bertness *et al.* (1992) carried out experiments on a New England (USA) saltmarsh to determine: (i) what conditions were like in bare patches and (ii) whether the patch pioneers tolerated or grew better under patch conditions than did the zonal dominants.

To investigate (i) they measured, for natural vegetation and for bare patches created with non-persistent herbicides, surface temperature, water loss over 4 hours on a cloudless sunny day from a sponge saturated with marsh water, and salinity of water remaining in the sponge. Results are shown in Figure 4.31. They also measured salinity of the surface soil under natural perennial vegetation and in the middle of artificial bare patches six weeks after creation and found that salinity was 2.5-fold higher for the bare patches (over 50 g kg^{-1} of NaCl). If patches were artificially shaded or covered by plastic straws inserted vertically into the mud (to mimic plant cover), there was no difference in salinity between bare and vegetated patches.

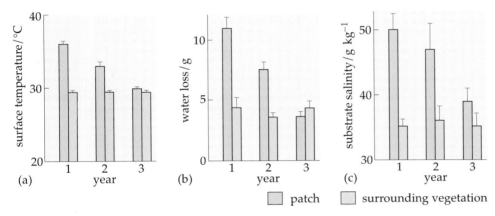

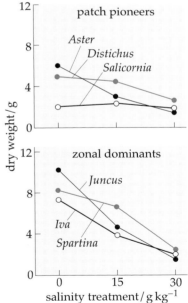

Figure 4.31 For use with Question 4.10. Conditions in high marsh bare patches and in surrounding undisturbed vegetation: (a) temperature at the soil surface; (b) water loss over 4 hours from sponges soaked in marsh water; (c) salinity of water remaining in the sponges. (The thin vertical bars denote SEM.)

(a) From the above information and the data in Figure 4.31, describe how conditions change when bare patches arise and explain why they change.

To investigate (ii), they grew patch pioneers and zonal dominant species at a range of salinities in a greenhouse for three months and then measured the dry weights of replicate batches (Figure 4.32). Some replicates of the zonal dominants died at the highest salinity treatment and these were not included in the dry weight measurements. The survival and growth of patch pioneers and zonal dominants were also measured in artificial bare batches. Eleven months after applying herbicide, seedlings that had colonized the patches were marked and their survival and growth measured after four months. During this time half the patches were flushed weekly with freshwater and half were left as untouched controls. Results are shown in Table 4.17.

Figure 4.32 For use with Question 4.10. Dry weights of replicate batches of patch pioneers and zonal dominants raised for three months in a greenhouse at three salinities. Values are means of 10–15 replicates ± SEM.

Table 4.17 For use with Question 4.10. Survival and final biomass of marked seedlings in watered and control bare patches in the *Juncus gerardi* zone of the upper marsh. Measurements were made four months after marking 26–50 seedlings of each species. * Significantly different from the control, $P < 0.05$.

	% survival	Dry weight per plant/g
Juncus gerardi (zonal dominant)		
control patches	61	0.26
watered patches	94*	1.42*
Iva frutescens (dominant in adjacent zone)		
control patches	45	0.49
watered patches	82*	3.18*
Salicornia europaea (patch pioneer)		
control patches	86	7.33
watered patches	94	14.41*
Atriplex patula (patch pioneer)		
control patches	68	0.34
watered patches	89*	1.09

(b) Give evidence for or against (i)–(iii).

(i) Patch pioneers are more abundant on patches than in the surrounding vegetation because they grow and perform better on patches.

(ii) Zonal dominants grow and perform less well on patches than in vegetated areas.

(iii) Conditions on bare patches are initially more favourable for patch pioneers than for zonal dominants; patch pioneers are rare in the undisturbed marsh because of competition from the dominant species.

AQUATIC ENVIRONMENTS: LAKES, RIVERS AND OCEANS

PART III

4.7 General characteristics

In this and the previous Chapter, we have discussed particular abiotic resources or conditions and considered only three habitats as a whole: rocky shores, hot deserts and saltmarshes. For the rest of this Chapter, we switch to the habitat approach and look at the nature of aquatic environments and some of the main factors – biotic and abiotic – that influence the growth and survival of individual organisms.

Since water occupies 72% of the Earth's surface and an estimated 99% of the volume of the biosphere (mostly in the oceans), aquatic environments exert a major influence on climate and global processes such as carbon exchange (Book 4). The surprising thing – and in sharp contrast to terrestrial environments – is that this influence is exerted mainly by organisms invisible to the human eye: the plants (primary producers) that are at the base of aquatic food-chains in 98% of water bodies are single-celled algae or **phytoplankton**. Similarly, the commonest grazers are very small animals (**zooplankton**), ranging from protistans to crustacean 'water fleas' such as *Daphnia*. For these organisms, life cycle duration is linked to size (Book 2, Chapter 5) – with phytoplankton having cell doubling times of a few hours or, at most, days – so it becomes impossible to distinguish between effects on individuals and effects on populations: factors that affect growth inevitably affect population size (the subject of Book 2).

4.7.1 Compartments and organisms

Phytoplankton and zooplankton are representatives of one kind of aquatic community and, in order to get a broader perspective, it is useful to review the general types of aquatic organisms. These fall into three broad groups, each associated with a particular spatial compartment of aquatic habitats:

1 **Pelagic organisms** float or swim freely in the water column. These are discussed further below because they are functionally the most important and much of the discussion in later Sections is about this group.

2 **Benthic organisms** or the **benthos** live on the bottom, in or on sediments or rock. This community is particularly evident in shallower rivers and streams (Section 4.10) where the strong current excludes all but the strongest swimmers from the water column. It exists in all aquatic habitats, however, comprising mainly micro-organisms and animals that feed on and process dead material (**detritus**) that sinks out of the water column. Functionally, therefore, it resembles the surface layers of the soil and is a decomposer community (Book 4). In shallow water, where light reaches the bottom, plants may occur – usually microscopic algae attached to surfaces.

3 **Fringing organisms** include large, rooted plants (*macrophytes*, such as pondweeds and water-lilies), larger filamentous algae and seaweeds

that grow round the edges (the *littoral zone*) of larger lakes and the oceans and may occur anywhere in shallow lakes and rivers. Growing on the surface of these large plants are numerous micro-algae (**periphyton**) which are often the main food for grazers. Structurally, therefore, the fringing communities are more complex and more like terrestrial communites than those in other aquatic compartments. They exist wherever light can penetrate to the substratum or where tidal movements regularly expose rocks or sediments – so rocky, seaweed-dominated shores are a good example (Chapter 3, Section 3.1). Other marine examples, which occur just below the seashore and in the *sub-littoral* zone, are 'meadows' of sea-grass (e.g. *Zostera*, Section 4.6.3), kelp 'forests' and coral reefs. This last is based on an animal–plant symbiosis (Chapter 1) but it resembles the communities dominated by large plants in being extremely productive, i.e. much carbon is fixed. The biggest contrast with the pelagic communities is that fringing communities dominated by macrophytes accumulate living material (**biomass**) in the perennial plants.

Pelagic organisms: plankton and nekton

The **plankton** comprises all the smaller pelagic organisms that cannot swim against the current: they drift passively and have no or only weak swimming ability which, at most, allows them to move up and down the water column. Table 4.18 shows how plankton may be classified according to size and function, using the system of Barnes (1980): only names shown in bold type need to be remembered.

Table 4.18 Classification of plankton according to size and functional type.

Size range /μm	Name	Organisms	Functional types
< 5	**ultra-** or **picoplankton**	mostly heterotrophic bacteria and cyano-bacteria; some true algae	decomposers (**bacterioplankton**) and primary producers (**phytoplankton**)
5–50	**nanoplankton**	autotrophic: mostly true algae; some cyanobacteria	primary producers
50–500	microplankton	true algae and animals	primary producers and consumers (**zooplankton**)
500–2000	macroplankton	mostly animals; a few algae	consumers and primary producers
> 2000	megaplankton	animals	consumers

There is clearly a wide range of sizes among plankton and an equally wide range of functions. The *picoplankton* contain a substantial proportion of heterotrophic bacteria which can decompose dead organic matter and are largely confined to the soil in terrestrial systems. *Phytoplankton*, the primary producers, are all autotrophs, i.e. they can photosynthesize, and include prokaryotic cyanobacteria (blue–green algae) and 'true' algae from eight phyla. The taxonomic diversity is thus enormous and because cyanobacteria are not even algae, some authors (e.g. Reynolds, 1984) use quotes – 'algae' – to indicate the wider meaning. Figure 4.33 illustrates the range of size and form among freshwater phytoplankton, where cyanobacteria, green algae and diatoms are all common. In contrast, green algae are rare in marine phytoplankton and cyanobacteria are confined mostly to the picoplankton; dinoflagellates (Figure 4.46) are more common than in freshwater, however. Notice that the larger types in Figure 4.33 are described as **net plankton**, a general term applied to to organisms that can be collected quantitatively by towing a net with the finest possible mesh size (about 70 μm). Smaller types are much more difficult to sample and are usually centrifuged or precipitated from water samples: even the existence of picoplankton was unknown until the early 1980s. We discuss later (Section 4.9) the factors that influence the distribution and abundance of phytoplankton which, in general, is the main determinant of zooplankton numbers.

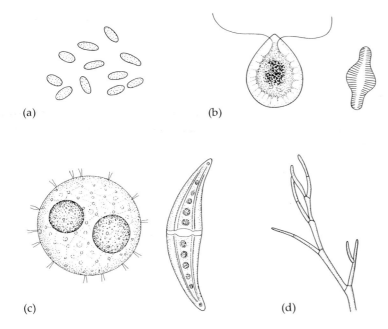

Figure 4.33 Examples of freshwater phytoplankton.
(a) Picoplankton, *Synechococcus elongatus* ($\times$ 1400).
(b) Nanoplankton: (left) *Haematococcus* sp., a green alga ($\times$ 700); (right) *Fragilaria construens*, a diatom ($\times$ 1000).
(c) Net plankton: (left) *Volvox* sp., a green alga ($\times$ 30); (right) *Closterium*, a green alga ($\times$ 50).
(d) Filamentous alga, *Cladophora glomerata* (blanketweed) ($\times$ 45).

Zooplankton, the heterotrophic consumers, range from unicellular protistans in the nanoplankton, which eat detritus and picoplankton, to macro- and megaplankton which include mainly herbivores and some carnivores. In both marine and freshwaters, small crustaceans (e.g. 'water fleas' and 'krill', a larger, shrimp-like marine type) are important grazers of phytoplankton, either filtering particles from the water or grasping prey. The only thing which separates zooplankton from **nekton** is that the latter can swim actively and maintain position in strong currents. Nekton includes most adult fish and, in the oceans, larger invertebrates such as squid, and mammals such as whales and seals. Nearly all of these are *carnivores*, and good swimming ability is necessary not only to reach feeding or breeding grounds but often also to catch prey and/or avoid being caught.

These are the main types of aquatic organisms. We next survey briefly the types of water bodies in which they live before considering the physical and chemical characteristics of water bodies that influence the survival and abundance of organisms.

4.7.2 Types of water bodies

For ecologists, the most useful way of grouping water bodies should reflect characteristics that have a major influence on organisms.

❏ Think of at least two such characteristics and then read on.

Chemical conditions is an obvious one: marine and freshwater environments have high and low concentrations of ions, respectively (although there are some saline lakes). These environments are, therefore, inhabited by different species with different physiologies but their modes of life – as phytoplankton, zooplankton, etc. – are basically similar. More relevant is the stability of chemical conditions. These are exceptionally stable in both space and time in the oceans (where pH is about 8) but highly variable in freshwaters (Table 4.15). Both pH and the major ions present in freshwaters depend on the geology, vegetation and land use of the drainage basin or **catchment**.

A second characteristic which profoundly affects aquatic organisms is the type and degree of *water movement*. At one extreme are streams and rivers with unidirectional flow which, if sufficiently strong and turbulent, excludes all plankton; rivers and current conditions are discussed as a separate case study in Section 4.10. At the other extreme are pools and smaller lakes where water movement occurs largely through wind and convection and varies on a scale of minutes to hours. As lakes get bigger and deeper, larger-scale water movements (eddies and currents) arise and there may be seasonal differences in turbulence. At the scale of ocean basins, these internal motions interact with motions induced by the Earth's rotation to create complex deep and surface currents (Figure 4.34) and tidal movements. Thus, dense cold water sinks in the polar regions and moves in deep currents towards the Equator, whilst water with salinity increased by evaporation sinks in the tropics, causing further circulation. So there are different types of water movement, from the smallest and shallowest to the largest and deepest, with the shape of the water body having a strong modifying effect. The organisms most affected by such movements are the plankton (Section 4.9).

A third way of thinking about water bodies is in terms of their *isolation and permanence*. Oceans are the most permanent (in geological terms) and the least isolated, being a series of linked habitats. Many marine species are cosmopolitan and it is factors such as temperature and ocean currents that are most likely to restrict geographical distribution. In contrast, both lakes and rivers are relatively impermanent, appearing and disappearing over geological time and undergoing slow but constant change. Each river system from headwaters to estuary is a series of linked habitats but the whole system is isolated from other rivers except via the sea. And lakes are the most uniform and *dispersed* habitats, isolated (except for inflows and outflows) rather like islands in the sea. The relative isolation of lakes and, to a lesser extent, rivers has some impact on the organisms present. It means, for example, that in large ancient lakes, such as Lake Baikal in central Asia or some of the African rift lakes, organisms such as fish which, by chance, were present when the lake formed – or colonized it soon after – may have persisted and evolved in isolation. **Endemic species** which evolved in isolated lakes and large, ancient rivers such as the Amazon are quite common. At the opposite end of the lake–river spectrum, however, are a wide array of **ephemeral habitats** which dry up during droughts or seasonally: in desert areas, sizeable lakes and rivers may appear briefly

Figure 4.34 (a) Section of the Atlantic Ocean showing the deep currents. (b) The main surface currents in the Atlantic Ocean.

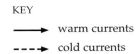

KEY

→ warm currents
⇢ cold currents

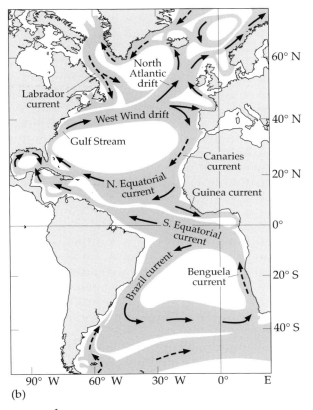

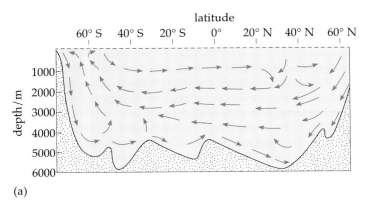

(a)

(b)

after rains. Such impermanence, an extreme form of disturbance, is the dominant influence on the organisms present and we consider this unique type of aquatic habitat briefly below.

4.7.3 Ephemeral habitats

A considerable number of aquatic organisms, especially among the invertebrates, are specialists of temporary water bodies. Notable are microcrustaceans (e.g. water fleas, cladocerans and copepods), certain insects (e.g. midges, Chironomidae, which have flying adults and aquatic larvae) and mites (Acari). Common characteristics are a *short life cycle*, possession of some *resistant stage* which can survive when the water dries up and/or good ability to *disperse widely* (e.g. as a flying insect or seed) and colonize other water bodies. In a study of midge larvae in tropical rain pools, McLachlan and Cantrell (1980) found that variation in these characters correlated with pool duration (Table 4.19).

Table 4.19 Characteristics of midge larvae found in temporary tropical pools of varying size.

Type of pool	Dominant midge species	Larval lifespan /days	Larval drought resistance	Ability to invade new pools
large, lasting several weeks	*Chironomus imicola*	12	nil (no resistant stage)	good, by egg-laying females
medium-sized, lasting for 1–2 weeks	*Dasyhelea thompsoni*	< 12	moderate: larvae intolerant of complete drying out	good
small, lasting for < 1 week	*Polypedilum vanderplanki*	<< 12	very good: larvae tolerate complete desiccation	poor

❑ From Table 4.19, why do you think *Polypedilum* commonly dominates the smallest pools?

■ These pools are likely to dry out before any of the three species has completed larval development. *Polypedilum* is the only species able to persist as a larva in completely dry mud. If the same pools refill, *Polypedilum* larvae will be the first there, giving them a competitive advantage.

However, the poor dispersal ability of *Polypedilum* means that it will only rarely colonize new pools and would decline if, for any reason, its small pools did *not* refill. Clearly there are risks associated with being a specialist in ephemeral habitats and, apart from drying up, conditions are often severe: temperatures may be high and, consequently, oxygen levels low (since gas solubility *decreases* with rising temperature); some pools, especially in desert areas, also become quite saline as the water level falls. There are some advantages, however. First, it is unlikely that any aquatic predators (fish or larger invertebrates with a long life cycle) will be present; and secondly, there is a fair chance of avoiding competitors in at least some habitats. Avoiding predation appears to be the main reason why some amphibians commonly deposit spawn in pools that may dry up.

Ephemeral water bodies have, therefore, a rich fauna and, to a lesser extent, flora (mainly of algae) and this richness is *because* of their impermanence rather than despite it. Most research, however, has concentrated on permanent water bodies and it is their physical and chemical characteristics that are considered next.

Summary of Section 4.7

- One of the chief differences between aquatic and terrestrial environments is that the major aquatic producers and consumers are very small organisms with life cycles of hours or days rather than weeks to years.
- The main types of aquatic organisms are pelagic, living in the water column; benthic, living on the bottom; and fringing, living on the edges of water bodies (littoral and sub-littoral) or in shallow water where light penetrates to the bottom, as large attached plants or, for corals, as an animal–plant symbiosis.
- The **pelagic organisms** comprise **plankton**, with weak or no swimming ability, and larger **nekton** (fish, marine mammals) which swim actively. The plankton range in size from a few μm to cm (Table 4.18) and include bacteria, small algae (**phytoplankton**) and **zooplankton** consumers.
- Water bodies may be grouped for ecological studies in terms of:
  (a) water chemistry (constant high ionic conentrations (marine) and variable, mostly low ionic concentrations (freshwaters));
  (b) water movement (direction and type of flow);
  (c) isolation and permanence.
- Ephemeral streams, pools and lakes are a highly specialized habitat with a rich fauna characterized by short life cycles, possession of desiccation-resistant stages and/or good dispersal ability.

4.8 Physical and chemical characteristics

Three properties of water have a particularly strong influence on the character of aquatic environments:

1   Water *absorbs radiation* (including light) very effectively and the longer the wavelength, the greater the effectiveness.

2   It has a *high density* relative to other, comparable liquids and the density of freshwater increases as temperature falls but reaches a maximum at 4 °C (not at the freezing point, 0 °C). The density of seawater increases uniformly down to the freezing point of around −2 °C.

3   It is a *poor conductor of heat*.

4.8.1 Light supply and temperature conditions

Because of property 1, both the amount and quality of light changes with depth. The surface layer where photosynthesis is possible is termed the **euphotic zone** and varies from a few centimetres to several hundred metres, depending on the amount of suspended sediment or phytoplankton or dissolved, coloured substances (such as the humic acids) which are present.

Because of properties 2 and 3, water bodies that are sufficently deep and are warmed at the surface (e.g. in a temperate spring) tend to *stratify*. As the surface layers warm up, they become less dense but are mixed by wind with the colder water below. The mixed-up warm layer extends downward until eventually there is insufficient energy to overcome the density gradient and two separate layers become established: a uniformly warm upper layer (**epilimnion**) and a uniformly cold lower layer (**hypolimnion**), which may vary by only a few degrees above the winter temperature.

❑   For deep lakes, what will this winter temperature be?

■   4 °C – the temperature of maximum density. Water freezes from the top down rather than the bottom up and poor heat conduction means that the deepest water virtually never freezes, so there is always a refuge habitat from ice.

Between the epilimnion and the hypolimnion is a region where density and temperature change rapidly, the **thermocline**, the whole process being termed **thermal stratification** (Figure 4.35). Water circulates quite separately within the hypolimnion and the epilimnion.

Figure 4.35 Diagrams to illustrate thermal stratification: (a) vertical section of a stratified lake; (b) typical temperature profile of a stratified temperate lake in summer.

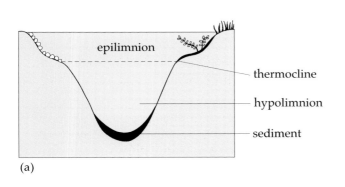

(a)

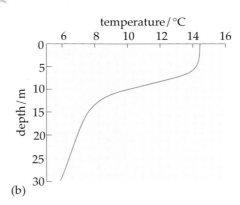

(b)

Themoclines may form in deeper lakes and even in the lower reaches of very large rivers during periods of low flow; they are a permanent feature of all oceans except those in polar regions, where surface water does not warm up sufficiently for stratification to occur. Figure 4.36 shows how the temperature profile in the oceans varies with latitude.

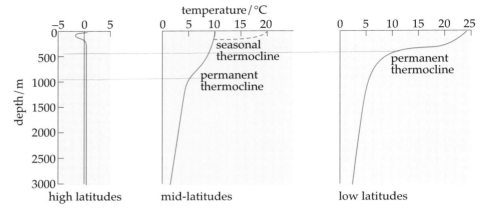

Figure 4.36 Temperature profiles at different latitudes in the open ocean. (The 'blip' near the surface of polar oceans occurs in summer when ice melts and cold, low salinity water floats over denser more saline water.)

❑ From Figure 4.36, at what depth(s) do thermoclines form in the oceans?

■ In tropical waters, at about 500 m; in temperate waters, at about 1000 m – but notice that in summer an abrupt seasonal thermocline forms at around 100–200 m. For shallow temperate seas over continental shelf regions, only the seasonal thermocline usually occurs.

The depth at which a thermocline forms and, if temporary, the time at which it breaks down can greatly affect aquatic organisms.

❑ Think of two reasons why a thermocline might break down.

■ (a) Extra strong wind turbulence and (b) cooling of the surface waters to a temperature below that in the hypolimnion or just above the permanent oceanic thermocline. This occurs in autumn for temperate lakes and seas but may happen *every night* in shallow tropical lakes.

Unless water is especially turbid (e.g. because of suspended sediment or dense phytoplankton populations), the depth of temporary thermoclines corresponds fairly well with that of the euphotic zones. So despite being carried vertically by surface turbulence, phytoplankton will still be able to photosynthesize throughout the day. But once the thermocline breaks down or is very deep and permanent, phytoplankton must spend considerable time below the euphotic zone. They may survive by entering a resting stage (common in temperate waters) or by a strategy of 'stuffing and starving': rapid photosynthesis and carbon storage in the light followed by periods in the dark living on stored carbon. Stratification thus affects both temperature conditions for surface organisms and the supply of light for phytoplankton. It also has a profound effect on the supply of chemical resources, which are discussed in Section 4.8.3.

4.8.2 Chemical conditions

Chemical conditions that vary in aquatic environments include: pH (and closely linked to this, *alkalinity*, which relates to all the ions that reduce acidity); the nature and concentrations of major ions (i.e. ions present in large amounts); levels of potentialy toxic ions, such as aluminium, and of organic 'pollutants'; and the levels of naturally derived dissolved organic substances such as humic acids, which often stain peaty water brown. Our main concern here is the impact of the first two items from this list on freshwater organisms.

❑ Why are these chemical conditions in the oceans of little ecological relevance?

■ Recall that the pH and ionic composition of oceans hardly vary (Section 4.7.3), so they do not affect the distribution and abundance of organisms.

The situation is very different in freshwaters, where pH is mostly in the range 4 to 9 and as high as 10–11 in the African soda lakes. There is a corresponding variability in the content of dissolved ions and indeed the two are related because the major ions of freshwaters – calcium (Ca^{2+}) and hydrogen carbonate (HCO_3^-) – influence pH. Soft rocks such as chalk or limestone are easily dissolved and water flowing over them may have high levels of calcium hydrogen carbonate which tends to raise pH and buffer the water against acid inputs. Calcium-rich water is also 'hard', producing an insoluble scum with soap. With rocks such as granite which dissolve slowly, the water may have very low concentrations of ions and be poorly buffered; CO_2 from the air dissolves to give a weak solution of carbonic acid, H_2CO_3, so the water is acidic and described as 'soft'. For both lakes and rivers, therefore, pH and ionic concentrations are closely linked to *catchment geology*.

Vegetation and land use also influence pH and chemical conditions. For example, water which drains through the acid leaf litter produced by conifers or through *Sphagnum* mosses or peat contains organic acids which are described as **humic substances**. These contribute to water acidity, colour the water brown (reducing light penetration) and bind metal ions such as iron(II) and aluminium. Al-binding is a useful feature of humic acids because Al is particularly toxic to aquatic organisms (especially fish). As described for soils (Section 4.4.3), Al is more soluble at low pH and tends to be leached from soils derived from acid rocks.

The important question is whether variations in the pH and ionic composition of freshwaters significantly affect aquatic life and the composition of communities. The simple answer is yes, they do. In a study of temperate rivers, for example, pH, alkalinity and water hardness emerged as a group of factors that explained much of the variation among invertebrate communities (see Section 4.10). But, when studied in isolation, many freshwater organisms tolerate a wide range of pH values from around 5.5 to 8: if pH appears to influence distribution in this range it is likely to act through biotic interactions (competitive ability), rather than acting directly to affect survival and reproduction. Thus, species may compete better over a narrower range of pH values than the range tolerated. However, there is clear evidence that more extreme values of pH or raised levels of aluminium associated with low pH affect survival and reproduction directly and a particular concern has been the widespread **acidification of lakes and rivers** observed in the 20th century in parts of Europe and North America.

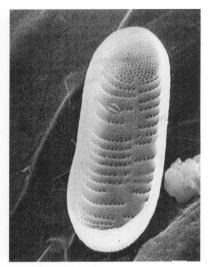

Figure 4.37 A silicaceous half cell wall of the diatom *Tabellaria binalis* (× 8000). Species can be recognized from the shape and pattern of the wall.

Acid deposition and the effects of low pH

Lakes and rivers may, of course, be naturally acidic but the acidification referred to above is *anthropogenic* (caused by human activities). It results from **acid deposition**, i.e. atmospheric inputs of strong acids such as sulphuric (H_2SO_4) and nitric (HNO_3), which derive from industrial emissions of sulphur dioxide (SO_2) and nitrogen oxides and are deposited mainly in rain. The effects of such '**acid rain**' depend very much on the nature of the catchment which drains into a lake or river: if the soil is rich in calcium, for example, and rain penetrates deeply rather than running off the surface, acidity can be reduced – the catchment buffers lakes against acidification. But if the catchment is on granite rocks with acid, calcium-poor soil, it has minimal buffering capacity; the presence of conifer trees, whose needle litter is acidic, exacerbates acidification. *Land use* in a catchment is thus a major factor influencing acidification.

❑   Would you expect upland or lowland lakes and rivers to be more affected by acid deposition?

■   Upland: these are more likely to have calcium-poor acid catchments, often with conifer plantations.

Anthropogenic acidification seems to have begun in some lakes around the middle of the 19th century in northern Europe. Evidence for this comes from *studies of diatoms* in cores of lake sediments (Battarbee *et al.*, 1988). **Diatoms** are abundant in lakes and are a type of alga with resistant silica cell walls which persist in sediments (Figure 4.37). The pH optima of different species range from > 7 to < 5 and the pH of lakes today can be predicted on the basis of the ratios of species with different pH optima. By using this method coupled with lead isotope (^{210}Pb) dating of sediments, it is possible to trace the pH 'history' of a lake (Figure 4.38).

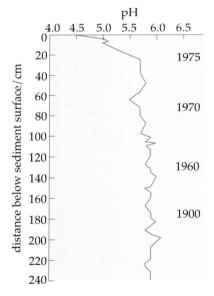

Figure 4.38 The pH history of Loch Fleet, Galloway (south-west Scotland), reconstructed from diatom assemblages in dated sediment cores.

The effects of acidification on organisms arise from interactions between low pH and other chemical conditions, which include raised aluminium levels, low calcium and high sulphate levels. In addition, effects on animals may arise from changes in food availability, and biotic interactions (competition and predation) may be altered in acid waters just as much as at higher pH values. The sorts of changes that have been attributed to recent acidification are listed below.

1   *Fish* are usually absent below pH 4.5 and salmonid (salmon and trout) eggs are affected below pH 5. The main cause seems to be aluminium toxicity when calcium levels are low.

2   The *macrophytes* rooted in shallow water or floating often become dominated by *Sphagnum* mosses and the rush *Juncus bulbosus* in acidified lakes, with considerable reduction in species number.

3   Both the number of species and the biomass of lake phytoplankton decrease with pH. Periphyton rather than plankton tend to dominate below pH 5 with thick mats of green algae and cyanobacteria forming.

4   Invertebrate responses vary widely but species that require high calcium levels for their shells (molluscs) or outer coat (crustaceans such as the isopod *Asellus aquaticus*, the freshwater louse) are usually absent below pH 5. Zooplankton numbers decrease. Among the insects, most mayflies (Ephemeroptera) are lost but there are some very acid-tolerant species of dragonflies and damselflies (Odonata) and, in streams, of stoneflies (Plecoptera). In streams where fish have been lost through

acidification, there have been increases in acid-tolerant prey species such as the large net-spinning caddis *Plectrocnemia*.

5   Dippers (*Cinclus cinclus* – Figure 4.39), a bird that feeds on invertebrates in upland streams, have declined, largely because of a reduction in food supply (Ormerod *et al.*, 1991).

The last two items on the list above provide examples of two contrasting ways in which the numbers of an organism or the species present may be influenced. If the important thing is the supply of nutrients or food or some other resource or abiotic condition that acts from 'below', this is described as **bottom-up** control; if the level of grazing or predation acting from 'above' in the food-chain is more important, this is described as **top-down** control.

Figure 4.39   Dipper *Cinclus cinclus* (× 0.3), a bird that feeds on invertebrates in upland streams and has declined because of loss of food following acidification.

❏   Identify an example of each type of control from the list.

■   The effect of stream acidification on dippers illustrates bottom-up control: loss of prey translates up the food-chain into loss of dippers. The effect on *Plectrocnemia* illustrates top-down control: loss of predators translates down the food-chain into an increase in prey.

These are important ideas, and other examples of top-down and bottom-up control are described in this Book with fuller discussion in later Books. To illustrate further the complex interactions through which water pH affects organisms and to illustrate also how experiments are carried out, we describe next a study of *naturally acid* bog lakes in Wisconsin, USA (Arnott and Vanni, 1993). These lakes (pH 4.3–5) are rich in humic substances, contain no fish and the sparse zooplankton are dominated by large species (mostly *Daphnia* spp.), whereas in nearby lakes containing fish and having a higher pH, small species of herbivorous zooplankton dominate and *Daphnia* are rare.

❏   Suggest two alternative explanations for this difference in size of the dominant zooplankton, one based on abiotic and one on biotic interactions.

■   (a) Small zooplankton may be less tolerant of low pH than large species (abiotic explanation). (b) Small zooplankton may be less abundant in acid lakes because of competition or predation from large species; large species are rare in lakes with fish because of selective predation by fish (biotic explanation).

Notice that (a) and (b) are also examples of bottom-up and top-down control, respectively. Considering just one species of small herbivorous zooplankton, *Bosmina longirostris* (Figure 4.40d) out of the 10 types studied, Figure 4.40a–c shows the results of three experiments. *Bosmina* were collected from a pH-neutral control lake (with fish) and introduced into large polythene exclosures in three acid bog lakes and the control lake. Water in these bags was filtered to remove large zooplankton, both herbivores and predators, but allowed small phytoplankton through.

•   Figure 4.40a shows the numbers of *Bosmina* after 2.5 weeks.
•   Figure 4.40b shows the results of another, similar experiment in which the pH inside acid lake exclosures (1–3) was raised to that of the control lake.

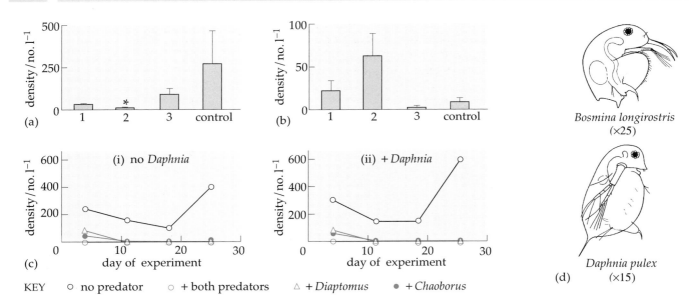

KEY   ○ no predator   ○ + both predators   △ + *Diaptomus*   ● + *Chaoborus*

Bosmina longirostris
(×25)

Daphnia pulex –
(×15)

Figure 4.40 (a) and (b) Results of experiments using *Bosmina longirostris* from a neutral control lake which was introduced into exclosures in three acid bog lakes, 1–3, as described in the text, (a) without treatments and (b) after increasing pH in exclosures in lakes 1–3 to that of the control. (c) Effect on *Bosmina* of adding two large predaceous species of zooplankton in the presence and absence of the large herbivore, *Daphnia*, when placed in exclosures in the neutral control lake. * Significantly different from the control, $P < 0.05$. (d) Drawings of *Bosmina longirostris* and *Daphnia pulex* – note the size difference.

- Figure 4.40c shows the results of an experiment in the control lake lasting 3.5 weeks; the effects of adding two types of large *predaceous* zooplankton were tested in the presence and absence of large herbaceous *Daphnia* (Figure 4.40d).

❏ Study Figure 4.40a–c and decide whether the data support the abiotic or the biotic explanations, or both, and then read on.

Figure 4.40a and b provide support for the abiotic hypothesis: *Bosmina* multiplied less well in exclosures in the acid lakes (although significantly so in only one case) but this effect disappeared when the pH was raised. Figure 4.40c indicates that predation but not competition from large zooplankton significantly affects *Bosmina*: *Daphnia*, a potential competitor, had no effect on numbers (compare (i) and (ii)) but the two predaceous species strongly reduced them. So the rarity of *Bosmina* in acid bog lakes seems to be caused by a combination of abiotic conditions (low pH) and predation. For other species of small zooplankton, the relative importance of pH and predation varied widely and the overall conclusion was that unfavourable abiotic conditions and predation by large zooplankton species (which were abundant because of the absence of fish at low pH) interacted to reduce the numbers of small zooplankton. In this example, therefore, elements of both bottom-up and top-down control operated simultaneously on the small zooplankton; low pH was the overriding control factor, acting both directly on the small zooplankton and indirectly via fish and larger zooplankton.

4.8.3 Chemical resources and trophic status

The distinction between oceans, with constant chemical conditions, and lakes and rivers, where they vary, does not apply to chemical resources. We consider here how and why three kinds of resources vary: oxygen, carbon (for autotrophs) and mineral nutrients.

Oxygen

Bearing in mind that gases diffuse thousands of time more slowly in water than in air, oxygen levels in water reflect the balance between oxygen-using and oxygen-supplying processes. The former depends on the oxygen requirements of organisms (the **biological oxygen demand** or **BOD**) and of chemical reactions such as the formation of iron oxides (the **chemical oxygen demand** or **COD**). The latter, oxygen replenishment processes, depend mainly on the amounts released by photosynthesis and the extent to which water movements (surface turbulence, convection and currents) facilitate diffusion from the air. Temperature is another factor affecting oxygen levels, especially when still and/or shallow water warms up in summer.

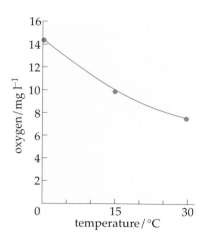

Figure 4.41 Effect of temperature on the oxygen content of pure water.

❑   From Figure 4.41, what is the effect on oxygen content of increasing water temperature?

■   Oxygen content goes down: gas solubility in water *decreases* with temperature.

Taking all these factors into account, the ways in which oxygen levels vary in different water bodies can be worked out.

❑   Explain the following observations: (a) For turbulent rivers and turbulent surface water of lakes and oceans, water is usually saturated with dissolved oxygen; (b) when rivers have smooth, non-turbulent flow, there is often an oxygen gradient with a minimum at the bottom sediments.

■   (a) Turbulent mixing of air and water constantly replenishes O_2. (b) Water at the surface can take up oxygen, but in the absence of turbulent mixing diffusion to the sediments is too slow to keep up with demand: the BOD of sediments that contain much dead organic matter will be high because of microbial respiration.

Low flow rates in rivers in summer, caused by drought or water abstraction, may result in drastic oxygen reduction, not only because of low velocity and turbulence but also because water becomes warmer. The other major cause of low oxygen in rivers is **organic pollution** from, for example, raw sewage or silage effluent. The BOD of these pollutants is enormous as they are decomposed by micro-organisms and reducing BOD is one of the major tasks of sewage treatment works.

In lakes, turbulence and *stratification* (Section 4.8.1) strongly affect oxygen supply. Diffusion from the air and photosynthesis usually maintain reasonably high levels in the epilimnion but not in the hypolimnion, which is outside the euphotic zone and cut off from the epilimnion circulation. If much dead organic matter settles out from the epilimnion, increasing the BOD of the hypolimnion, oxygen levels in the latter can fall to low levels or even complete anoxia can result. Perhaps surprisingly, this does *not* happen in the oceans.

❑   Look back at Figure 4.34a and try to explain this observation.

■   It is because of the deep ocean currents, which carry cold, well-oxygenated water from polar regions into the hypolimnion of other latitudes. In addition, BOD is low in most mid-ocean regions (for reasons discussed later).

Oxygen supply is thus a major factor influencing river organisms; it may influence survival in the hypolimnion of lakes but is not a significant factor in oceans.

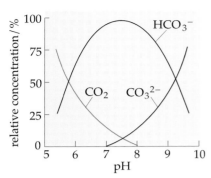

Figure 4.42 Effect of pH on concentrations of CO_2, hydrogen carbonate and carbonate in seawater.

$$CO_3^{2-} + 2H^+ \rightleftharpoons HCO_3^- + H^+$$

$$\Updownarrow$$

$$H_2CO_3$$

$$\Updownarrow$$

$$H_2O + CO_2$$

Carbon

Carbon dioxide is about 200 times more soluble in water than is oxygen, so autotrophs photosynthesizing near the water surface ought to be almost as well supplied with carbon as are terrestrial plants. The complicating factor is that the pH–CO_2–HCO_3^- system operates in water, just as in soil (Section 4.4.3), and so the supply of free CO_2 depends on water pH (Figure 4.42).

In seawater (pH 8) and in freshwaters where pH exceeds 7.5, the levels of free CO_2 are very low. The usual solution to this 'problem' for aquatic algae and macrophytes is to use hydrogen carbonate as a carbon source – taking up HCO_3^- and releasing CO_2 inside the cell. However, not all are able to do this and some algae and the majority of aquatic mosses are confined to relatively acid water for this reason. In addition, surface pH in lakes may vary on a seasonal or daily basis between values of 7–10 *because* of photosynthesis: when there are dense populations of phytoplankton or submerged macrophytes, high rates of photosynthesis remove CO_2 and then HCO_3^- ions so rapidly that the pH rises.

❑   So at what times of the year and day would you expect to find the highest pH values?

■   In midsummer (June–July) when days are longest and temperatures high and at midday or early afternoon on a sunny day when photosynthesis has been brisk.

It has been argued that the varying ability of phytoplankton to function and obtain sufficient carbon at high pH influences considerably the species present in some lakes. Supplies of other mineral nutrients, however, are of far wider significance.

Mineral nutrients

Changes in the availability of mineral nutrients, chiefly N and P, are among the most important factors determining the numbers and types of aquatic plants in a system. We discuss this in relation to phytoplankton in Section 4.9 and concentrate here on the general processes that cause variation in nutrient levels over space and time.

The situation is simplest for rivers because nutrient levels here depend largely on inputs from the catchment rather than on processes within the water body. Catchment geology, vegetation and, increasingly, land use are what matter, together with inputs from industry and sewage treatment plants. Thus, inputs to rivers and streams increase when land is disturbed by forestry or agriculture, and heavy applications of nitrogenous fertilizer can lead to very large N inputs. The most commonly observed effects of nutrient enrichment in rivers, especially of the shallower reaches, is greatly increased growth of filamentous algae attached to stones and sometimes of rooted, fringing macrophytes. Many nutrients that enter rivers simply end up in the sea, reflecting the open nature of river systems.

In lakes and oceans, by contrast, internal processing of nutrients is of equal or much greater importance than external inputs. Leaving aside fringing macrophytes, the first step is that phytoplankton take up mineral nutrients. When they die, large algal cells are likely to sink to the bottom sediments where they are decomposed by micro-organisms; but small pico- or nanoplankton may be decomposed actually in the water column. Alternatively, algal cells may end up inside grazing zooplankton, whose nitrogenous wastes and faeces are usually decomposed rapidly in the water column but whose bodies often sediment to the bottom. Now imagine what happens in stratified conditions. Nutrients are mostly *used*, i.e. taken up by growing phytoplankton, in the epilimnion; some are regenerated here, especially if grazing intensity is high with rapid recycling of zooplankton wastes, but any organic matter that sinks below the thermocline will be decomposed in the hypolimnion and its nutrients will become largely unavailable to the epilimnion.

Nutrient levels are thus always higher in the hypolimnion compared with the epilimnion, where they commonly fall to undetectable levels. This does not mean that all the phytoplankton are dead but simply that they are taking up very rapidly any nutrients regenerated within the epilimnion or diffusing upwards across the thermocline. Typically, it is the smallest pico- and nanophytoplankton that operate in this fashion, having remarkable abilities to absorb nutrients present at very low concentrations and having high rates of cell division that allow some individuals to escape being eaten. For large areas of the oceans this situation is permanent, notably the mid-ocean regions where there are no surface currents (Figure 4.34b) and a permanent thermocline. Here 90% of algal production depends on nutrients regenerated within the epilimnion, with the remaining 10% supported by nutrients that diffuse across the thermocline. However, there are parts of the oceans where surface waters become enriched with nutrients; for example, upwelling areas (discussed in Section 4.10), the coastal regions where inputs from rivers are substantial, and shallow waters over continental shelves where there is only a temporary thermocline which breaks down in autumn. Periodic nutrient enrichment of the epilimnion is also the norm for lakes.

❑   When and why will this occur?

■   It will occur whenever the thermocline breaks down, allowing complete mixing of the hypolimnion and epilimnion – this commonly happens in autumn for temperate lakes and erratically whenever there is enough turbulence or surface cooling for shallow lakes.

There are marked differences between lakes in which the total stock of circulating nutrients is high and those where it is low. The former are described as **eutrophic** ('well-feeding') and the latter as **oligotrophic** ('small-feeding'), terms which describe the overall productivity or **trophic status** of the lake. Eutrophic lakes can develop naturally from oligotrophic lakes because, over centuries, nutrients gradually accumulate from the catchment. But far more common today are lakes that have become rapidly eutrophic as a result of increased inputs from human activities, a process described as **cultural eutrophication**. Perhaps the most striking difference between temperate eutrophic and oligotrophic lakes is the state of the hypolimnion in summer. This difference arises mainly because algal growth in spring–early summer greatly outstrips zooplankton grazing in eutrophic lakes so that large numbers of phytoplankton sink into the hypolimnion.

❑   Compare Figure 4.43a and b: which relates to a eutrophic and which to an oligotrophic lake?

■   Figure 4.43a relates to a temperate eutrophic lake in midsummer and 4.43b to an oligotrophic lake. Label each part of the Figure with its trophic status, try to list the evidence for this answer and then read on.

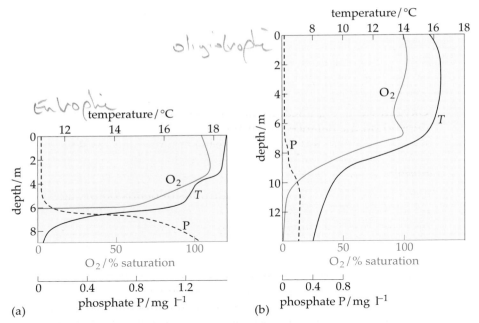

Figure 4.43 Vertical profiles of temperature (*T*), oxygen (O₂) and phosphate phosphorus (P) in midsummer for temperate lakes of different trophic status.

The very large difference in phosphate concentration in the hypolimnion is one indicator – but notice that *both* types of lakes have low P in the epilimnion. Once nutrients here have been depleted by algal growth, they remain low until the thermocline breaks down, although eutrophic lakes with, for example, large constant inputs of P from sewage treatment plants, may not show such extreme surface depletion. You should also have noticed that oxygen levels fall rapidly to zero in the eutrophic hypolimnion but fall more gradually in the oligotrophic lake and reach zero only close to the sediment surface. This reflects the higher microbial activity in the former, arising from the heavy deposit of dead algal cells. Anoxia in the hypolimnion is a classic symptom of a eutrophic lake. Although not formerly the practice, the terms eutrophic and oligotrophic are also now commonly applied to areas of oceans where there are high and low levels of nutrients in the epilimnion, even though anoxic hypolimnia exist only in virtually enclosed seas such as the Baltic and Black Sea.

From this brief account, you should begin to see how steep gradients of resources and conditions over both time and space characterize many aquatic habitats. We focus next on the ways in which these gradients affect one group of organisms – phytoplankton.

Summary of Section 4.8

- Arising from the physical properties of water, aquatic environments have a vertical gradient of light (with photosynthesis confined to the **euphotic zone**) and may develop a discontinuous temperature gradient by **thermal stratification**. A warm, mixed **epilimnion** overlies a cooler, mixed **hypolimnion**, separated by the **thermocline**.

- Chemical conditions vary significantly only in freshwaters and pH is often the most important factor influencing organisms. Acidification of lakes and rivers through acid deposition is, therefore, of major concern.

- Extreme pH values (or associated high aluminium levels) may influence organisms directly but many effects of pH are through altered biotic relationships (competition or predation). Loss of food organisms can influence predators (**bottom-up control**) and loss of predators (especially fish) can influence prey (**top-down control**).

- Excluding food, the three chemical resources that influence aquatic organisms most strongly are oxygen, carbon (CO_2 or HCO_3^- ions) and mineral nutrients.

  > Oxygen varies significantly only in freshwaters, where levels depend on the balance between supply (determined mainly by water turbulence and photosynthesis) and demand (**BOD** and **COD**): it varies greatly in rivers and between the epilimnion and hypolimnion of stratified (especially eutrophic) lakes.

  > Carbon availability to aquatic autotrophs depends on pH and relative ability to utilize HCO_3^- ions rather than free CO_2.

  > Nutrient supplies depend on external inputs (very important for rivers and to a lesser degree for coastal waters and eutrophic lakes) and on internal regeneration from decomposition in the water column and in bottom sediments. Nutrients regenerated in the hypolimnion become available to algae only if the thermocline breaks down, which occurs seasonally in temperate lakes and shallow seas. In the open oceans, where there is a permanent thermocline, virtually all nutrients are regenerated in the epilimnion and very small algae with rapid rates of nutrient uptake and cell division dominate. Nutrient-rich waters are commonly described as **eutrophic** and nutrient-poor waters as **oligotrophic**.

Question 4.11   *(Objectives 4.1, 4.8 & 4.9)*

Explain which of (a)–(e) are generally true, false or partly false.

(a) If living phytoplankton sink below the thermocline, they probably stop growing and die prematurely.

(b) River systems are dispersed and relatively isolated habitats with two compartments – benthic and fringing.

(c) The open oceans show virtually no variation in the chemical properties that affect organisms.

(d) Biological oxygen demand will be high in the hypolimnion of a eutrophic lake.

(e) Fish that are restricted to polar oceans are likely to be more stenothermal (Chapter 3, Section 3.4.3) than are fish from temperate oceans.

Question 4.12   *(Objective 4.10)*

Armoured catfish *Ancistrus spinosus* were studied in a Panamanian river (Central America). They feed on attached filamentous algae whose growth potential (determined in exclosures where grazing was prevented) varied by a factor of 17-fold between shaded and sunny pools. In all pools studied (> 50), a sparse uniform algal cover was present wherever catfish grazed. Catfish were abundant in sunny pools and sparse in shaded pools but growth rates and survival were the same in both. The well-protected catfish had no aquatic predators but were attacked by birds in shallow water (< 20 cm deep) and avoided these areas. Along the shallow margins of the river, dense bands of attached green algae occurred. Explain, giving reasons, whether this example illustrates top-down or bottom-up control.

Figure 4.44 A group of elongated diatom cells (*Asterionella formosa*) arranged in a star shape (× 250).

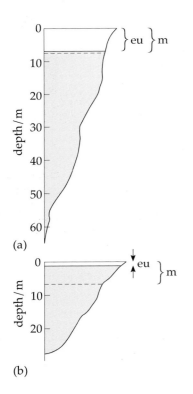

Figure 4.45 Relative depths in June–July of the euphotic zone (eu) and the mixed zone (m), i.e. the depth of the epilimnion through which water circulates, for (a) an oligotrophic lake and (b) a stratified eutrophic lake.

Question 4.13   *(Objectives 4.10 & 4.11)*

A species of herbivorous zooplankton lived mainly among rooted macrophytes in small, upland English lakes. In lakes where anthropogenic acidification had occurred, the species disappeared. Suggest at least three alternative hypotheses to explain its disappearance and, for each, state whether top-down or bottom-up control or neither is implied.

4.9 Phytoplankton

Because phytoplankton (Section 4.7.1) are at the base of food-chains wherever water is sufficiently deep, they play a central role in aquatic ecology. You have already seen that change in diatom species caused by lake acidification happens on a scale of decades in response to altered chemical conditions. However, phytoplankton undergo constant change at much shorter time-scales and a major concern of *limnology* (the study of lakes) and marine ecology is to understand the causes of these changes and be able to predict them. To do this, it is useful to describe briefly the characteristics of three broad types of phytoplankton that are often dominant in lakes and oceans.

4.9.1 Characteristics of dominant types

Diatoms (phylum Chrysophyta, family Bacillariophyceae) are very common in lake and ocean phytoplankton. They range in size from nano- to microplankton (Table 4.18, Figure 4.33c) and their most striking feature is that the cell walls contain a lot of silica (SiO_2). This is a dense material and, since diatoms are non-swimmers (although they can glide over surfaces), they sink readily and in still water rapidly fall out of the euphotic zone. The elongated shape seen in many diatoms and the habit of linking cells in chains or stars (illustrated for *Asterionella* in Figure 4.44) greatly reduce the rate of sinking compared with spherical cells of the same volume. If well-supplied with nutrients, diatoms can achieve high rates of growth (i.e. cell division), sometimes at quite low temperatures and light levels. Species abundant in winter or early spring have been called **winter diatoms**.

A second group of phytoplankton are taxonomically mixed but characterized by small size (in the pico–nanoplankton range) and the ability to divide rapidly over a wide range of nutrient levels provided that light supply is high. This group will be referred to as **small-type species**, although in other publications the names *r*-strategists or *r*-species may be used. They have a rapid rate of growth and are 'opportunists', whose numbers may increase rapidly and may just as rapidly decrease, usually because of grazing.

The third group of phytoplankton are the **large-type species** (which are often called *K*-strategists or *K*-species). Apart from large size, the main characteristics of these species are that they grow relatively slowly and can achieve high densities. In this situation, the resources needed for growth are in very short supply and large-type species (unlike small-type species) are strong competitors. In a stable, stratified water column, the resources in shortest supply in the epilimnion will be nutrients and *light*: look at Figure 4.45.

❑   Why is the euphotic zone so much shallower in the eutrophic lake and what effect will the altered ratio of eu : m have on phytoplankton growth?

■ Remember that nutrient levels are higher in the eutrophic lake and consequently phytoplankton numbers reach a much higher value. The result is self-shading by algal cells – the water becomes an algal soup, light penetrates only a short distance and the eu : m ratio falls. Since phytoplankton circulate throughout the mixed zone, they inevitably spend much of their time *outside* the euphotic zone, which will slow down growth.

Large-type phytoplankton include green species such as *Volvox* (Figure 4.33c), larger cyanobacteria such as *Microcystis* or *Anabaena* which form colonies or filaments and, especially in the oceans, large dinoflagellates (Figure 4.46a). These species are usually too large or unpalatable to be grazed by zooplankton; they are often favoured by high temperatures and they possess characteristics that aid survival in conditions when nutrients and/or light are in very short supply. Many cyanobacteria, for example, supplement their nitrogen supplies by *nitrogen fixation*, carried out in special cells called **heterocysts** (Figure 4.46b) and they also have a unique mechanism for moving up or down in still water by regulating buoyancy using **gas vacuoles** (protein-bounded structures which can accumulate or lose gases). Flagellated species may also carry out vertical migrations and – coupled with an ability to take up and *store* nutrients – this means that large-type species can move up to the euphotic zone in very still conditions (when all other cells tend to sink), they can move below the thermocline at night to stock up with nutrients, and they can live off stored nutrients during more turbulent periods when cells circulate out of the euphotic zone.

4.9.2   Seasonal patterns of change

With some information about the main groups of phytoplankton, we can now discuss how and why the species present change at different seasons. Figure 4.47a shows an idealized pattern of change or **seasonal succession** of phytoplankton in temperate waters – both lakes and shallow seas – under moderately eutrophic conditions. Notice that stratification is stable from spring to autumn; the line labelled 'envelope' indicates the upper limit of algal biomass set by the availability of light and nutrients.

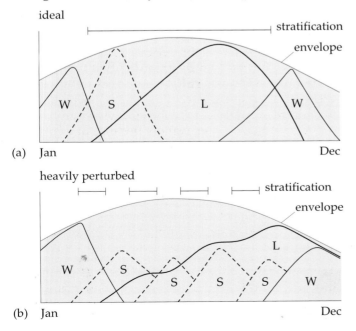

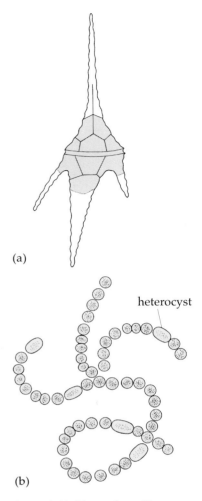

Figure 4.46 Examples of large-type phytoplankton: (a) *Ceratium hirundinella,* a freshwater dinoflagellate (× 800) (similar species occur in temperate seas); (b) *Anabaena flos-aquae,* a freshwater filamentous cyanobacterium (× 800).

Figure 4.47 Seasonal successions of phytoplankton in temperate waters: (a) idealized pattern with a stable water column and constant stratification through the growing season; (b) 'perturbed' pattern where the water column is less stable and stratification frequently breaks down because of strong turbulence. W = winter diatoms; S = small-type species (mostly small flagellates); L = large-type species.

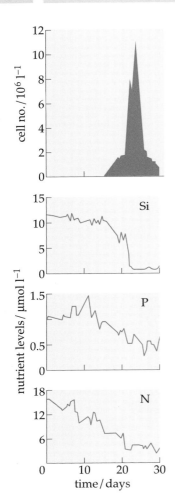

Figure 4.48 Spring diatom bloom in marine, temperate waters (Firth of Clyde, Scotland) and change of dissolved silica (Si), phosphate (P) and nitrate (N).

❑ Starting in January, think through the *reasons* for the idealized succession (Figure 4.47a) and then read on.

At the beginning of the year, nutrient levels are high in the fully mixed water column and, as temperature begins to rise, the winter diatoms increase rapidly. This is the **spring bloom** of diatoms, which may end for a variety of reasons. One reason is shortage of dissolved silica (Figure 4.48); a second is increasing grazing by zooplankton; and a third, particularly in shallow lakes with a shallow mixed zone, is that heavy diatom cells simply sink below the thermocline. Several different species of diatoms may contribute to the spring bloom, each starting and peaking at slightly different times.

As the diatoms decline, opportunist small-type species increase. Zooplankton grazing on nanoplankton is one of the commonest reasons for their decline but, in eutrophic water, shading out by the steadily increasing large-type species – which are *not* grazed by zooplankton – is another contributing factor. If conditions are stable, the latter reach a peak in the warm conditions of late summer, declining only after the autumn overturn (breakdown of the thermocline) when diatoms show a second peak in abundance. A problem which afflicts eutrophic lakes is the formation of **water-blooms** of cyanobacteria: when summer populations are high and most cells are positively buoyant, a sudden reduction in wind, so that surface turbulence falls, leads to a mass upward movement and formation of a dense surface scum – the bloom. Scums tend to pile up at one side of the lake as soon as the wind picks up and dying cells may release powerful toxins that can be harmful to mammals who enter or drink the water. Similar blooms caused by dinoflagellates may also occur in eutrophic coastal waters, the most infamous being the '*red tides*' which release toxins that kill fish, contaminate shellfish and cause havoc in holiday areas.

Look now at Figure 4.47b which illustrates seasonal succession when the water column is much less stable (e.g. because of strong winds) and stratification breaks down, injecting pulses of nutrients from the hypolimnion, several times during the growth season. This pattern is most likely in relatively shallow lakes and least likely in large, deep lakes and the oceans.

❑ List the differences between the perturbed and the ideal seasonal succession and suggest explanations for them.

■ 1   The spring diatom peak is prolonged (because disturbance keeps diatoms suspended in the water column).

2   Instead of a single large peak of small-type species, there are several smaller peaks right through the summer (each period of stratification is an 'opportunity' but each breakdown reduces the population, probably because too much time must be spent out of the surface euphotic zone and there is increasing competition with large-type species as well as zooplankton grazing; different species are likely to peak at different times under the slightly different conditions).

3   The large-type species show a slower build-up with a peak later in the season (these species can survive temporary destratification but growth rates slow down).

4   The autumn diatom peak occurs later and declines in parallel with large-type species, both showing maximum use of available resources (reaching the envelope) and possibly, therefore, competing for resources.

Conditions in the water column vary in unpredictable ways depending on the weather: temperature varies and wind affects turbulence, the amount of mixing in the epilimnion and temporary destratification episodes. Thus, it is not possible to predict exactly which types of algae, let alone which species, will be most abundant at a particular time. What Figure 4.47 indicates, however, is that only rarely do algal populations reach the envelope – the limit set by resource supply – and so for most of the time they are not short of resources and cannot be thought of as competing strongly. This applies especially in the perturbed conditions (Figure 4.47b).

Overall, temperature, nutrient supply and light (which is affected by stability of the water column) are the three abiotic factors that influence most strongly phytoplankton numbers and the species present. In addition, levels and selectivity of zooplankton grazing certainly have an important influence. This applies to all water bodies but the relative importance of different factors varies with latitude and with size and depth of the water body. In polar seas, for example, conditions are cold, turbulent and nutrient-rich during the period when light is available.

❑   Recall from Section 4.8.1 why conditions are turbulent and nutrient-rich in summer and predict the types of phytoplankton likely to dominate.

■   The water never warms up sufficiently to stratify (Figure 4.36) and upwelling maintains high nutrient levels: the conditions described favour rapidly growing diatoms and other net plankton (large-type species) co-exist.

In polar seas, therefore, there is generally a single massive summer peak of algal biomass, followed by a zooplankton peak that in turn supports a rich diversity of animal life ranging from fish and whales to penguins and seals. The decline in algal numbers is probably little affected by grazing here, since algal division rates far outstrip grazing capacity, and falling light levels as autumn approaches are more important. Patterns of algal production, i.e. the amounts of carbon fixed per unit time and area in different water bodies, will be discussed in more detail in Book 4. It is relevant to point out here, however, that *average annual algal production* matches surprisingly well the *average annual concentration* of phosphate in freshwaters and, often, of nitrate in seawater. The influence of nutrients on the types and numbers of phytoplankton is considered next.

4.9.3 Nutrient supply and the question of top-down or bottom-up control

Nutrient supply is one of the factors that determine the limit to algal biomass – the envelope – in a system (Section 4.9.2). The limits will be higher in eutrophic compared with oligotrophic water and if you compare the crystal-clear water of an oligotrophic lake in late summer with the green algal soup of a eutrophic lake, the obvious conclusion is that nutrients – in this case phosphate – are what really matter: the more nutrients there are the more algal growth and biomass, and bottom-up control is, therefore, paramount. Obvious conclusions are often the most dangerous so we need to re-examine this one.

The first point to bear in mind is that phytoplankton, particularly the small types, have short life cycles and the biomass present at any time depends not only on their rates of cell division but also on rates of *loss*, through sinking or being eaten. Even in eutrophic lakes there is often a period in early summer, before the numbers of large-type species have built up and when small-type species have been heavily grazed, when the water is quite clear (see Figure 4.47). This is primarily a function of high loss rates of small species to zooplankton and so represents a period of top-down control.

Now consider the situation in temperate oligotrophic lakes.

❑   Predict what will happen (a) in early spring and (b) in early summer.

■   (a) The pattern will be the same as in eutrophic lakes (Figure 4.47), but with a lower peak of algal biomass because nutrient levels, and therefore the limiting envelope, will be lower. (b) Following the spring bloom, the nanoplankton will be grazed out by zooplankton, but large-type species will not take over to the same extent as in eutrophic lakes because nutrient levels are lower in the epilimnion, and lower levels in the hypolimnion (Figure 4.43) mean that periodic pulses during disturbance events (Figure 4.47b) will be smaller. Oligotrophic lakes also tend to be deeper than eutrophic lakes, with a deeper thermocline and less likelihood of destratification during summer.

What seems to happen subsequently is that the very small picoplankton become the dominant type of algae together with small numbers of large-type species that can obtain nutrients by vertical migrations below the thermocline. How the picoplankton function in this situation is still a matter of debate (they are extraordinarily hard to study) but, as described in Section 4.8.3, it is generally agreed that they can take up nutrients very efficiently when concentrations are low and that they have the potential for rapid cell division. One scenario – but still only a hypothesis – is as follows:

1   Autotrophic and heterotrophic picoplankton associate on particles of dead organic matter, usually zooplankton faeces, which will be abundant after the early production peak, and as these particles are decomposed, the autotrophs immediately take up released nutrients and may grow quite rapidly. There is also evidence that autotrophs in nutrient-poor conditions possess phosphatase enzymes in their cell walls which allow them to release phosphate from organic compounds (as described for the roots of nutrient-unresponsive plants, Section 4.3.1).

2   Picoplankton on particles are consumed by unicellular protistan grazers, whose dead remains will be decomposed *in situ*, and the process continues until either the whole particle is consumed and picoplankton disperse to new particles or until small zooplankton come along and consume the particle – resulting in more faeces (new particles).

Here, therefore, you have islands of nutrients (the **island hypothesis**) in what is otherwise a nutrient desert with *very tight coupling* between nutrient release, nutrient uptake and zooplankton grazing; growth would be sustained almost entirely by nutrients recycled in the epilimnion and top-down and bottom-up control are of about equal importance. The alternative scenario, in extreme form, is that picoplankton spend most of their time floating free and growing at a very slow rate under extremely nutrient-poor conditions; the slow upward diffusion of nutrients across the thermocline is here the main factor that sustains growth and bottom-up control dominates. The situation in the oligotrophic waters of open oceans is basically similar but even more tightly constrained because here there is no spring burst of production to set things going.

Problems of cultural eutrophication

The question of whether top-down or bottom-up control is more important in determining algal biomass has real practical significance in the context of cultural eutrophication. This now occurs world-wide and, as well as frequently reducing wild-life value, causes increasing problems to water and leisure industries: summer blooms of cyanobacteria in lakes and slow-flowing, large rivers block pumps, taint drinking water, may be a toxic hazard and interfere with leisure pursuits. Much effort and large sums of money are being spent in trying to restore or improve eutrophic freshwaters and, assuming primacy of bottom-up control, all attention will be focused on reducing nutrient inputs; assuming primacy of top-down control, attention will be focused on manipulating conditions so as to favour algal grazers (e.g. by removing or reducing predatory fish and providing refuges for zooplankton grazers where they can escape predation). Even quite subtle effects on predators can be magnified in a downward cascade of effects on grazers and plants – the so-called **trophic cascade**. This idea is incorporated into a top-down approach to algal control which is termed **biomanipulation** and has been used with some apparent success in, for example, certain of the Norfolk Broads and some reservoirs in the Thames Valley. However, earlier discussion in this Section should convince you that biomanipulation cannot work in all situations.

❑   Look at Figure 4.47, think about grazer selectivity and predict when biomanipulation could not work.

■   If there are high nutrient levels and a stable water column with an early build-up of large cyanobacteria – which are not grazed – then biomanipulation will be ineffective.

Grazers will have a significant impact only on small-type species or diatoms. So only if nutrient levels are sufficiently low and/or if conditions are sufficiently unstable as to suppress growth of cyanobacteria, or the equally inedible filamentous algae, can biomanipulation significantly improve water quality (the goal). An example of a trophic cascade which illustrates how biomanipulation can reduce phytoplankton biomass in a lake where phosphate levels were only moderate is shown in Figure 4.49.

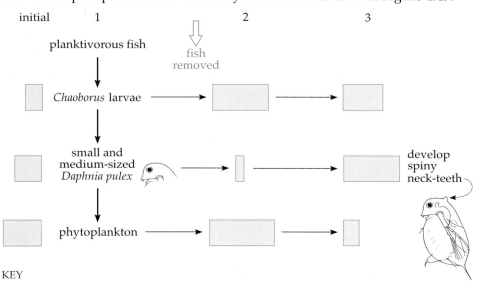

Figure 4.49 Events occurring in Gråfenhain, Germany – a lake with moderate phosphate levels – following removal of planktivorous fish during a biomanipulation experiment. The initial situation is shown in column 1 and columns 2 and 3 show what happened initially and subsequently after biomanipulation. (Benndorf, 1992)

KEY

▢   relative population size

❑   Study Figure 4.49 and describe the situation at stages 1, 2 and 3.

■   Initially (stage 1), predation by planktivorous fish maintained low numbers of *Chaoborus* larvae, whose main prey – the water flea *Daphnia pulex* – therefore had moderate populations, as did the phytoplankton on which *Daphnia* grazed. At stage 2, after removal of planktivorous fish, *Chaoborus* numbers increased, *Daphnia* numbers declined and phytoplankton increased! By stage 3, the *Daphnia* had developed spiny neck-teeth which prevented their being eaten by *Chaoborus* (an example of an induced defence – Chapter 2); *Daphnia* numbers increased and phytoplankton populations fell.

Clearly, biomanipulation may work in rather unpredictable ways and requires sound knowledge about the relationships between grazers and predators; but it can be helpful in certain situations. Other techniques which have been used include constant or intermittent destratification of reservoirs (usually by means of large pumps), which delays and reduces the peak of large-type algae (Figure 4.47b); and the addition of barley straw to smaller lakes and rivers where algal inhibitors are released as the straw undergoes aerobic decomposition. In the long term, however, the real solution to cultural eutrophication is reduction of nutrient inputs and, even then, if nutrients stored in sediments are continually released, mechanisms to block this release (including sediment removal) must be considered.

Summary of Section 4.9

*   The three main types of phytoplankton are: **winter diatoms**, which sink easily but can grow rapidly and flourish in turbulent conditions that keep the cells suspended in the water column; **small-type** species, which are very small, opportunist species with potentially high rates of increase, efficient nutrient uptake and high susceptibility to grazing; and **large-type species**, which are large, slow-growing species that are unpalatable to most grazers, tend to dominate in stable, high-nutrient conditions and have a variety of mechanisms that allow them to compete strongly when algal numbers are high.

*   In water bodies that stratify seasonally, a succession of different algal types occurs which can be represented for relatively eutrophic waters as: winter diatoms → small-type species → large-type species → winter diatoms. The sequence can be modified by temperature and water column stability (which affects light supply), both of which interact with nutrient supply to determine the maximum potential algal biomass (envelope). Grazing pressure and the frequency of destratification events commonly determine whether algae reach the envelope (limit of resources or maximum potential biomass) and – usually – they rarely do.

*   Nutrient supply is probably the most important factor influencing the envelope, usually phosphates in freshwater and nitrates in seawater. Overall, bottom-up control operating through nutrient supply dominates in eutrophic conditions.

*   The situation is less clear in oligotrophic conditions: if the island hypothesis is accepted (aggregates of picoplankton and protistan grazers on detritus particles acting as tightly coupled systems of nutrient release and uptake), then top-down and bottom-up control are

of equal significance. If picoplankton are regarded as mainly free-floating, then bottom-up control dominates.

- Improvements in water quality in culturally eutrophic lakes or rivers have involved nutrient reductions (bottom-up approach) and **biomanipulation** (top-down approach based on the idea of **trophic** cascades). Biomanipulation can be successful only if nutrient levels are relatively moderate and if unpalatable cyanobacteria are not dominant.

Question 4.14   *(Objectives 4.1 & 4.12)*

Complete statements (a)–(c) by choosing one or more items from (i)–(x).

(i) Winter diatoms; (ii) zooplankton; (iii) small-type species; (iv) large-type species; (v) cyanobacteria; (vi) dinoflagellates; (vii) unstable; (viii) stable; (ix) nutrient; (x) light.

(a) ...✓...... such as certain ...✓...... are most likely to be the dominant type of phytoplankton in eutrophic lakes in late summer.

(b) The spring bloom of ...✓...... may be prolonged if the water column is very ...✓...... .

(c) ...✗..... supply is likely to limit phytoplankton growth if the euphotic zone is much shallower than the mixed zone.

Question 4.15   *(Objectives 4.12 & 4.15)*

Figure 4.50 shows the change in cell numbers for two types of phytoplankton when cultured together under high and low light intensities.

(a) Explain, giving reasons, which is the green nanoplankton and which the cyanobacterium.

(b) If both species occurred together in a temperate eutrophic lake, when would you expect the green nanoplankton species to be most abundant?

Question 4.16   *(Objectives 4.10 & 4.15)*

This question is about top-down/bottom-up control in Lake Erie, one of the North American Great Lakes. Between 1972 and 1991, US$7.5 billion was invested in a phosphate reduction programme to reduce eutrophication in this lake. Figure 4.51a and b show changes in annual average phosphate concentrations and average summer phytoplankton biomass, respectively. In addition:

(i) The maximum biomass of several species of phytoplankton (measured by counting cells) that are indicators of eutrophic conditions declined by 85–94% between 1972 and 1983/4.

(a) From the information given so far, do you think that the data support bottom-up control of phytoplankton through phosphate supply? Justify your answer and point out inadequacies in the data.

Figure 4.51c and d show changes in the numbers of two types of fish: the piscivorous (fish-eating) walleye (*Stizosedion vitreum vitreum*), a top predator; and the planktivorous (plankton-eating) alewife (*Alosa pseudoharengus*) which forms a major part of the walleye's diet. Figure 4.51e shows changes in two types of zooplankton: the large herbivorous cladocerans (such as *Daphnia*) and small herbivorous copepods. No data were available for the 1970s. In addition:

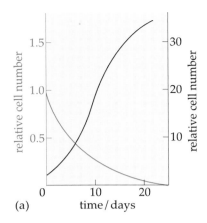

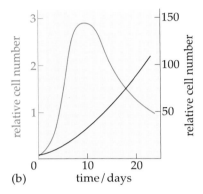

Figure 4.50 For use with Question 4.15. Relative increases in cell number of a green nanoplankton *Scenedesmus protuberans* (green curves) and a cyanobacterium *Oscillatoria agardhii* (black curves) when cultured together under (a) low light intensity and (b) high light intensity.

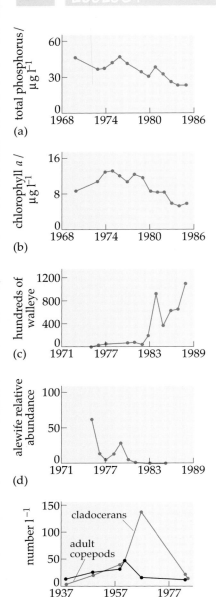

Figure 4.51 For use with Question 4.16. Changes in the western basin of Lake Erie in: (a) average annual total phosphorus; (b) phytoplankton (measured as the average summer chlorophyll-*a* concentration); (c) walleye (a top predator fish) – data based on sport angler harvest; (d) alewife (a typical plankton-eating fish); (e) zooplankton (adult copepods and cladocerans).

(ii) A large herbivorous cladoceran, *Daphnia pulicaria*, first appeared in Lake Erie in 1983, was the dominant zooplankton in 1984 but crashed (90% reduction) in 1985. In October 1985, a very large (> 10 mm) predaceous cladoceran was observed in the lake for the first time and in considerable numbers.

(b) From Figure 4.51c–e and item (ii), identify any examples of top-down or bottom-up control.

4.10 Flowing waters: rivers and streams

Only a tiny fraction, 0.004%, of the Earth's freshwater occurs in rivers but they are extremely important as a habitat and a functioning part of the biosphere. Rivers and streams are the arteries of the land drainage system, carrying water from land to sea and it is their unidirectional *flow* that defines them. In this Section we look in more detail at the ways in which water flow affects conditions in rivers and at the distribution and types of river organisms

4.10.1 Water flow and river habitats

Current velocity (U, usually measured in m s^{-1}) is a characteristic of major significance for river organisms so it is useful to consider how and why it varies. This has to be done at both large and small scales. At any point along a river, current velocity varies with depth (Figure 4.52). The *mean* velocity is denoted by the symbol $\overline{U}$.

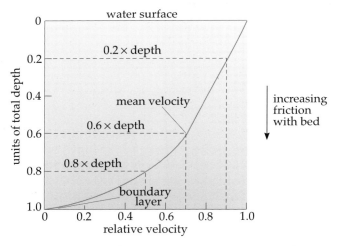

Figure 4.52 Water velocity as a function of depth in an open channel.

❏ Where in the channel is velocity greatest and least and why?

■ Velocity is greatest at the surface (where friction due to the bed is least) and least adjacent to the bed. You can see that mean velocity (i.e. $\overline{U}$) could be measured at 0.6 × the total channel depth.

These relationships hold only near the centre of the channel because friction from banks decreases velocity at the edges. Obstructions such as boulders also influence velocity, creating areas of turbulent flow and, in general, there is much variation in velocity along shallow streams and greater uniformity along deep rivers. On a larger scale, velocity is affected by **discharge**, Q, the volume of water moving past a point in unit time. Assuming an approximately rectangular cross-section,

$$Q = WD\overline{U} \text{ m}^3 \text{ s}^{-1} \tag{4.2}$$

where W and D are channel width and depth, respectively. If discharge increases after heavy rain, then any or all of W, D and $\overline{U}$ may increase – streams get wider, deeper and/or flow faster: in small streams, faster flow usually dominates. Q also increases as small streams merge to give larger rivers and although this causes some rise in velocity, it is mostly accommodated by increases in channel width. Finally, and still on a large scale, mean velocity increases with *gradient* and mean depth and decreases with the 'roughness' of the bed. Considering all these large-scale factors together, we can get a picture of how flow conditions change from the smallest (*first-order*) streams going downstream.

❑   Where do you think velocity will be greater – in small streams or large rivers?

■   Most people think it is small streams but, despite their lower gradient, current velocity in large rivers is usually the greater – because of their much larger discharges.

Typically, small streams are shallow, have a rough bed, much turbulent flow and current velocity is very sensitive to changes in discharge. Large rivers are wider, deeper, with a smoother bed, less turbulent flow and lower sensitivity of velocity to discharge.

Water discharge and velocity profoundly affect the channel bed (**substratum**) which in turn affects the organisms that live there. Turbulent water erodes banks and stream bottoms, moving solid particles downstream. The greater the velocity and discharge, the larger the particles that can be moved and the further they are carried. Extreme floods can move huge boulders a surprising distance so that, on timescales varying from months to years, virtually all stream and river beds can be regarded as disturbed habitats. Channel erosion is generally greatest in the upper reaches of rivers and here a substratum of gravel and boulders is common. Silt and detritus tend to be deposited in deeper **pools** where the velocity is lower and there is commonly an alternating sequence of pools and **riffles**, which are shallow, fast-flowing zones with a gravel bed. Deep but fast-flowing stretches with a gravel bed are referred to as **runs**. Large boulders and obstructions such as fallen trees slow down the flow so that silt and detritus settle out and these sites (**flow refugia**) are often of considerable importance as refuges for animals and plants in times of flood. The availability of flow refugia influences strongly the numbers of organisms lost from an area through being washed downstream (**drift**). In the lower reaches of rivers, where water is deeper and the flow smoother, silt tends to accumulate and sediments out to give muddy/silty substrata. Figure 4.53 summarizes the main physical characteristics of streams, and if you add a time dimension to this picture – seasonal variation in flow and temperature, occasional severe floods and droughts – then you gain some idea of the great variability in time and space of stream or river habitats. Bear in mind, however, that much natural variation in rivers, including pool–riffle sequences, has been substantially altered by engineering, for example straightening, channelization and dam and weir construction.

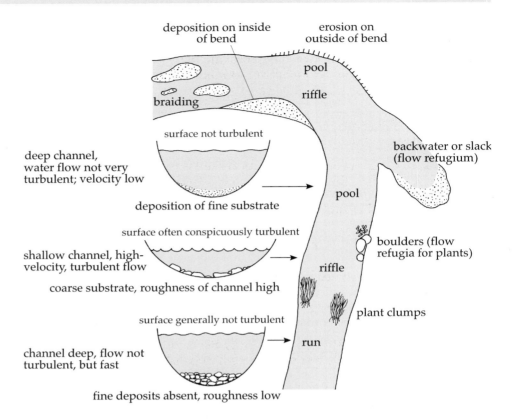

Figure 4.53 Physical habitat features of a stream with cross-sections of pool, riffle and run.

❑   Recall from Section 4.8.3 how flow conditions in rivers affect oxygen supply for river animals.

■   High velocity and turbulence are both associated with good oxygenation.

Both these features are commonly associated with stony or gravel-bed rivers and a characteristic of these rivers, only recently appreciated, is that a surprisingly large proportion of the river habitat is actually invisible underground. The **hyporheic zone** is a region below the river bed and the adjacent floodplain in which oxygenated water is connected to the river channel (Figure 4.54).

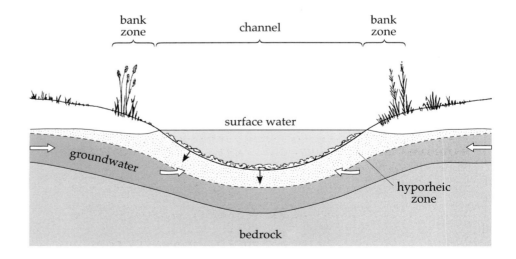

Figure 4.54 The hyporheic zone (stippled). Water is oxygenated and lies above relatively anoxic groundwater or directly above bedrock.

The hyporheic zone is inhabited by aquatic animals and although extending for only a few centimetres from silt-bed rivers, Stanford and Ward (1988) found that the hyporheic zone under a gravel-bed river extended to an average depth of 10 m and had an average width of 3 km around a channel 50 m wide! Oxygen diffuses in from channel water, temperature varies little and water may flow in either direction through the hyporheic zone between the channel and aquifers in the floodplain. Fine particulate organic matter (detritus) is carried into the zone where it is decomposed by micro-organisms and is the main food source for aquatic invertebrates. An important point to appreciate is that where an extensive hyporheic zone exists, the river is especially sensitive to human activities in the catchment: toxic chemicals dumped 2–3 km from a river may pass directly to the river via the zone.

4.10.2 The distribution of organisms in streams and rivers

In shallow, turbulent rivers, benthic and to a lesser extent fringing organisms tend to predominate, with strong-swimming nekton (mostly fish) the only occupants of the water column. Only in deeper, less turbulent zones is there a substantial planktonic component. Clearly, therefore, flow conditions and their variability – the frequency of *spates* (high flow) and severe disturbance – profoundly influence the distribution of river organisms. On a local scale, the other influential factors are mostly chemical: oxygen supply, pH (and, closely related to this, alkalinity and hardness of water), ionic conditions, the supply of dissolved nutrients and, for animals, the nature and quantity of food resources.

Plants

Plants are an easy place to begin because there are relatively few different types. In shallow streams or rivers with fast, turbulent flow and a stony bottom, the only plants present are those that attach firmly to rocks. These include certain aquatic mosses and a range of surface (periphytic) algae – including many diatoms which often form a slimy film over rocks. Where sediment accumulates, rooted plants such as *Ranunculus fluitans*, a submerged species of water crowfoot, may occur, its finely divided leaves offering little resistance to the current (Figure 4.55). This species is typical of nutrient-poor water up to a maximum of 1 m depth and with a coarse, gravelly substratum, often over limestone. If nutrient levels increase, usually through anthropogenic inputs, filamentous algae tend to increase markedly.

❑   From the above description, identify at least three factors which influence the distribution of plants in shallow rivers.

■   Current velocity, depth, nature of the substratum, nutrient status of the water and, probably, ionic status (since *R. fluitans* occurs mainly over limestone). You might also guess (correctly) that mosses and benthic algae (on the bottom) will be shaded out by water crowfoot or filamentous algae, so that competition is also a significant interaction.

In the deeper, lower reaches of rivers, phytoplankton are usually the dominant primary producers, even though they are on a one-way journey to the sea. Current velocity, the concentration of nutrients, sediment load (which affects light penetration) and the numbers of grazing zooplankton all influence phytoplankton numbers but the dominant types (diatoms and small-celled green algae) are typically fast-growing small-type species:

Figure 4.55 Finely divided leaves of *Ranunculus fluitans* (× 0.3), a submerged species of water crowfoot found in moderately fast to swiftly flowing rivers.

populations of large-type species (often cyanobacteria) build up only in warm, low flow rate conditions. Despite their high rates of cell division, however, the persistence and numbers of small-type species often appeared to be too great for the known rates of wash-out. The puzzle was resolved for the River Severn by five years of careful study, summarized in Reynolds (1995), and the answer turned out to be *storage zones*: low or zero flow patches adjacent to bank cavities or behind projecting bars where algae could multiply and pass gradually into the main flow. Storage zones are another example of *flow refugia*.

Animals

The situation is more complex for animals because there are so many different types. The majority of these are invertebrates and the most common groups are molluscs, crustaceans, insects (many of which spend only the larval or nymph stage in water, e.g. mayflies and stoneflies) and assorted 'worms' (platyhelminths, nematodes and annelids). Fish are, of course, the main kind of vertebrate in rivers.

A useful way of grouping the larger invertebrates is in terms of their food and method of feeding (**functional feeding groups**), summarized in Table 4.20 and with illustrations of some types in Figure 4.56.

Table 4.20 Functional feeding groups of invertebrates in rivers and streams. CPOM, FPOM = coarse and fine particulate organic matter, respectively.

Functional feeding group	Food consumed	Method of feeding	Examples
shredders	leaf litter and other CPOM + associated micro-organisms	chewing and scraping	Crustacea, e.g. *Gammarus*, (freshwater shrimp), crayfish; some insect larvae or nymphs, e.g. caddis flies (Trichoptera), crane flies (Tipulidae), stoneflies (Plecoptera)
suspension-feeders or collectors	FPOM + associated micro-organisms	filter or collect particles floating in the water	net-spinning caddis (Trichoptera); black-flies (Simulidae) and other dipteran larvae; some mayflies (Ephemeroptera)
collector-gatherers	FPOM + micro-organisms on surfaces and in sediments	collect or scrape off surface deposits or burrow in sediments	some mayflies (e.g. burrowing types, *Ephemera*); some midges; many 'worms', e.g. *Tubifex* (an oligochaete)
grazers	periphyton	scrape surfaces	snails (gastropod molluscs); many insect larvae/nymphs
	macrophytes	pierce or mine	some caddis larvae; weevils
	phytoplankton	filter	'water-fleas' (Crustacea)
predators	other animals	bite or pierce and suck	dragonflies (Odonata); bugs (e.g. water boatman *Notonecta*); some beetles (Coleoptera) and larvae of several other insects

❑ Use information from Table 4.20 to link (a)–(d) in Figure 4.56 with the correct name and functional feeding group.

■ (a) is *Simulium*; notice the basal attachment and collecting fans which trap fine particles (FPOM). It is a suspension-feeder.

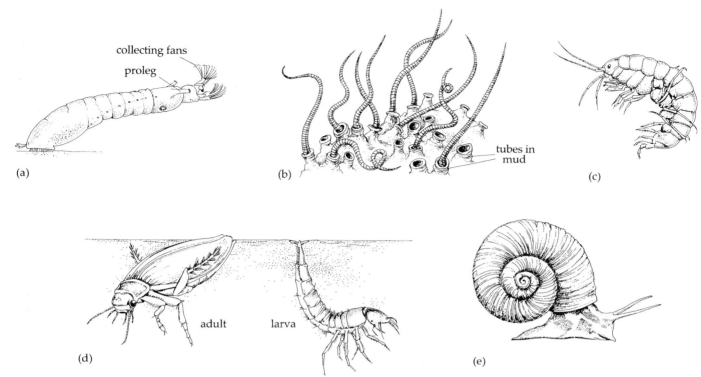

Figure 4.56 Examples of river invertebrates from different functional feeding groups. Organisms (not in order) are: *Dytiscus marginalis* (great diving beetle) (Insecta, order Coleoptera) (× 1.2); *Tubifex* (tubificid worms) (Annelida, order Oligochaeta) (× 0.9); *Simulium* (black-fly larva) (Insecta, order Diptera) (× 8); *Planorbarius corneus* (ramshorn snail) (Mollusca, order Pulmonata) (× 1.4); *Gammarus pulex* (freshwater shrimp) (Crustacea, suborder Amphipoda)(× 2).

(b) are tubificid worms, collector-gatherers which burrow in sediments and emerge from their tubes.

(c) is *Gammarus*; it has no obvious large jaws or filtering apparatus but the small jaws and thoracic legs are used to grasp and shred large particles (CPOM). It is a shredder.

(d) is the great diving beetle and both adult and larva are predators; note the large jaws and powerful legs used for swimming.

(e) is the ramshorn snail, a grazer on periphyton that scrapes surfaces with its rough 'tongue' or radula.

The availability of different kinds of food varies from source to mouth of rivers and thus influences the distribution of river animals. If the upper reaches of a river are heavily wooded so that there are large inputs of leaf litter, shredders feeding on CPOM are likely to dominate. If the upper reaches are not wooded, grazers on periphyton are more likely to dominate. In lower reaches, fine particles of organic matter, including faeces, provide food for suspension-feeders and collector-gatherers, with benthic collectors occurring wherever organic sediments build up. Herbivorous zooplankton occur only in the lower reaches, along with phytoplankton.

Characteristics that are often closely correlated with feeding type are tolerance of fast, turbulent flow and a requirement for high oxygen levels. For example, larvae of *Simulium* (black-flies or buffalo gnats, see Figure 4.56a) are suspension-feeders with a high oxygen requirement and live attached firmly to rocks in shallow, turbulent, well-oxygenated water: lake outflows, which may carry many phytoplankton, are a favourite habitat. Living in the low-velocity microhabitat on or among stones and gravel is another way of persisting in fast-flowing streams for organisms such as *Gammarus* (freshwater shrimp) that lack attachment organs.

Mayfly nymphs (order Ephemeroptera) all have moderate or high oxygen requirements which are met in three ways. *Flattened mayflies* such as *Ecdyonurus* (Figure 4.57a) cling to stones in fast-flowing water and depend on this constant supply of O_2; *swimming mayflies* such as *Baetis* (Figure 4.57b) swim vigorously in slow-flowing or still water and provide their own oxygenating current; *burrowing* or *creeping mayflies* (e.g. *Ephemera*, *Ephemerella*, Figure 4.57c, d) live in burrows or move slowly among plants but can ventilate their gills by moving them about.

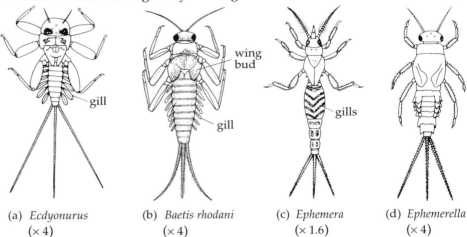

Figure 4.57 Four British mayfly nymphs.

(a) *Ecdyonurus* (× 4)   (b) *Baetis rhodani* (× 4)   (c) *Ephemera* (× 1.6)   (d) *Ephemerella* (× 4)

❑ In Figure 4.58, one curve relates to a creeping mayfly nymph and one to a flattened type: identify which is which.

■ Curve A relates to a flattened mayfly (*Rhithrogena*); it can survive at low current velocities only if the water is very well oxygenated and depends on high flow rates for O_2 delivery. Curve B relates to the creeping mayfly *Ephemerella* which, by fluttering its gills, obtains enough O_2 almost independent of current velocity.

The animals that are most tolerant of low oxygen conditions are tubificid worms and larvae of some non-biting midges (chironomids), which live in sediments and contain a red, haemoglobin-like pigment that helps in the uptake and storage of O_2. These organisms are often abundant immediately downstream of raw sewage outfalls (Figure 4.59d) where there is plenty of food but also very low oxygen.

❑ From Figure 4.59, (a) why is there a low-oxygen zone? (b) Why does *Cladophora* increase some distance downstream of the outfall?

■ (a) Because the large numbers of bacteria, protistans and sewage fungi that decompose the raw sewage immediately below the outfall use up much of the available O_2, hence the high BOD (Section 4.8.3). (b) *Cladophora* increases once water turbidity and sewage fungus growth have decreased and sufficient light penetrates the water. Its growth is stimulated by high nutrient levels as described earlier – but it also requires light!

Out of this plethora of factors affecting river organisms, is it possible to say which are the most important? One approach developed during the 1980s was to use statistical techniques (*multivariate analysis*). Based on

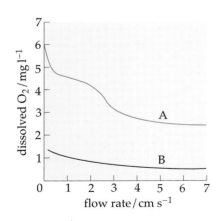

Figure 4.58 Minimum dissolved oxygen contents at which two kinds of mayfly nymphs survive at different current velocities.

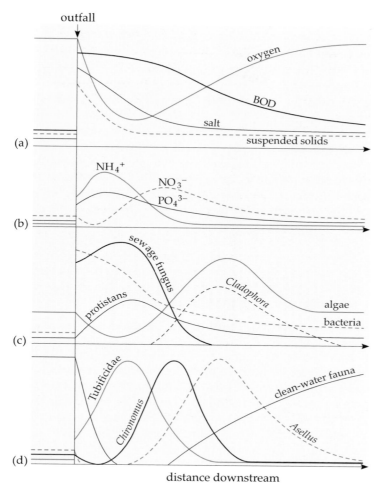

Figure 4.59 The effects of discharge of organic (sewage) effluent on a river and the changes observed downstream from the outfall: (a) and (b) chemical changes (BOD is biological oxygen demand); (c) micro-organisms and algae (sewage fungus forms dense filamentous growths and *Cladophora* is a filamentous alga); (d) macro-invertebrates.

invertebrate surveys, sample sites can be grouped on the basis of similarities in their most abundant species, each group identifying a community type. You can then determine whether there is any relationship between community groupings and environmental variables – the physical and chemical conditions at the sample sites. Do all the sites in community X, for example, have a low pH and all those in community Y a high pH? If so, communities will show a distinct ordering along a pH axis and pH will have been identified as an important element that determines community composition and the distribution of species. This technique, therefore, identifies communities in rivers and then indicates what factors determine where they occur.

Hildrew and Townsend (1987) compared three studies which all showed that pH or water hardness or alkalinity (all of which are closely correlated) was, indeed, a major factor influencing the species present at a site and thus the position of communities along an axis. Communities could also be arranged along a second axis which corresponded to a variety of factors (e.g. slope, discharge, distance from source) that changed from source to mouth. Notice that the emphasis here is on abiotic factors but this does not mean that biotic interactions are not important in rivers. Recall (Section 4.8.3) that top predators such as fish which are exerting top-down control may be especially sensitive to factors such as pH.

Monitoring and restoration

Rivers in many parts of the world, especially in industrialized countries, have been grossly affected by chemical pollution and by engineering works. In Europe, restrictions on chemical inputs and penalties for pollution incidents have increased and there is a strong movement towards improving or 'restoring' rivers. Both to detect chronic, low-level pollution (which is the commonest sort) and to decide what state rivers should be restored *to* require much information from surveys and monitoring. Increases in acidification of upland rivers or nutrient inputs to lowland rivers, for example, may be detected by changes in river communities – provided there have been regular surveys. The standard surveys of invertebrates and fish carried out regularly by the National Rivers Authority in the UK aims to identify, at quite a fine scale, the effects of natural disasters and human activities on river organisms and then decide what, if any, steps to take. An alternative approach where regular surveys are not possible is to use information about catchment geology and land use, and discharge and slope of the river channel to predict which organisms *should* be present: if a survey indicates that they are not, then something has gone wrong and needs investigating.

Apart from chemically cleaning up rivers, the other area strongly emphasized in river restoration relates to the importance of flow refugia. The primary aim of drainage and flood control engineers is to keep water moving and prevent overflow, so there has been much straightening of river channels with removal of obstructions such as gravel bars or fallen trees. Meanders and obstructions, however, play a major role in providing flow refugia and a diverse array of habitats, so there is conflict between engineering and conservation aims. In the early 1990s, there were limited attempts to restore small sections of straightened rivers by *re*-introducing meanders and gravel bars (at considerable cost). Perhaps the most that can be expected for the future is that conservation interests will be taken into account when assessing the suitability of any river engineering project.

Summary of Section 4.10

- Unidirectional flow is the unique characteristic of rivers. On a small scale, flow conditions depend on **current velocity** and turbulence; on a larger scale, they depend on water **discharge** (determined by channel width and depth and current velocity), gradient and bed roughness.

- Water discharge and velocity affect the **substratum** (e.g. gravel and boulders or silt and detritus). Local areas of low flow, often associated with obstructions, provide **flow refugia** which are important in times of flood.

- An extensive **hyporheic zone** may extend below and laterally from stony-bedded rivers, providing an oxygenated link habitat between the river channel and groundwater in the floodplain.

- The distribution of river organisms is influenced by the nature and variability of flow conditions (which in turn affect oxygen supply) and by a range of chemical conditions.

- Plants are also influenced by nutrient supply and, if attached or rooted, by the nature of rocks and substratum. Phytoplankton occur mainly in lower, less turbulent zones and their persistence may depend on the presence of flow refugia.

- Animals are also influenced by the supply of food of an appropriate nature and size. They may be divided into **functional feeding groups** on the basis of their food and method of feeding. Multivariate analysis has been used to correlate type of invertebrate community with environmental variables; pH emerged as a major influence with physical conditions (slope and discharge) and distance from river source as a second important influence.

- Regular surveys and monitoring of river organisms are useful tools for detecting change caused by human activities. The restoration of degraded rivers involves not only changing chemical conditions but also the physical nature of the habitat; reinstating or encouraging the development of flow refugia is particularly important.

Question 4.17   *(Objectives 4.8 & 4.9)*

(a) List the factors that may affect current velocity in rivers.

(b) Two streams, (i) and (ii), have the following characteristics at a certain time: discharge, 2 and 18 m^3 s^{-1}; width, 2 and 6 m; average depth, 0.5 and 1 m. In which stream is mean current velocity greatest?

(c) If a river flows through a deep rocky gorge and discharge increases by 50%, which other physical characteristic of the river is most likely to change?

Question 4.18   *(Objectives 4.1 & 4.12)*

(a) Define and give examples of flow refugia. What is their significance for river organisms?

(b) What sorts of organisms and with what characteristics would you expect to find in riffles in a river?

Question 4.19   *(Objective 4.13)*

(a) A species of water crowfoot (*Ranunculus*) occurs in rivers and has the following characteristics: it has lobed, rounded floating leaves and the stem usually grows horizontally over the substrate, often with roots at the nodes ('joints'). In which one of habitats (i)–(iii) is this species likely to occur and for what reasons?

(i) On the edges of a shallow, turbulent, swift-flowing stream with a stony bottom; (ii) in a shallow, slow-flowing stream with a muddy bottom; (iii) in a deep pool (1 m) with a silty substratum and moderate current velocity for most of the time.

(b) The aquatic nymph (juvenile stage) of a stonefly (Plecoptera) has the following characteristics: (i) body strongly flattened; (ii) legs with stout claws on the ends; (iii) weak swimming ability; (iv) no external gills; (v) does not usually survive long in a static tank of water; (vi) moderately sized jaws. (1) Is this species more likely to occur in slow-flowing parts of rivers on a muddy substratum or among rooted plants; or among stones on the bed of turbulent, fast-flowing streams? (2) To what functional feeding group is this species most likely to belong?

Question 4.20   *(Objective 4.15)*

Experiments were carried out to investigate the effect of different factors on the drift (movement downstream) of bottom-living invertebrates. Drift was measured from laboratory channels (flumes) in which the force acting per unit area of flume bottom (shear stress, dyn cm^{-2}) could be varied by

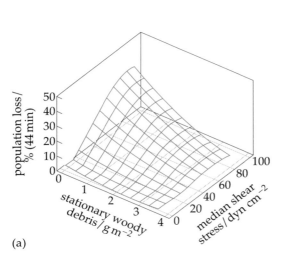

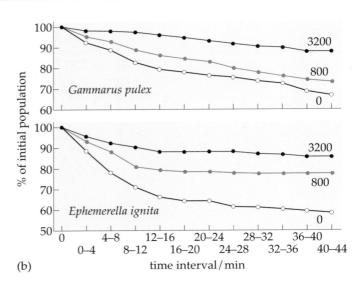

(a)

(b)

Figure 4.60 (a) Drift loss of *Gammarus pulex* as related to shear stress and abundance of woody debris. (b) Pattern of drift loss over 44 minutes for *G. pulex* and *Ephemerella ignita* with three levels of fixed woody debris present (0, 800 and 3200 g m^{-2}) and a median shear stress of 89.5 dyn cm^{-2} (1 dyn = 10^{-5} N).

altering water velocity and depth. Figure 4.60a shows how drift losses for *Gammarus pulex* varied when shear stress and the amount of woody debris fixed to the flume bottom were varied.

(a) From Figure 4.60a, what two factors influence drift of *Gammarus*, in what ways do they do so and why?

Figure 4.60b shows how drift losses in the flume changed with time for *Gammarus* and a creeping mayfly, *Ephemerella ignita*, with different amounts of fixed woody debris. *Ephemerella* is a less mobile species than *Gammarus*.

(b) In Figure 4.60b, what is the main difference between *Gammarus* and *Ephemerella* in their temporal patterns of drift loss? Suggest an explanation for this difference.

(c) One aim of river restoration projects is to improve the habitat for invertebrates and, to this end, flow refugia such as backwaters or bankside debris are often incorporated. On the basis of his flume experiments (Figure 4.60), Borchardt (1993) suggested that flow refugia were essential also in the middle of the channel. What is the basis for this suggestion?

4.11 Special features of oceans

We have already described general features of the oceans – chemical stability, currents, patterns of stratification and phytoplankton. In this Section, we examine one of the unique ocean habitats – the deep sea (Section 4.11.2); and consider briefly a number of ways in which special features of oceanic environments influence marine animals (Section 4.11.1)*.

Figure 4.61 provides background information and shows the main habitat divisions of oceans. The **littoral zone** is the transitional area between land and sea and includes rocky shore environments (discussed in Chapter 3); relatively shallow waters (up to about 130 m deep) over the continental shelf are described as **neritic**; and deeper **oceanic waters** overlie continental slopes, the flat abyssal plains and mid-ocean ridges and trenches.

* For a full discussion of the biology, chemistry and physics of oceans, see Course S330 Oceanography.

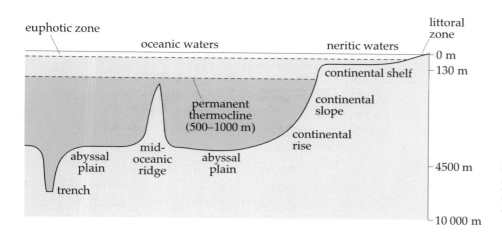

Figure 4.61 Section of the ocean showing the major divisions of habitat. (Modified from Barnes and Hughes, 1988.)

4.11.1 Behaviour of zooplankton and nekton

As a broad rule of thumb, the numbers of marine animals in the water column are highest where phytoplankton production, i.e. food supply, is greatest.

❏ What sort of control does this imply? Where in oceans is algal production greatest?

■ Bottom-up control (Sections 4.8.2 and 4.9.3). Areas of high production occur mainly where the surface waters are enriched with nutrients (Section 4.8.3); these include neritic waters, especially around estuaries, and upwelling areas.

Figure 4.62 shows the main areas of upwelling (mentioned in Section 4.7.3), which occurs when deep, nutrient-rich currents upwell to the surface. The reason why upwelling occurs along the Equator is that wind-driven surface currents move apart here and deep water is sucked up; it occurs off the coasts of Peru, California, north-west and south-west Africa and the Arabian peninsula because winds here blow surface water offshore and –

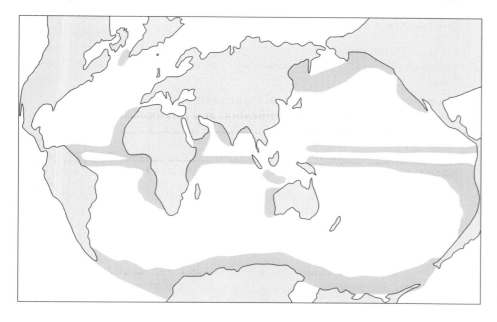

Figure 4.62 The main areas of upwelling in the world's oceans.

provided the continental shelf is fairly narrow – deep water is drawn in to replace it. An important point to appreciate is that upwelling is often sporadic and rarely continuous, just as nutrient enrichment of neritic waters may vary seasonally. Food supply for marine animals thus shows global patchiness in space and time. It may also show small-scale patchiness arising from local turbulence and a variety of other causes, the patches ranging from less than a metre to hundreds of metres across. How is the behaviour of marine animals influenced by this patchy food supply?

Zooplankton

The larger herbivorous zooplankton are stronger swimmers than, for example, flagellated algae but because of the turbulence and currents of oceanic surface water, they cannot swim horizontally from one food patch to another. Neither is staying still, or rather drifting passively with their algal food a viable option, because – usually – the algae are quite rapidly grazed out. The answer to this puzzle can be obtained if you study Figure 4.63 in conjunction with Figure 4.34; note that deep currents often move in the opposite direction to surface currents.

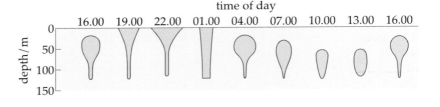

Figure 4.63 Vertical diurnal movements undertaken by a zooplankton species such as adults of the copepod *Calanus* (× 6).

❑ (a) Describe the pattern of movement and feeding of adult *Calanus*. (b) How could the movement shown transfer *Calanus* to a new patch of food?

■ (a) This copepod undergoes **vertical migration**, swimming up to the surface during the evening (19.00–22.00 h), dispersing through the water column around 1 a.m. (01.00 h) and spending the day at a depth of 50–100 m. Since it is a surface feeder, feeding occurs at night. (b) The vertical movements exploit the fact that deeper currents commonly move in the opposite direction to surface currents. During the day, the copepods are carried along sideways and swim upwards at night to a 'new' patch.

Such vertical movements are very common among marine zooplankton and may extend down to a depth of 1000 m for larger species. Apart from helping to solve the feeding problem, there is evidence that by moving to cooler, deeper water in the day, less energy is used during the period of digestion.

Seasonal vertical migrations between surface and deep currents are also used as a means of reaching feeding grounds – notably those in the Antarctic – at just the right time (Figure 4.64). Krill (*Euphausia* sp.) is a large zooplankton that feeds continuously in the food-rich surface waters of the Antarctic during summer, drifting constantly southwards. At the end of the season, it releases eggs which sink into deep water and drift north during winter, the larvae hatching in spring/early summer and swimming up to the surface to begin the whole cycle again. It is possible that this migration pattern helps to reduce predation on larvae by surface predators: vertical *daily* movements also occur on a much smaller scale for freshwater zooplankton such as *Daphnia* and there is good evidence that this does indeed reduce predation risk and may even do so to a small degree for the marine zooplankton.

Nekton

Seasonal migrations are also very common among nekton (e.g. fish, whales and large squid) of neritic waters and the open oceans (but excluding the numerous species that live permanently around coral reefs or in the sub-littoral zone). Figure 4.65 shows a triangular migration circuit that occurs at higher latitudes for fish such as cod (*Gadus morhua*) and plaice (*Pleuronectes platessa*). Seasonally patchy food supplies and the existence of ocean currents are again important driving forces for these migrations. Other factors relate to the small size and, therefore, planktonic nature of young. Eggs are released at a site from which the planktonic young will drift passively to a good and well-protected nursery area where they can grow rapidly without heavy predation, sheltered lagoons or sea-grass meadows in estuaries being typical examples. The adults then return from the spawning site to their feeding grounds, which might be polar waters or another rich upwelling area; they rarely swim directly against the current but will often move vertically to an appropriate depth where the current is flowing in the direction they want to go. There are numerous variations of this triangular migration pattern: some fish, for example, spawn directly at nursery areas if currents will not carry eggs away.

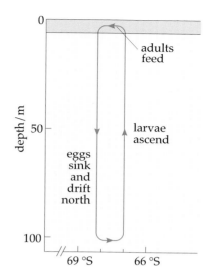

Figure 4.64 Seasonal migrations of Antarctic krill (*Euphausia* sp.).

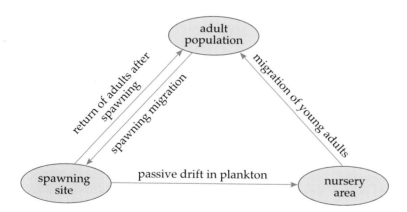

Figure 4.65 The migration circuit of migratory nekton typical of higher latitudes.

Fish that undergo extensive migrations usually tolerate a broad range of temperatures (i.e. they are eurythermal – Chapter 3). The fish of polar seas, especially the Antarctic, are different in that their migrations are vertical and a high proportion of these fish are *stenothermal*: their muscles operate best, for example, at a temperature close to 0 °C (Robinson, 1994). This appears to be the price paid for having a highly specialized physiology that allows them to move rapidly at freezing temperatures. Despite their specialized physiology, poikilothermic fish cannot compete with homeotherms such as whales and seals in seawater at 0 °C and are more likely to serve as easy prey. Thus, the majority of Antarctic fish species are *benthic* and remain in polar waters throughout the year whilst their young, which are too small to be eaten by large mammals, may migrate upwards and feed nearer to the surface in summer.

Antarctic mammals, on the other hand, migrate to warmer places to breed after the brief feeding bonanza of the polar summer, having laid down abundant stores of food as body fat. The highly specialized baleen whales, which filter krill via massive brush-like filtering systems, do not appear to feed whilst in the tropics and the rationale for their epic journeys seems to be energetic: they expend so much less energy to maintain body temperature in warm water that this more than compensates for the energetic cost of migration.

4.11.2 The deep sea

For marine organisms, the main features of surface water environments can be summarized as: patchy in space and time with seasonal variation increasing with latitude; turbulent and with a strong vertical light gradient. By contrast, the main features of the deep ocean environment are: cold, dark, rich in mineral nutrients and with little variation with season or latitude. With one exception, food is generally scarce here, the only source being dead material that sinks below the thermocline. The exception is around **hydrothermal vents**, places on the ocean bed where hot water saturated with minerals gushes out. Sulphides in the water provide a source of energy for chemosynthetic bacteria (a type of primary producer, discussed in Book 4) which support a rich invertebrate fauna.

The **deep sea** can be defined as water below the permanent thermocline and covering the area over the continental slopes, abyssal plains and deep trenches (Figure 4.61). It is widely regarded as the largest remaining wilderness area on Earth. Knowledge of the deep-sea habitat and organisms has increased dramatically since the 1970s, mainly as a result of greater research effort and the use of deep-sea submersibles to take photographs, make observations and collect samples. Nevertheless, we still know very little about deep-sea ecology and a brief review of current knowledge and of how ecologists are starting to investigate this strange environment is a fitting ending to this Book.

Considering the deep-sea environment in more detail: it is dark, cold (mostly −1 to +4 °C, apart from the Mediterranean and Red Sea depths, which are much warmer), with little variation in salinity and oxygenated via the deep currents − although O_2 levels are often quite low (around 5% of air saturation). The only significant abiotic gradient is in pressure, which increases by 10^5 Pa (1 atmosphere) for every 10 m increase in depth. Over the abyssal plains, the sea-bed is covered in a monotonous blanket of oozy mud, which has a low organic content and derives mainly from the calcareous or siliceous skeletons of plankton. Food resources in this heterotrophic realm are scarce. Since they derive from dead organisms and detritus raining down from the surface, there is some seasonal and spatial variation with occasional rich patches from large carcasses (e.g. fish, whales). The warm, mineral-rich waters of hydrothermal vents with their autotrophic communities are the only oases in what is otherwise a marine desert.

You can gain some idea of how knowledge and views about *deep-sea organisms* have changed over the last 20–30 years from Table 4.21. Perhaps the biggest surprise has been the finding that numbers of species in the deep sea are high, approaching the total of those in tropical rainforest (although spread over a much greater area). The main types are scavengers (equivalent to freshwater shredders), suspension-feeders, sediment collectors and predators. Highly specialized feeding mechanisms are rare: most organisms are opportunistic, flexible feeders and consume anything organic, dead or alive, that is of a suitable size.

Table 4.21 Changing views of deep-sea organisms. ? = no consensus view.

	pre-1970s	1990s
number of species	low	high
biomass/numbers of individuals	low	low, decreasing with depth
rates of metabolism and growth	uniformly low	variable
distribution patterns	?, probably uniform	some degree of clumping for most types
timing of reproduction	non-seasonal	seasonal in some invertebrates and many fish

Among the larger animals (megafauna), echinoderms are well-represented, especially brittle stars (Ophiuroidea) (Figure 4.66a) living on or near the sea-bed. Whereas many shallow-water echinoderms are predators, the scarcity of prey has led to the evolution of deposit-feeding among deep-sea echinoderms and a shift in feeding mechanism is common among other deep-sea invertebrates. Many bivalve molluscs, tunicates and sponges, all of which are filter-feeders in shallow water, are macrophagous carnivores in the deep sea (capturing and then digesting large particles or organisms, as do sea-anemones).

Other deep-sea megafauna include giant amphipods (Crustacea) up to 18 cm long, giant squid (cephalopod molluscs), and a range of fishes that live on or swim near the bottom. Predaceous fish have evolved some bizarre forms (Figure 4.66b), largely, it is thought, because of the need to minimize weight and energy requirements when living at such high pressures and with scarce prey.

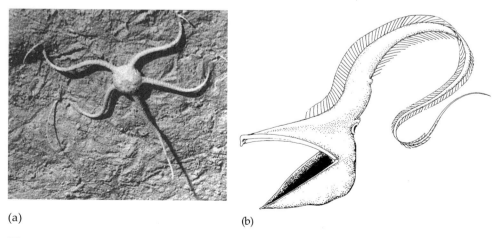

(a)   (b)

Figure 4.66 Larger deep-sea animals (megafauna): (a) a deposit-feeding brittle star (echinoderm), *Ophiomusium lymani* ($\times 0.3$); (b) a predaceous fish, the deep-sea gulper *Eupharynx pelecanoides* ($\times 0.25$).

Predaceous fish commonly adopt the low energy, sit-and-wait lifestyle of ambush predators, often using bioluminescent lures to attract prey; some have huge jaws capable of snapping up any prey that does come within range and sufficient energy reserves for one rapid lunge (Figure 4.66b). Among fish that swim actively, there seems to have been convergent evolution towards the slender, elongated body shape seen in Figure 4.66b and semi-transparent muscles with a high water content are characteristic of megafauna generally, probably as an adaptation that reduces weight and aids survival at high pressure.

Species richness is even greater among the smaller deep-sea animals (*macrofauna*), which mostly live in or on the sediments and include bristle worms (polychaete annelids), crustaceans, molluscs and various worms, in order of decreasing abundance (Figure 4.67). Compared with similar surface-dwelling species, body size in the macrofauna is generally smaller, a trend towards dwarfism that is the direct opposite of the trend to gigantism seen in the megafauna.

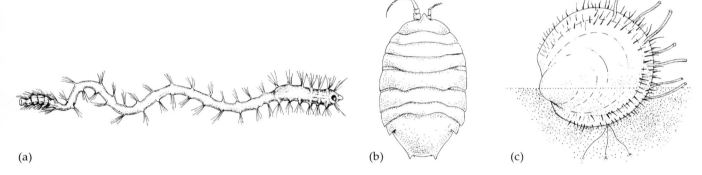

(a)   (b)   (c)

Figure 4.67 Smaller deep-sea animals (macrofauna): (a) a deposit-feeding polychaete worm (Annelida), *Aparaonis abyssalis* (× 1.3); (b) an isopod (Crustacea), *Haploniscus* (× 14); (c) a bivalve mollusc, *Limopsis cristata* (× 4) anchored by byssal strands (three shown here) to clumps of sediment.

Information about the physiology, life histories and general ecology of these deep-sea animals is still rudimentary but ecologists have still asked the usual questions about what determines their distribution and abundance. Unlike rocky shores (Chapter 3), there is no clear vertical zonation with depth in the deep sea.

❏   What factor does this rule out as a major determinant of distribution?

■   Pressure, which increases with depth.

This implies that many species are **eurybathic** (tolerating a wide range of pressures) and some of the megafauna certainly are whilst others (particularly among the echinoderms) are relatively **stenobathic** (tolerating only a small range of pressures). There is, however, a general decrease in total biomass with depth, for which there is still no clear explanation, although it might be related to a decrease in available food: lowest biomass occurs under the centres of oceanic gyres, which are areas of very low surface production. Other relevant factors are the low average rate of metabolism (respiration), slow reproduction and longevity of many deep-sea organisms. Respiration rate may be constantly low (e.g. bacteria and some pelagic fishes) or vary between 'normal' (similar to surface species) and very low, corresponding to periods of movement and torpor (e.g. some fish and scavenging amphipod crustaceans), or show no decline relative to surface species (some echinoderms, gelatinous pelagic groups and hydrothermal vent organisms).

The general picture is that many deep-sea animals are globally cosmopolitan: you can find the same deep-sea species in the tropics and in polar waters. Some local differentiation has been detected, for example the fauna of very deep trenches (Figure 4.61) often differs from that of the surrounding abyssal plain and may contain many **endemic species** that occur nowhere else. There is also good evidence of clumped distributions for some species on scales ranging from centimetres to hundreds of metres (Gage and Tyler, 1991) but only speculation as to why.

❑ Suggest at least three possible explanations for clumped distribution.
■ 1 Resources, e.g. food, might be patchily distributed.
2 Conditions, e.g. physical nature of the sea-bed, might vary.
3 It might reduce the risk of predation (i.e. biotic interactions).
4 It might increase the chances of finding a mate.

The strength of biotic interactions such as predation and competition can still only be guessed for the deep sea. There are important practical reasons for finding out more because human interference – even in the deep seas – is increasing. Deep-sea mining of metal-rich sediments and nodules on the sea-bed is already being explored; the deep sea has long been used as a disposal site for sewage, dredging spoil, industrial waste and low-level radioactive waste; and there is growing presssure from commercial fishermen to harvest deep-sea fish. To assess fully the impact of these activities requires much more knowledge of deep-sea ecology than we currently have and, in particular, information about the regulation of population size and community structure (the topics of Books 2 and 3) and about how deep-sea ecosystems function (the topic of Book 4).

Summary of Section 4.11

• The food supply of open-water marine animals is patchy. Areas of high algal production, on which most food-chains depend, occur in scattered **upwelling** areas and in neritic waters around continents; upwelling is discontinuous and algal production is often highly seasonal. There is also much small-scale patchiness.

• Zooplankton often show diurnal **vertical migrations**, commonly feeding at the surface by night. They are moved laterally by deep currents during the day, which minimizes energy requirements because of the lower temperature, and ascend to a new patch of food at night.

• Polar zooplankton, such as the Antarctic krill, may show seasonal migrations with adults drifting polewards at the surface during summer and eggs and larvae drifting in the opposite direction near the sea-bed in winter.

• Nekton often undergo migrations that exploit patchy food supplies and ocean currents. Adult fish may move between rich feeding grounds and spawning sites from which planktonic young drift to a third, protected nursery site. Antarctic fish are mostly benthic with a specialized stenothermal physiology; they do not migrate to lower latitudes but their young may migrate vertically to the surface.

• In contrast to the patchy nature of the surface waters, the **deep sea** shows little variation with latitude or season. Increasing pressure with depth is the only physical gradient and food (detritus from the surface) is scarce except around **hydrothermal vents**, where chemosynthetic producers support food-chains.

- In the deep sea, the number of species is higher, distributions are more clumped and reproduction more often seasonal than previously thought. Little is known about the strength of biotic interactions. Gigantism is common in the invertebrate megafauna and dwarfism in the macrofauna. Fish are often ambush predators. Biomass correlates roughly with surface production and decreases with increasing depth. Many deep-sea species are cosmopolitan with endemic species apparently confined to deep trenches.

Question 4.21 (Objectives 4.1 & 4.12)

(a) Identify two factors that may have led to the evolution of diurnal vertical migrations in marine zooplankton.

(b) Sponges were discovered in shallow Mediterranean marine caves that were macrophagous predators, capturing small crustaceans by means of filaments covered in hooked spicules. The only other sponges that feed in this way live in the deep sea and are closely related to the cave sponges. What does this indicate about the selective pressures that have led to the evolution of a different feeding method in the deep sea?

Question 4.22 (Objective 4.12)

Mammalian nekton, such as whales and seals, often feed in polar waters during summer and then produce young at a nursery site in temperate or tropical latitudes. Which arm of the migration triangle (Figure 4.65) do these migrations correspond to? Why are the other two arms omitted?

Question 4.23 (Objectives 4.1 & 4.14)

Explain which of statements (a)–(d) about the deep sea are true, partially true or false.

(a) The deep sea is a heterotrophic habitat where there is a universal shortage of food.

(b) Deep-sea organisms are all relatively sluggish compared with surface-living species and tend to have low metabolic rates.

(c) Deep-sea predatory megafauna show a range of characteristics that may be related to shortage of prey and high pressure; these include possession of bioluminescent lures, huge jaws and watery tissues.

(d) Lack of knowledge about life histories and the strength of biotic interactions between deep-sea organisms mean that it is very difficult to predict the effects of human interference.

Objectives for Chapter 4

After completing Chapter 4 (and the associated video 'Soils' and AC programme 'Understanding soils'), you should be able to:

4.1   Recall and use in their correct context the terms shown in **bold** in the text. (*Questions 4.1, 4.4, 4.8, 4.11, 4.14, 4.18, 4.21 & 4.23*)

4.2   Describe or recognize the main components of soil, the characteristics of different soil types and the processes that affect soil development and properties. (*Questions 4.1, 4.5 & 4.7*)

4.3   Recognize the ways in which the chemical and physical properties of soils affect the supply of mineral nutrients and living conditions for plants and animals. (*Questions 4.1, 4.5 & 4.10*)

4.4   Describe and identify from relevant data the three types of mechanism by which plants obtain adequate mineral nutrients when these are in very short supply. (*Question 4.3*)

4.5   Given relevant information about the structure, growth or physiological tolerance of named organisms, predict whether they are likely to be found on a defined type of soil, saline habitat or zone in a saltmarsh. (*Questions 4.2, 4.5, 4.6 & 4.8*)

4.6   Identify some of the major factors that affect survival and distribution in estuaries and saltmarshes. (*Question 4.10*)

4.7   Assess the suitability of or recommend management practices for defined purposes, including the reclamation of saline areas, increasing species richness in grassland and revegetating spoil heaps. (*Question 4.9*)

4.8   Describe and compare the major physical and chemical characteristics, including water movements, thermal stratification and chemical composition, of rivers, lakes and oceans. (*Questions 4.11 & 4.17*)

4.9   Describe how and for what reasons physical and chemical resources and conditions may vary in space and time for different types of water body. (*Questions 4.11 & 4.17*)

4.10 Describe and recognize examples of top-down and bottom-up control in aquatic habitats. (*Questions 4.12, 4.13 & 4.16*)

4.11 Explain and give examples to illustrate the effects in lakes and/or rivers of organic pollution, cultural eutrophication and anthropogenic acidification; assess critically management strategies intended to alleviate these effects. (*Question 4.13*)

4.12 Given relevant information about a water body, predict: (a) the types of phytoplankton that are likely to dominate at different times of year; (b) the characteristics of river organisms likely to occur there; and (c) the type of migration that may occur in zooplankton or nekton. (*Questions 4.14, 4.15, 4.18, 4.21 & 4.22*)

4.13 Given relevant information about aquatic organisms, predict: (a) the type of habitat or compartment of a water body where they are most likely to occur; (b) for animals, their functional feeding group. (*Question 4.19*)

4.14 Describe the main characteristics of the deep-sea environment and deep-sea organisms. (*Question 4.23*)

4.15 Interpret experimental data in ways that are consistent with the principles described in this Chapter. (*Questions 4.2, 4.6, 4.7, 4.10, 4.15, 4.16 & 4.20*)

References for Chapter 4

Adam, P. (1990) *Saltmarsh Ecology*, Cambridge University Press.

Arnott, S. E. and Vanni, M. J. (1993) Zooplankton assemblages in fishless bog lakes: influence of biotic and abiotic factors, *Ecology*, **74**, 2361–80.

Barnes, R. S. K. (1980) The unity and diversity of aquatic systems, in R. S. K. Barnes and K. H. Mann (eds) *Fundamentals of Aquatic Ecosystems*, pp. 5–24, Blackwell Scientific Publications.

Barnes, R. S. K. and Hughes, R. N. (1988) *An Introduction to Marine Ecology*, 2nd edn, Blackwell Scientific Publications.

Battarbee, R. W. *et al.* (1988) *Lake Acidification in the United Kingdom 1800–1986, ENSIS* Publishing, London.

Benndorf, J. (1992) The control of the indirect effects of biomanipulation, in D. W. Sutcliffe and J. G. Jones (eds), *Eutrophication: Research and Application to Water Supply*, Freshwater Biological Association, UK.

Bertness, M. D., Gough, L. and Shumway, S. W. (1992) Salt-tolerance and the distribution of fugitive salt marsh plants, *Ecology*, **73**, 1842–51.

Borchardt, D. (1993) Effects of flow and refugia on drift loss of benthic macroinvertebrates: implications for habitat restoration in lowland streams, *Freshwater Biology*, **29**, 221–27.

Castellanos, E. M., Figueroa, M. E. and Davy, A. J. (1994) Nucleation and facilitation in saltmarsh succession: interactions between *Spartina maritima* and *Arthrocnemum perenne*, *Journal of Ecology*, **82**, 239–48.

Choong, M. F., Lucas, P. W., Ong, J. S. Y., Pereira, B., Tan, H. T. W. and Turner, L. M. (1992) Leaf fracture toughness and sclerophylly: their correlations and ecological implications, *New Phytologist*, **121**, 597–610.

Crawford, R. M. M. (1992) Oxygen availability as an ecological limit to plant distribution, *Advances in Ecological Research*, **23**, 93–185.

Enright, N. J. and Lamont, B. B. (1992) Survival, growth and water relations of *Banksia* seedlings on a sand mine rehabilitation site and adjacent scrub-heath sites, *Journal of Applied Ecology*, **29**, 663–71.

Gage, J. D. and Tyler, P. A. (1991) *Deep-Sea Biology: A Natural History of Organisms at the Deep-Sea Floor*, Cambridge University Press.

Grundon, N. J. (1972) Mineral nutrition of some Queensland heath plants, *Journal of Ecology*, **60**, 171–81.

Hildrew, A. G. and Townsend, C. R. (1987) Organization in freshwater benthic communities, in J. H. R. Gee and P. S. Giller (eds), *Organization of Communities Past and Present*, pp. 347–71, Blackwell Scientific Publications.

Kazda, M. (1995) Changes in alder fens following a decrease in the ground water table: results of a geographical information system application, *Journal of Applied Ecology*, **32**, 100–10.

Marrs, R. H. and Bannister, P. (1978) Response of several members of the Ericaceae to soils of contrasting pH and base-status, *Journal of Ecology*, **66**, 829–34.

McLachlan, A. J. and Cantrell, M. A. (1980) Survival strategies in tropical rain pools, *Oecologia*, **47**, 344–51.

McNaughton, S. J. (1988) Mineral nutrition and spatial concentrations of African ungulates, *Nature*, **334**, 343–45.

Ormerod, S. J., O'Halloran, J., Gribbin, S. D. and Tyler, S. J. (1991) The ecology of dippers *Cinclus cinclus* in relation to stream acidity in upland Wales: breeding performance, calcium physiology and nestling growth, *Journal of Applied Ecology*, **28**, 419–33.

Pigott, C. D. and Taylor, K. (1964) The distribution of some woodland herbs in relation to the supply of nitrogen and phosphorus in the soil, *Journal of Ecology*, **52** (Supplement), 175–85.

Port, G. R. and Thomson, J. R. (1981) Outbreaks of insect herbivores in plants along motorways in the United Kingdom, *Journal of Applied Ecology*, **17**, 649–56.

Proctor, J., Johnston, W. R., Cottam, D. A. and Wilson, A. B. (1981) Field capacity water-extracts from serpentine soils, *Nature*, **294**, 245–46.

Reynolds, C. S. (1984) *The Ecology of Freshwater Phytoplankton*, Cambridge University Press.

Reynolds, C. S. (1995) River plankton: the paradigm regained, in D. M. Harper and A. J. D. Ferguson (eds), *The Ecological Basis for River Management*, pp. 161–74, John Wiley & Sons Ltd.

Robinson, D. (1994) Temperature and Exercise, in S324 *Animal Physiology*, Book 2, Ch. 5, The Open University.

Rorison, I. H. (1960a) Some experimental aspects of the calcicole–calcifuge problem: I The effects of competition and mineral nutrition upon seedling growth in the field, *Journal of Ecology*, **48**, 585–99.

Rorison, I. H. (1960b) Some experimental aspects of the calcicole–calcifuge problem: II The effects of mineral nutrition on seedling growth in solution culture, *Journal of Ecology*, **48**, 679–88.

Snow, A. A. and Vince, S. W. (1984) Plant zonation in an Alaskan salt marsh. II An experimental study of the role of edaphic conditions, *Journal of Ecology*, **72**, 669–84.

Snowdon, R. E. D. and Wheeler, B. D. (1993) Iron toxicity to fen plant species, *Journal of Ecology*, **81**, 35–46.

Stanford, J. A. and Ward, J. V. (1988) The hyporheic habitat of river ecosystems, *Nature*, **335**, 64–66.

Treseder, K. K., Davidson, D. W. and Ehrleringer, J. R. (1995) Absorption of ant-provided carbon dioxide and nitrogen by a tropical epiphyte, *Nature*, **375**, 137–39.

Tyler, G. (1994) A new approach to understanding the calcifuge habit of plants, *Annals of Botany*, **73**, 327–30.

Answers to Questions

CHAPTER 1

Question 1.1

1   Table A1 is a completed version of Table 1.3.

Table A1   Completed version of Table 1.3.

| | Effect of the rabbit: | | |
	Positive/negative	Direct/indirect	Evidence
Plants:			
Anthoxanthum odoratum	negative	direct	observation
Festuca ovina	positive	indirect	observation/ experiment
Cynosurus cristatus	positive	indirect	observation
Carex spp.	positive	indirect	observation
Sanguisorba minor	positive	indirect	observation
Rubus fruticosus	negative	direct	observation
Cirsium acaule & C. palustre	positive	indirect	experiment
Cornus sanguinea	negative	direct	observation
Crataegus monogyna	negative	direct	observation
Animals:			
detritivores:			
Ceratophyus typhoeus	positive	direct	natural history
Armadillidium vulgare	negative	indirect	experiment
herbivores:			
brown hare *Lepus europaeus*	negative or positive	indirect	observation
mountain hare	negative	indirect	natural history
red deer *Cervus elaphus*	negative	indirect	natural history
Chorthippus brunneus	negative	indirect	experiment
roe deer	negative	indirect	natural history
short-tailed field vole	negative	indirect	natural history
carnivores:			
fox	positive	direct	natural history
buzzard *Buteo buteo*	positive	direct	observation
badger	positive	direct	natural history
stoat	positive	direct	natural history
weazel	positive	direct	natural history
other animals:			
rabbit flea *Spilopsyllus cuniculi*	positive	direct	natural history
stone curlew	positive	direct	natural history
wheatear	positive	direct	natural history
large blue *Maculinea arion*	positive	indirect	observation
Myrmica sabuleti	positive	indirect	observation

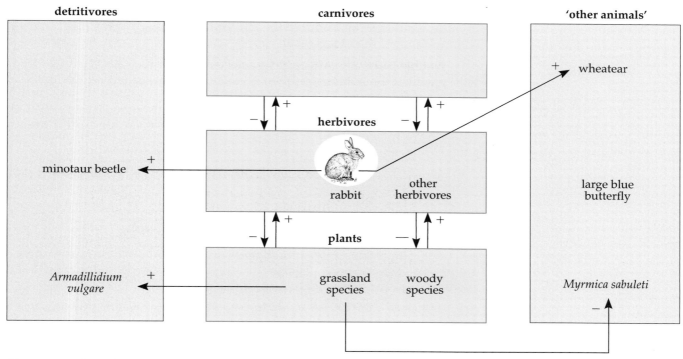

Figure A1   Completed version of Figure 1.4.

2   Figure A1 is a completed version of Figure 1.4.

(a)   Carnivores and herbivores each have negative effects upon their food organisms, while plants and herbivores both have positive effects upon the organisms that eat them.

(b)   Unlike the relationships between carnivores and herbivores and between herbivores and plants, the direct relationships between the rabbit and the minotaur beetle (a detritivore) and the rabbit and the wheatear (an 'other animal') are one-way and both positive.

(c)   Unlike the relationships between carnivores and herbivores and between herbivores and plants, all the indirect relationships are one-way. The indirect effect of the rabbit on *Myrmica sabuleti* is positive, and in the longer term so is the indirect effect of the rabbit on grassland plants whose competitors (particularly woody plants) are kept in check by rabbit grazing. The indirect effect of the rabbit on *Armadillidium vulgare* is negative.

3   The relationships between species *within* the groups shown in Figure 1.4 should be either neutral, or negative arising from competition (except among the miscellaneous 'other animals'). For example, there seems to be no evidence of *direct* interaction between the rabbit and brown hare, but on the other hand the plant species appear to compete with each other.

Question 1.2

(a)   Ecology is the study of interactions between organisms and their environment, as exemplified by the effect of the rabbit on other animals and plants when rabbit numbers were reduced by myxomatosis. (Many other examples of ecological interaction may, of course, have been given.)

(b)   Levels of organization refer to the different scales at which ecological investigations may be directed: the individual, the population, the community and the ecosystem.

(c)   Trophic levels are groups of species defined by their source of food, with plants at the bottom, herbivores consuming plants and carnivores consuming herbivores.

(d)   The term 'habitat' describes where a named species is to be found, while the term 'landscape' refers to a larger area that may contain some patches suitable and some unsuitable as habitat for various species.

Question 1.3

(a) $-/+$. The botflies are parasites of oropendolas.

(b) $-/+$. The cowbird chicks are predators of the botflies.

(c) $-/+$. The cowbirds are brood parasites of the oropendolas when these are protected from botflies by bees or wasps. When oropendolas reject cowbird eggs, the relationship is competitive ($-/-$).

(d) $+/+$. The cowbirds and oropendolas are mutualists when botflies are present.

(e) $+/+$, $+/0$ or even perhaps $+/-$. It is fairly clear that the oropendolas benefit from the association but the benefit or loss to the insects is not known. Perhaps it is most likely that they are predators on botflies, in which case they will benefit from the presence of oropendolas if these attract more botflies to the area.

(f) $+/-$, or $0/-$ depending upon whether wasps and bees prey upon botflies or just keep them away for some reason.

(g) $-/0$. When protected from botflies by wasps or bees, oropendolas scare off cowbirds.

(h) $+/-$, because the trees provide the discriminators with a habitat but the birds cause the limbs of the trees to snap off. On the other hand, the oropendolas probably provide useful nutrients to the tree in their droppings so the relationship could be a mutualistic one.

Question 1.4

(a)   Lichen symbioses contain a fungus and a green alga or a cyanobacterium (sometimes both). The fungus is the larger member and acquires carbon fixed by photosynthesis from the alga or nitrogen and carbon from the cyanobacterium.

(b)   Ectomycorrhizas mainly form between the roots of temperate forest trees and basidiomycete fungi. The trees obtain minerals, particularly P, from the fungi.

(c)   Vesicular–arbuscular mycorrhizas form between phycomycete fungi and a very wide variety of plants which obtain minerals, particularly P, from the fungi.

(d)   Root nodules form on the roots of plants in the legume family (and less commonly in some other species) and contain symbiotic bacteria that fix atmospheric nitrogen which is supplied to the plant.

(e)   Green *Hydra* spp. are cnidarians that contain symbiotic green algae which provide the animal with carbon fixed by photosynthesis.

(f)   Reef-building corals are cnidarians that contain symbiotic dinoflagellate algae in the genus *Symbiodinium*. These provide the animal with carbon fixed by photosynthesis and possibly organic N.

(g)   Ruminants are mammalian herbivores with a fermentation chamber called the rumen in their guts. The rumen contains bacteria and flagellated protistans that are able to digest cellulose. The short-chain fatty acids produced by this are used by the animal.

(h)   Wood-eating lower termites have flagellates or a mixture of bacteria, amoebae and ciliates in the hind part of the intestine that are essential to their ability to digest the wood on which they feed.

(i)   Blood-feeding insects such as bed-bugs possess special groups of cells called mycetocytes which contain bacteria that provide the animal with vitamins.

Question 1.5

(a)   Hypothesis (1), that VAM infection improved the phosphorus nutrition of plants, is not supported by the data because Figure 1.23 shows no positive effect of VAM infection upon P inflow. Furthermore, if VAM affected plant nutrition, you would expect the experimental removal of VAM infection to cause a significant decrease in the dry weight of shoots, but this did not occur (Table 1.6).

(b)   Hypothesis (2), that VAM protected plants' roots from infection by *Fusarium oxysporum*, is supported, though only indirectly. The application of benomyl reduced the abundance of *F. oxysporum*, but this apparently had no effect upon plant growth. This could mean that neither VAM infection (also reduced by benomyl) nor the presence of *F. oxysporum* had any effect on plants, or that the presence of the two fungi had opposite effects so that the removal of both cancelled out any overall effect of benomyl on *V. ciliata*.

(c)   Many other experiments and observations could be useful, for example:

(i)   P inflow could be compared between controls and plants treated with benomyl. In order to test for possible nutritional benefit alone, this should be done in the absence of *Fusarium* infection.

(ii)   Mycorrhizal and non-mycorrhizal plants could be exposed to infection by *Fusarium oxysporum* and their growth compared.

Question 1.6

(a)   Secondary hosts may harbour an essential developmental stage in the parasite's life cycle, and they also play an important role in the transmission of many endoparasites between primary hosts. For example, *Plasmodium* and *Trypanosoma* are transmitted by insect vectors and the dog tapeworm *Echinococcus granulosus* uses other mammals as a secondary host.

(b)   Dutch elm disease is transmitted by the bark-beetles and myxomatosis by the rabbit flea (in Britain).

(c)   Ectoparasitic animals mentioned are: crab, head and body lice, the sheep tick, feather mites, the cirripedes *Rhizolepus* and *Anelasma*. Ectoparasitic plants mentioned are common dodder, mistletoe, broomrape and the hemiparasites *Rhinanthus*, *Euphrasia* and *Striga*.

Question 1.7

(a)   (i) Flowers visited by bats tend to open at night, have exposed anthers and stamens, a strong odour and copious nectar. (ii) Flowers visited by birds open by day, tend to lack scent but are often scarlet in colour and tubular or trumpet-shaped. (iii) Butterflies particularly visit flowers that are tubular in shape and red, blue or yellow in colour. (iv) Bees are attracted to flowers that appear to us as blue, yellow or white. (v) Beetles, flies and wasps tend to visit drab-coloured or white flowers that are cup-shaped.

(b)   The relationship is mutualistic (+/ +) and highly specific, as it seems that neither species can reproduce without the other.

(c)   All three *Prodoxus* species are parasites of *Y. whipplei* (+/ −).

(d)   The pod-feeding species of *Prodoxus* is dependent upon *Tegiticula* because without its pollinating activities there would be no pods in which the larvae could feed. As this *Prodoxus* sp. and *Tegiticula* feed on different tissues in the pod, there is unlikely to be competition between them so the relationship is (+/0). The other *Prodoxus* spp. are not dependent upon pods so they have no direct relationship with *Tegiticula* (0/0). On the other hand, they are dependent upon *Yucca* so it may be considered that all the *Prodoxus* species have a (+/0) association with *Tegiticula*.

It is worth noting how like parasitism the feeding behaviour of *Tegiticula* larvae is and how small a difference in feeding site exists between the mutualist, and parasitic moths of yucca which feed in its pods This suggests that one of these symbioses may have evolved from the other.

(e)   Epibionts use other organisms for living space. They are not parasites because they do not obtain their nutrition at the expense of a host.

(f)   Epiphytes mentioned are: tropical orchids, lichens, mosses, ferns, bromeliads including *Tillandsia* spp. Epizoites mentioned are: the many that live on the jewelbox clam *Chama pellucida*, sea-anemones living on the shells of hermit crabs, and hydroids on the fish *Minous inermis*.

(g)   Defence and shelter associations mentioned are: the woodlouse *Platyarthrus hoffmannseggii* living in nests of the common yellow ant *Lasius flavus*; various ants associated with various trees; fish living among the tentacles of jellyfish, anemones or the spines of sea-urchins; gobies and other animals living in the burrows of the worm *Urechis caupo* and the pistol shrimp *Alpheus djiboutensis* (Figure 1.51); the hermit crab *Eupagurus bernhardus* that carries the sea-anemone *Calliactis parasitica* on its shell.

(h)   Cleaners mentioned are: tick-pickers on wildebeest, the Nile crocodile's cleaner bird, the cleaner wrasse, the Pederson shrimp.

CHAPTER 2

Question 2.1

(a)   The two mechanisms are interference and resource competition. In interference competition, species physically interfere with their competitor's access to resources. Examples mentioned are: three-spot damselfish removing sea-urchins from the algal lawn where fish and sea-urchins feed; encrusting organisms on rocks in the sea that grow over the top of each other or sting each other; carrion beetles whose phoretic mites attack eggs and small larvae of bluebottles which compete with the beetles for food; antibiotics produced by fungi against other fungi.

In exploitation competition, species compete with each other by depleting resources. Examples mentioned are: competition between tits and goldcrests, and competition between plants for light (though there may be an element of interference in this, too).

(b)   Apparent competition occurs when species have negative effects on each other because they share a common enemy such as a parasite. Examples are difficult to verify, but possibly include two leafhoppers (*Elegantula* spp.) sharing the parasitoid *Anagros epos* and the alternate hosts of white pine blister rust.

(c)   A limiting resource is anything whose limited supply can lead to one species negatively affecting another when the resource is consumed.

(d)   A species' niche is that part of a limiting resource that it is able to exploit better than its competitors.

(e)   A species' fundamental niche is the set of resources that a species could potentially utilize in the absence of competing species. The realized niche is the set of resources actually used when competing species are present.

Question 2.2

(a)   The observation that clams and snails are negatively associated (the former preferring boulders and the latter cobbles) could itself be interpreted as circumstantial evidence of competition, but in fact there is no direct evidence that this is the cause. On the contrary, the fact that snails and clams feed in quite different ways and occur in different places suggests that competition between them is most unlikely.

(b)   As with the evidence for the competition hypothesis, the observation that clams 'prefer' boulders and snails 'prefer' cobbles could itself be interpreted as circumstantial evidence that the two groups select different habitats, but in fact there is no independent evidence that this happens.

(c)   Apparent competition can occur when two or more prey share a common predator. Although the predators prefer clams to snails (i), they do attack both and may find snails easier prey (iii). On boulder substrata, there were more predators and fewer snails where clams were present than where they were absent (iv). This suggests that predators attracted to areas where their preferred prey (clams) are available produce lower snail densities in these areas. The experiment (v) showed that this also occurred on cobble substrata, and taken together (iv + v) this evidence suggests that bivalves have a negative effect on snails because they raise the local density of a common predator to which snails are more vulnerable (iii). This is good though only partial evidence for the apparent competition hypothesis.

(d)   The apparent competition hypothesis is the best supported of the three.

However, experiments did not demonstrate that snails have a negative effect on clams, and that this was due to predation. Two experiments would test this: (1) a snail-addition experiment, reciprocal to the bivalve-addition experiment; (2) repetition of both the bivalve- and snail-addition experiments with and without the presence of predators which could be excluded by cages.

Question 2.3

(a) When ants were present, aphids were much more heavily parasitized by the more important of the two parasitoids (Ti). This refutes the hypothesis.

(b) Parasitized aphids produced more honeydew (i), and therefore *L. cardui* benefited *L. niger*. The fact that more ants were attracted to parasitized colonies (ii) supports this.

Ants also protected *L. cardui* from its hyperparasitoids (Tiv), and *L. cardui* females foraged more in ant-attended colonies (Tiii). Therefore, the overall relationship was a *mutualistic* one.

(c) Ants attacked foraging females of *Trioxys angelicae* (iii), which therefore parasitized very few ant-attended colonies (Tii). In ant-free colonies,
T. angelicae experienced very high rates of hyperparasitization (Tv). Unlike the relationship between ants and *L. cardui*, this relationship was therefore *not mutualistic*.

In fact, *L. niger* and *T. angelicae* were competitors for a common resource (aphids) and the fact that ants attacked *T. angelicae* (iii) indicates that there was some interference competition, too.

(d) Hyperparasitoids were apparently less common in ant-attended colonies (Tiv), but there is no *direct* evidence that hyperparasitoids had any effects on ants. This would suggest a −/0 relationship. However, hyperparasitoids kill *L. cardui* parasitoids which are mutualistic with ants. One might therefore regard the ant–hyperparasitoid relationship as −/−, though this is a rather tenuous conclusion.

Question 2.4

(a) (i) Parasitoids are insects whose parasitic larvae develop in the bodies of other insects. Examples mentioned in the text are *Anagrus epos* attacking leafhoppers in California; chalcid wasps in the genus *Trichogramma* that parasitize insect eggs; *Apanteles* attacking cabbage white caterpillars; and braconid wasps in the genus *Aphidius* that attack aphids.

(ii) Top carnivores are predators in the highest trophic level of a community. Examples mentioned in Section 2.2 are wolves, bears, foxes, lynx, lions, leopards, cheetahs and wild dogs.

(iii) Raptors are birds of prey. Examples mentioned in the text are marsh harrier, honey buzzard, goshawk, osprey, white-tailed eagle, buzzard, hen harrier, red kite, golden eagle, merlin, peregrine and sparrowhawk.

(b) The terms monophagous, oligophagous and polyphagous refer to the breadth of diet in animals that feed respectively upon prey belonging respectively to one, several or many species (or sometimes families, depending on the author).

Question 2.5

(a) Lynx are the main predators of snowshoe hares in mainland Canada. Snowshoe hares probably increased their numbers rapidly when introduced into Newfoundland because there were very few lynx there.

(b)   The increase in lynx followed the introduction of snowshoe hares which elsewhere are their main prey species. Lynx numbers increased because the population of their prey was large.

(c)   The most likely explanations are that the snowshoe hares ate out their food supply and/or lynx numbers had built up to such a level by 1915 that their predation on snowshoe hares meant that few hares survived to breed.

(d)   Lynx eat arctic hares in Newfoundland but elsewhere in Canada the two species do not occur together (ii). The experimental introductions of arctic hares onto islands (vii–ix) show that predation by lynx is much more important to arctic hare numbers than is competition from snowshoe hares. Hence, the rarity of the arctic hare since 1915 is probably due to lynx predation. These predators had subsisted on snowshoe hares but when that population crashed the lynx probably *switched* to arctic hares as prey. The recent rarity of arctic hares is therefore an indirect result of the introduction of snowshoe hares in 1864 and of *predator switching* by lynx when snowshoe hare numbers crashed. Some additional information indirectly corroborates this explanation. Snowshoe hares and lynx both have feet that enable them to run fast on loose-packed snow while arctic hares have feet which exert more pressure on the ground; this enables them to run fast in the arctic tundra but not in southern Canada and Newfoundland where winter snow is soft. So arctic hares may be confined to areas where they can escape lynx because of a better turn of speed.

(e)   The fall in caribou numbers in about 1920 was probably also a consequence of lynx *switching* from snowshoe hares and then arctic hares to caribou. Experiment (vi) shows that lynx can seriously affect recruitment to caribou populations. The slight delay between the fall in snowshoe hare numbers and the fall in caribou numbers may reflect the longer generation time of caribou and the fact that lynx only prey on young caribou calves. Some of the fall in caribou numbers around 1920 was caused by Newfoundlanders hunting adult caribou. The predation on calves by lynx would have reduced or prevented the replacement through recruitment of adults killed by hunters.

(f)   The cycling of lynx and snowshoe hares in Newfoundland today (v) suggests that these populations have reached the same state as in the rest of Canada where coupled oscillations between them are found. These oscillations are probably the result of time-lags between changes in prey density and its effects upon the predator which produce delayed density-dependence. This tends to destabilize populations.

(g)   Variation from year to year in the survival of caribou calves is probably due to variation in lynx predation because this is the most important source of mortality for caribou in Newfoundland (iv). Variation in the predation rate by lynx is in turn probably due to predator switching between snowshoe hares (which cycle in abundance) and caribou calves and the cycles in abundance of lynx will also cause variation in predation upon caribou from year to year.

Question 2.6

(a)   Non-chemical defences against herbivores include: spines, stinging hairs, tough bark, waxy leaves with thick cuticle, latex, the presence of lignified fibres and silica, symbiotic associations with ants and mimicking inedible plants or stones.

(b)   Secondary compounds may reduce the digestibility of plants or they may be toxic to herbivores. Tannins and other phenolic compounds are examples of digestibility-reducers. Alkaloids, cyanogenic glycosides, glucosinolates and other nitrogen-containing chemicals are examples of toxic secondary compounds.

(c)   The fungal endophytes of grasses produce alkaloids that are toxic to herbivores.

Induced defences increase a plant's defences against further attack after a plant is damaged.

Phytoalexins are defensive chemicals induced by fungal or bacterial infection.

Compensatory regrowth is the chief means by which plants tolerate herbivory.

(d)   Specialist insect herbivores may use the smell of their host plant's defensive compounds to locate the plant; they may use plant secondary compounds as pheromones to attract other insects; and they may use secondary compounds in their own defence.

Question 2.7

(a)   In the experiment (Table 2.6), *Pieris rapae* and *P. melete* both grew significantly better on their usual food plants than on the plants they normally avoid. This supports hypothesis (i) that diet breadth is determined by larval performance. By contrast, *P. napi* avoided some plant species (*Brassica pekinensis* and *Rorippa indica*) on which its larvae grew significantly better than on its chosen host plant and also avoided others on which it grew just as well.

(b)   Rates of parasitism were quite high on all or several of the food plants chosen by *Pieris rapae* and *P. melete* (Table 2.7), suggesting that hypothesis (ii) is wrong. Compared to the other butterflies, *P. napi* had a very low rate of parasitism (5%) on its food plant, suggesting that in this instance hypothesis (ii) might be correct.

(c)   The data suggest that hypothesis (i) is correct for two species and that hypothesis (ii) is correct for the third. However, a proper test of hypothesis (ii) would require the experimental measurement of parasitism in larvae feeding on non-host plants, so these could be compared with rates on host plants. Although this experiment was not performed, Ohsaki and Sato (1994) did carry out another experiment on the cause of low parasitism in *P. napi*. *Arabis gemmifera*, this butterfly's food plant, is a perennial which grows in habitats where it is concealed among other plant species. Ohsaki and Sato cleared the neighbours from around some *A. gemmifera* in the field and compared the parasitism rate on larvae of *P. napi* feeding on the exposed plants with unmanipulated controls. The parasitism rate increased from 5% in controls to 85% in exposed plants. This value is at the upper end of the range of parasitism rates found in the other species of *Pieris* (Table 2.7), whose food plants tend to grow in the open, not concealed by vegetation.

CHAPTER 3

Question 3.1

(a)   Common limpets occur from lower to upper shore in temperate regions but are most abundant in the middle shore (Figure 3.7). Density is highest on semi-exposed to exposed shores and decreases on the most exposed and most sheltered shores (Section 3.2.4).

(b)   Kelps (*Laminaria* spp.) are large brown seaweeds that occur in the sub-littoral and the lowest parts of the lower shore (Sections 3.2.2, 3.2.4). The species present vary with exposure (Figures 3.6 and 3.12), *Alaria esculenta*, for example, occurring on very exposed shores and *Laminaria saccharina* on very sheltered shores.

(c)   *Pelvetia canaliculata* (channelled wrack) is usually the dominant alga on the upper shore for all except the most exposed sites (Figure 3.5). It may be sparse on exposed/semi-exposed shores because of grazing (Figure 3.12).

(d)   Two or three species of barnacle occur on British rocky shores. On very exposed sites, *Chthamalus* spp. occupy the middle and upper shore, extending into the splash zone; on moderately exposed shores (where barnacle cover can be very high), *Semibalanus* occurs mainly in the middle shore and *Chthamalus* spp. mainly in the upper shore; and on very sheltered shores (where barnacle cover is low and patchy), either *Chthamalus* or *Semibalanus* may be found from lower to upper shore with the latter more likely to be found in the lower part of this range. See Sections 3.2.3(2), 3.2.4 and Figures 3.7 and 3.12.

Question 3.2

(a) Two biotic interactions are important in determining this boundary (Section 3.2.3(1)) :susceptibility to fungal infection when submerged for too long; and shading (competition for light) by faster-growing fucoids.

(b) In Britain, this boundary is determined primarily by physical factors that affect the risk of desiccation (Section 3.2.3(2)). You might, therefore, have guessed that degree of shore exposure and aspect may influence its position: the boundary tends to be higher on north-facing shores and on exposed shores.

Question 3.3

(a)   Barnacles colonized the plots soon after clearance. *Chthamalus* was dominant for the first year but declined after 20 months as *Balanus* cover increased. Switchover between barnacles was most probably due to interspecific competition, as found for *Semibalanus balanoides* and *Chthamalus montagui* (Section 3.2.3(2)). The implication is that C colonizes and/or grows faster than B, but is then ousted by interference competition and other experiments by Farrell supported this view.

(b)   Figure 3.14a shows that algal cover appeared only after 20 months and increased once *Balanus* was the dominant barnacle, suggesting a positive effect of *Balanus* on algae. Observations (i) and (ii) support this idea because very few algae became established on cleared plots if they were kept free of *Balanus*, whereas removing *Chthamalus* but not *Balanus* had no effect on algal density.

(c)   Relevant information is mostly in Figure 3.14c, which provides some support for hypothesis 1 (in the absence of limpets, more algae settled and grew in the presence of barnacles (−L+B) than in their absence (−L−B)); and some for hypothesis 2 (in the presence of barnacles, more algae grew when limpets were absent (−L+B) than when they were present (+L+B)). So both mechanisms contribute to barnacle facilitation of algal growth and the data do not allow you to decide which mechanism is more important.

(d)   Figure 3.14b shows that limpets decrease the overall rate of change. *Balanus* cover (and linked with this, algal cover) increased more rapidly in limpet removal plots and *Chthamalus* cover decreased more rapidly. The implication is that limpet grazing interfered more severely with *Balanus* settlement than with *Chthamalus* settlement – but there is no information about this.

Question 3.4

(a) and (e) are true (Section 3.4.3) and the rest are false.

(b)   Desiccation or lack of oxygen is more likely to cause death at high temperatures than is enzyme inactivation (Section 3.4.2).

(c)   Stenothermal organisms can persist in a dormant state during periods of unfavourable temperatures, or they can migrate.

(d)   Endothermy is a physiological (not a behavioural) mechanism that maintains body temperature (Sections 3.4.3 and 3.4.4).

Question 3.5

Since *Geum turbinatum* reproduces by seeds and does not flower above 15 °C, temperature acting on reproduction is likely to limit distribution. Its habitat must, therefore, have low summer temperatures and at least five weeks below freezing in winter (to break seed dormancy). The only habitat which meets these requirements *and* has a wide latitudinal distribution is an *alpine or sub-alpine site* (recall that temperature falls with altitude, Section 3.4.1): the common name for this species is alpine avens!

Question 3.6

(a)   From Figure 3.32a, two differences are clear. First, the length of the growing season is shorter at the late-melting site. This is an effect of temperature acting as a *physical condition* (Table 3.4) and influencing the day-degrees available for growth. Secondly, fewer soil nutrients are available at the late-melting site, so there is a difference in *nutrient resources*. Other differences not shown in Figure 3.32a may exist. For example, the late-melting site is at the bottom of a hollow so there could be more water available in summer (a chemical resource) and/or the soil might be waterlogged here or differ in some other way as a chemical or physical condition.

(b) (i)   At the late-melting site, temperature through its effects on the length of the growing season appears to be the dominant factor influencing plant cover. Artificially increasing the time for growth by 12 days at the late-melting site increased plant cover by about threefold to a value that was not significantly different from that at the early-melting site (although it was still somewhat lower). However, the converse did not hold: decreasing time for growth by 12 days had no significant effect on plant cover at the early-melting site (although there was some decrease) and temperature cannot, therefore, be the dominant factor influencing growth here. The most likely explanation is that temperature (length of growing season) determines growth made when nutrient supply is low (as at the late-melting site) but not when it is high (as at the early-melting site).

(ii)   From Figure 3.32c, it appears that temperature (time for growth) is the major factor influencing seed weight at both early- and late-melting sites. There is a perfect reciprocal effect: increasing the growing season by 12 days results in an increase in seed weight at the late-melting site to the same value as at the early-melting site control plot; decreasing the growing season by 12 days decreases seed weight at the early-melting site to the same value as at the late-melting control plot.

(c)   Temperature conditions, specifically the length of the growing season, interact with other factors, probably the supply of soil nutrients, to influence the growth of snow buttercups. Temperature appears to be the dominant factor when nutrient supply is low. Temperature also influences seed weight and, since this affects the chance of seedling establishment, the fecundity of snow buttercups. Since heavy seeds are less easily dispersed, temperature also affects the chance of dispersal to new sites. Snow buttercups are probably unable to persist in colder sites at higher altitude because of poor growth and low seed weight. There appears to be no abiotic explanation for their absence outside snowbeds (they clearly grow better and produce heavier seeds at the outer edges of the beds), so biotic interactions are probably involved: inability to compete or heavier grazing pressure are possible explanations.

Question 3.7

The second statement is true for all of (a)–(c) and for (b) and (c) it adequately explains the first statement. For (a), however, statement 2 is unlikely to be the full explanation of statement 1. Besides reducing water loss, conditions of temperature and humidity are changed in the lee of a shelter belt and these could also contribute to increased crop yields.

Question 3.8

(i)   The direction and strength of wind will influence distribution of trees with wind-blown seeds, such as birch and willow.

(ii) Tree growth may be influenced downwind of pollution sources (e.g. factory chimneys) or sources of nutrients (e.g. large manure heaps).

(iii) Catastrophic windthrow can influence all trees in an area or act selectively so that less susceptible species become relatively more abundant.

(iv) Trees may be dwarfed by wind alone (e.g. in exposed coastal sites) or winter desiccation may result from the interaction of wind and low winter temperatures (e.g. krummholz, Section 3.4.6). Since krummholz cannot produce seeds, wind also influences fecundity and the potential to spread.

(v) Wind interacting with water supply or winter temperatures may determine forest boundaries as at the treeline (Section 3.4.6).

Question 3.9

(a) Ferns and bryophytes require free water in order to reproduce sexually (Section 3.6.4) and many bryophytes grow well only in moist, humid conditions. These conditions are more commonly provided in northern and western woodlands because water supply (precipitation) is higher here than in the south and east.

(b) Water supply will be similar for epiphytes in these two sites (recall that epiphytes have no roots and absorb rainwater through shoots), but climatic factors influencing water loss will be very different. Canopy epiphytes will be exposed to full sunlight and, therefore, become hotter than shaded epiphytes near the forest floor; air temperatures and wind speed will also be higher in the canopy and the potential for water loss here will be much higher – hence the desert-like characteristics.

(c) This observation suggests that water supply for germination or early seedling growth critically affects distribution of such plants with mature plants being better able to restrict water loss or forage for water and, therefore, less dependent on moist soil.

Question 3.10

(a) They show that although *Quercus* and *Tilia* desiccate to the same extent during the daytime, *Quercus* was not irreversibly damaged whereas *Tilia* was and *Quercus* also recovers to a much greater extent, i.e. rehydrates at night. Growth is possible only when shoots are well hydrated.

(b) The data are consistent with this hypothesis in that lime trees on northern slopes do not desiccate below the critical value of −2 MPa whereas stunted seedlings on southern slopes do. However, an underlying assumption is that water supply differs on the two slopes and there are no data on this point.

(c) An alternative hypothesis is that water supply does not differ on north- and south-facing slopes but the potential for water loss does (because of higher temperatures on the latter). Greater water loss by lime trees rather than lower water supply would then be the critical factor.

Question 3.11

Desert environments are extreme with respect to: (1) the maximum temperature of the air and surfaces; (2) the daily or seasonal range of temperature; (3) low rainfall; (4) the uneven distribution of rain over time; (5) the uneven availability of water in the soil in different places (run-off and drainage habitats); (6) absence or instability of soil in some areas.

You might also have mentioned the dry, abrasive winds and the saline soils of salt pans in some areas (Section 3.7.1).

Question 3.12

Species A is one of the deep-rooted plants of drainage habitats and only (viii) and (x) apply. Recall that these species are not resistant to desiccation and are not strictly drought-tolerant because they always have access to water (Section 3.7.3).

Species B is clearly an avoider (Section 3.7.2) which may grow in any habitat and is more likely to be an ephemeral because of its C4 character; so (i) and (ix) apply.

Species C is a cactus described by (iii), (v), (vii) and (x) (Section 3.7.3).

Question 3.13

(a)   Food and shelter are the most likely factors. Small rodents survive in deserts only if they can retreat to sheltered microhabitats during the day. Food availability (seeds) would depend on whether there was sufficient soil and rain at least to support ephemeral plants.

(b)   The existence of waterholes that were not too far apart and the presence of sufficient vegetation to graze would probably be the two most critical factors (Section 3.7.3). Predation from desert lions (or humans) is another possible factor.

Question 3.14

(a)   In each year, there was a period in summer when the soil became very dry (water potential very low). No germination would be possible during this period and seedlings would not survive if germination occurred in early summer *before* the drought. Germination would not occur after a long drought extending through autumn into winter because temperatures would be too low. The unique feature of 1976 is that there was a short summer drought followed by a long period when rain fell intermittently and soil moisture remained high. Thus, seedlings were able to germinate in warm conditions in late summer (one critical condition) and then had a long period during which growth was possible before the next drought in summer 1977 (probable second critical requirement for successful establishment).

(b) (i)   For eight years, i.e. all the years shown as coming below the tolerance curve. In all these years, the length of the growing season (high soil moisture) was followed by a drought period that did *not* result in seedlings losing more than 84% of tissue volume and dying. For all the other years (above the curve), the growing season was too short to allow seedlings to survive the following drought.

(ii)   Yes, in part: Figure 3.54 shows that 1976 (but not 1978) falls below the critical survival curve and, therefore, confirms the suggestion in (a) that length of growing season was critical. What was not clear for (a) was that length of drought in 1977 was also critical.

Question 3.15

(a)   Partially false: the first part of the statement is true but the second part is not. The low light compensation point of shade plants arises mainly from their low respiration rate and not their efficient photosynthesis (Section 3.8.1).

(b)   This is true: the arrangement of leaves and branches (i.e. canopy shape) influences greatly the extent to which upper leaves shade those lower down and hence the LAI (Section 3.8.1).

(c)   This is true, the reason being that temperature falls with altitude and the competitive advantage of C4 plants is therefore lost (Section 3.8.1).

(d)   False: depth of the euphotic zone is also influenced by the numbers of algae (or other plants) in the water column, which often depends on levels of mineral nutrients (Section 3.8.1).

(e)   True: inability to germinate in shade cast by plants is due to the low R : FR ratio which prevents phytochrome-stimulated germination but there would be no such change in light conditions in shade cast by rocks (Section 3.8.2).

Question 3.16

Curve X relates to *Lamiastrum*: it illustrates a Type I shade-tolerant response because there is a large increase in leaf area in shade and, even at 3% of daylight, leaf area is only slightly less than in full light. Curve Y relates to *Liriodendron*; this is typical of shade-intolerant species because there is some increase in leaf area in shade (but less than for A) and this is not maintained in deep shade – leaf area is less than in full daylight. Curve Z relates to *Fagus*; here you see the paradoxical response of a Type II shade-tolerant species – leaf area is reduced in proportion to shading – so the plants do not maximize light interception as X and Y do.

Question 3.17

Species A is a Type I shade-tolerant herb and is *Stachys officinalis* wood betony. You can deduce this because it has comparatively large seeds (item 8 in Table 3.12); shows much less inhibition of weight increase in shade than species B (item 3); and shows a strong elongation response even in dense shade, although plants do not collapse (item 4b).

Species B is shade-intolerant and is *Arenaria serpyllifolia* thyme-leaved sandwort, a plant of bare ground, walls and arable fields. Seeds are small, there is strong inhibition of dry weight gain in shade, considerable elongation in full sunlight (item 4a in Table 3.12) but miserable growth in shade, mainly because of the very small seed reserves.

Question 3.18

(a)   This observation suggests that bluebells are shade-tolerant but actually grow better in more open habitats, so their restriction to woods probably results from poor competitive ability in the open. Woodland clearings, where open conditions have been created around existing bluebells or which could be invaded easily by adjacent populations before more distant competitors invaded, represent a golden opportunity for shade-suppressed populations.

(b)   C4 plants are normally competitively excluded from shaded habitats because C3 plants grow better when photosynthesis is light-limited. The presence of a C4 plant in shade suggests that there is little competition here and this is the accepted explanation.

(c)   In midsummer at higher latitudes, daylength would certainly be longer than 13 h and so the long-day plant could reproduce and establish (if other conditions were suitable). However, daylength could well be longer than 12 h in late spring at higher latitudes (and low temperatures would probably prevent earlier flowering) so that the short-day plant would be unable to flower. It is unlikely that a photoperiodic ecotype with different photoperiod requirements could have evolved in 100 years.

Question 3.19

Types (b) (c) and (e) are all likely to be common. Their characteristics would allow plants to survive, at least underground, or recolonize rapidly as seeds (Section 3.9.2). Serotinous pines, (a), are unlikely to be common given the very high fire frequency because few, if any, pines would be able to mature and produce cones. Type (d) are also unlikely to be common because few trees would mature sufficiently to develop the thick bark necessary to protect epicormic buds.

Question 3.20

(a)   Two short-term changes that would occur are: (i) the frequency of seed-reproducing shrubs would decline progressively between 20 and 40 years as shrubs died and were not replaced in the unburnt scrub; (ii) the cover of the 'sprouting' shrubs would increase as they continued to grow.

(b)   In the longer term, species with short-lived seeds would probably be eliminated but those with long-lived seeds in a seed bank would increase after fire, with frequency gradually falling over the next 20 years. The post-fire increase would be especially marked if sprouting ability of the other shrubs was reduced after 40 years (compared with after 2–10 years), as described for heather in Section 3.9.2. However, any shrubs with *enhanced* sprouting ability after 40 years would block this trend and such shrubs might eventually dominate the community. Overall, it is likely that reduced fire frequency would lead to fewer species being present; the dominant species might be entirely shrubs able to sprout rapidly or a cycle of 20 years of mainly seed-reproducing shrubs followed by 20 years of sprouting shrubs.

Question 3.21

(a)   The light fire would not kill heather plants and the rapid restoration of cover suggests that plants resprouted from unburnt tissues. The slower restoration of cover with a moderately severe fire suggests that plants were killed and regeneration occurred from seed. The long delay in cover restoration with very severe fires suggests that the seed bank had been destroyed, probably because the peat had ignited down to a considerable depth (Section 3.9.1). However, the 10-year delay is surprising because you might expect that seeds would be dispersed over the burnt area from adjacent unburnt sites.

(b) (i)   This observation is consistent with hypothesis A but does not actually support it. The observation confirms that after a severe fire the seed bank would indeed be almost completely destroyed so that restoration of cover would depend on new seeds being dispersed into the burnt area.

(ii)   This does not support hypothesis A because although primary dispersal is poor, secondary dispersal should have allowed faster recolonization than was observed. Shortage of heather seed may contribute to the slow restoration of cover but cannot explain it entirely, so there is indirect support for hypothesis B.

(iii)　This gives some support for hypothesis B because germination of heather seeds was reduced in the presence of a lichen–alga crust. However, germination is not 100% inhibited and the lichen–alga crust did not cover 100% of the severely burnt area, so it may contribute to poor recolonization but cannot explain it entirely.

Overall, it appears that a combination of few seeds (due to poor dispersal) coupled with unfavourable conditions for germination or seedling establishment (including the presence of a lichen–alga crust) is the best explanation. Legg *et al.* showed in further experiments that several other factors contributing to poor seedling growth were: surface instability of the peat, trampling by sheep and unfavourable nutrient supplies (a high ratio of nitrogen to potassium after a few years) leading to poor root development and high susceptibility of seedlings to drought.

Question 3.22

(a)　Dispersal ability will determine whether or not a species has been able to reach all climatically suitable habitats since the species evolved or the climate changed (Section 3.10.1). It will also determine whether a species can disperse to more suitable habitats (e.g. higher latitudes or altitudes) after future climate change (Section 3.10.3).

(b)　If a species migrates to new geographical areas in the face of climate change, so that conditions of temperature and water supply are still within tolerance limits, it is still likely to encounter other changes, such as different photoperiods, higher UV-B levels (at higher altitudes), different wind conditions or fire risk: the species will survive only if it has sufficient phenotypic plasticity or can adapt genetically to these altered conditions. In the same way, these two properties will determine whether a species can survive without migration.

Question 3.23

The brief answer is 'no' for two aspects of forest clearance. The impact on local precipitation and surface temperature will depend on local climate and forest type (cloudiness and the extent to which evapotranspiration contributes to local rainfall); the impact on water supply will depend on slope, the nature of soils and the type of vegetation and land use that replaces the forest. However, a third aspect of forest clearance – contribution to the increase in atmospheric CO_2 – is relatively independent of geographical area and depends only on the biomass removed, so the answer here is 'yes'.

Question 3.24

(a)　It is a 'learning from the past' study in which long-term monitoring data are correlated with climatic records from the immediate past and used to predict the effects of future climate change.

(b)　The temperature records were not for Norfolk (the ideal) but for Central England; and rainfall data were too broad (monthly records would have been better and revealed the effects of dry seasons) and were also not for Norfolk.

(c)   First, the same care and luck in recording the date of first appearance may not have applied for each year – care in searching and luck in finding the first snowdrop, swallow etc. Secondly, there is much variation between individual plants in flowering and leafing dates, irrespective of climate: was the earliest plant always observed or the same plant or clump each year?

(d)   There is no compelling reason why the climate in Norfolk should have any influence on arrival date for migratory birds; climate on their migration route or wintering site would be more relevant.

(e) (i)   It might mean that oak trees had a longer growing season and, therefore, fixed more carbon and grew more (e.g. revealed as wider annual rings). However, this assumes that leaves do not fall earlier in autumn and photosynthesis in summer is not inhibited by higher temperatures or altered rainfall, so the prediction is not sound.
(ii)   Bearing in mind that successful completion of development for insects such as winter moth depends critically on hatching very close to the date of leaf expansion (because tannins make older leaves inedible), this three-week shift would probably lead to complete failure *unless* hatching date of eggs also shifted to an earlier, matching date.

(f)   There are two major criticisms. (i) The global warming scenario used in the Marsham analysis could be wrong: it is impossible to predict accurately what changes will occur in localized areas (Section 3.10.3). (ii) The predicted effect on oak tree leafing was determined by extrapolating in a linear way from past correlations with climate: this is a massive extrapolation and not a firm prediction as the statement implies.

CHAPTER 4

Question 4.1

(a)   This relatively raw sandy soil will be very free-draining (and so likely to dry out quickly), contain little organic matter and have little capacity to bind and retain nutrients, so that nutrient levels will be very low. These features together with the low-growing nature of the plant colonists suggest that they would be nutrient-insensitive types (Section 4.3.1) and probably also tolerant of dry conditions.

(b)   The probable explanation is that as the dunes age and soil organic matter and nutrient levels increase, faster-growing, more nutrient-responsive species (usually grasses) can invade and displace the low-growing, nutrient-insensitive species. Displacement is prevented if heavy rabbit grazing prevents the former from growing so tall.

(c)   In the short term, heavy fertilizer application might allow some very nutrient-responsive species to invade – provided that they had good powers of dispersal and water supply was adequate for germination. In the long term, nutrients would be lost by leaching (even phosphate, on this type of soil) and the invaders would either die out or else persist but show slow, stunted growth.

Question 4.2

(a)   An increase in growth with addition of N or P alone suggests that the nutrient was in limiting supply; an increase or a greater increase with N+P addition suggests that both nutrients were limiting or one nutrient became limiting after addition of the other. Thus:

- for species 1–3, P alone is clearly in limiting supply; there is no significant response to added N and no strong interaction in the N+P treatment. These species show an increase in growth, relative to controls, of 10-fold or more, so they can be classified as strongly nutrient-responsive.

- for species 4, there is a clear response to P, a smaller response to N and no interaction between the two, so P is the more important limiting nutrient. The maximum responses are a 3–4-fold increase in growth above controls, so this species is moderately nutrient-responsive.

- for species 5, there is a small response to N, none to P and a strong interaction between the two, so both nutrients are about equally limiting. As for species 4, the data indicate only moderate nutrient-responsiveness.

- for species 6–8 there is no significant growth response to any nutrient additions and, therefore, neither N nor P is in limiting supply and the species can all be classed as nutrient-insensitive.

(b)   Species characteristic of nutrient-rich soils would be expected to show very large growth responses to added nutrients, whereas those characteristic of nutrient-poor soils would not. So, based solely on the *relative* response (maximum weight with nutrient addition/control weight), species 1–3 are very likely to grow on nutrient-rich soils – and you know from Section 4.3.1 that nettles indeed do. Note that as the initial weight of seeds or seedlings is unknown, the *actual* growth rate ((initial – final weight)/duration of experiment) cannot be calculated. However, even assuming an initial weight of zero, species 2 (rose-bay willow-herb) grew relatively slowly after addition of N and P, and some other nutrient (e.g. K) could be more important for this species.

The moderate response of species 4 and 5 to nutrient additions is too small to conclude that these would occur naturally on nutrient-rich soils. In fact they do – but other factors are important. Species 4 (red campion) grows on fertile soil in *shade*; and 5 (goosegrass, cleavers), whilst appearing to grow well without nutrient addition, produces no flowers or fruit unless nutrient levels are high.

Species 6–8 might all be expected to grow on nutrient-poor soils although 6 (nettle-leaved bellflower) and to a lesser extent 7 (slender false-brome) appear to have quite high growth rates. These two species actually grow most commonly in woodlands on moderately rich soil whereas 8, as its apparently low growth rate suggests, is widespread on nutrient-poor soils in a variety of habitats.

From the above, you can see that it is not possible to make firm predictions about a species' habitat solely on the basis of response to added nutrients. Even for a species like nettle which clearly grows on nutrient-rich soils, you know from Section 4.3.1 that it also requires a good light supply. Nutrient supply from soil is only one niche dimension and the supply of water and light and conditions such as temperature, exposure to wind, soil stability, etc. may all be important.

Question 4.3

Characteristics (ii)–(iv). (i) is unlikely since it involves highly modified leaves and (in England) occurs only for species (like sundew) of wet habitats (Section 4.3.1). (ii) is likely since this is a slow-growing species on a very nutrient-poor site where anti-herbivore mechanisms such as high tannin levels are common (Section 4.3.1). (iii) and (iv) are both related to nutrient supplementation, (iii) being a characteristic of legume species. (v) suggests a very low ratio of root : shoot biomass, which is the opposite of what you might expect for a plant colonizing nutrient-poor soil (Section 4.3.1).

Question 4.4

(a) defines an ecotype (Section 4.4.4). Although defined in this Section in relation to heavy metal tolerance, recall that thermal ecotypes were mentioned in Chapter 3 (Section 3.4.3); the term is a general one.

(b) defines impedance (Section 4.4.1).

(c) describes soil pH (Section 4.4.3).

Question 4.5

(a)   When $CaCO_3$ is added, the carbonate affects the CO_2–HCO_3^-–CO_3^{2-} system, shifting the equilibrium so that pH rises (Section 4.4.3); calcium nitrate and sulphate do not usually interact with this system and hence do not affect soil pH.

(b)   The rusty sheath of iron(III) hydroxide was mentioned in Section 4.4.2. It forms because oxygen diffusing out of roots possessing aerenchyma tissue oxidizes reduced iron (iron(II)) in the anaerobic mud and reduces root uptake of iron(II). This probably aids survival because iron(II) is potentially toxic and has been shown to influence the composition of fen vegetation (Section 4.4.4).

(c)   Zinc-rich soils will carry zinc-tolerant ecotypes of widely distributed species which are not visibly different from 'normal' plants; these are likely to be the only species present on very localized sites with medium to high nutrient levels. Distinctive vegetation that includes species confined to zinc-rich soils (e.g. alpine penny-cress) is more likely to arise if the site is larger and, especially, if macro-nutrient levels are low: recall that this was a major factor producing serpentine vegetation (Section 4.4.4).

Question 4.6

(a)   The strong inhibition of growth caused by increasing calcium levels is the clearest evidence that B. robur is a calcifuge. In Section 4.4.3, calcium toxicity was mentioned as one factor that may exclude calcifuges from calcareous soils, others being shortages of iron and phosphate at high pH. Neither of these last two factors can be assessed in this experiment.

(b)   The data do not provide conclusive evidence that B. robur is nutrient-insensitive but two results indicate that this is probably true (and, in fact, it is). First, the response to increasing N, although appearing large in percentage terms, is very small for the actual dry weight (i.e. the height of the green bar is small compared with the maximum response to P and Ca). However, you have to bear in mind that the response to level 5 N was measured at level 5 Ca (the amount in the standard nutrient solution) and this level of Ca is clearly very inhibitory. Secondly, there is a significant reduction in growth at the highest level of P, suggesting phosphate toxicity, which is most likely to occur in nutrient-insensitive species (Section 4.3.1). What is needed are comparative data showing the response to N and P of known, nutrient-responsive species so that the poor response of *B. robur* is clear.

Question 4.7

(a)   Table 4.14a shows a correlation between the fall in the water table (increasing distance from the surface) and a fall in pH (an increase in the area with low pH). Thus, the lowest pH (< 3.5) occurs mainly where the water table is >150 cm and the highest pH (> 5.5) occurs mainly where the water table is within 50 cm of the surface. As the fen dries out, pH falls. The most likely explanation is that oxidation of reduced ions (NH_4^+, Fe^{2+}, Mn^{2+}) and possibly of some organic components in the peat occurred as the peat dried out and became less waterlogged; such oxidation can cause acidification (Section 4.4.3).

(b)   Table 4.14b shows a correlation between low pH and high Al levels (or high pH and low Al levels). From Section 4.4.3, you know that decreasing pH is associated with increasing availability (i.e. solubilization) of aluminium ions (Figure 4.10b). Usually, however, the Al is released from aluminium complexes in clay, whereas this was a *peat* soil and Al here would be incorporated into organic compounds. This suggests that the peat must have been broken down (a process called mineralization which is discussed in Book 4) as it dried out.

(c)   (i) Good growth of nettles requires high levels of nitrate and phosphate (Section 4.3.1). This observation suggests, therefore, that when peat is moderately dry (pH > 4, Table 4.14a) and has a pH that is tolerated by nettles, high levels of these macro-nutrients occur. They must have derived largely from the oxidation of ammonium ions or organic compounds in the peat – which is further evidence that the peat mineralizes as it becomes drier.

(ii)   Rusty deposits suggest that the iron was oxidized (Fe^{3+}) and that, therefore, the waterlogged peat contained high levels of reduced iron (Fe^{2+}). This oxidation would have contributed to the fall in pH (see (a) above).

(iii)   Since the decrease in pH correlates with fall in the water table, this observation suggests that Ca must be lost by leaching as pH falls and also that organic nitrogen compounds are broken down. This process, together with the oxidation of ammonium ions released from organic N, would have *contributed* to the fall in pH.

(d)   As the water table fell, the peat became better oxygenated (less waterlogged) and this resulted in oxidation of Fe^{2+} (and possibly Mn^{2+} – there are no data about this) and decomposition of the peat itself, including organic nitrogen compounds. Peat decay and NH_4^+ oxidation gave high levels of nitrates and phosphates (hence the nettles) and also released aluminium. The oxidation reactions released hydrogen ions which caused a fall in pH which, in turn, caused leaching of calcium and maintained Al in a soluble form.

Question 4.8

(a)   Stenohaline organisms cannot tolerate large changes in salinity (and are all strictly marine). The swelling observed results from osmotic movement of water into the body and this, coupled with the outward movement of salts from the body, causes death. Euryhaline organisms tolerate large changes in salinity and are able to prevent excessive influx of water and loss of salts at low external salinities by mechanisms such as low permeability of the body surface to water and ions (as in the brine shrimp *Artemia*) (see Section 4.5.1).

(b)   Salty run-off water will accumulate in depressions in the salt desert (cf. run-off habitats in hot deserts, Chapter 3) and, after evaporation, very high salt levels are left, with a gradient from edge (lowest) to centre (highest). Thus, the most likely explanation for zonation is differences in salinity tolerance (and possibly competition) between species. On saltmarshes, zonation is not related primarily to variations in salinity and differences in submergence, substrate, aeration and competition are commonly involved (Section 4.5.4).

(c)   Mosses have no roots and are, therefore, unlikely to occur in the lower parts of saltmarshes because they would be washed away. In southern and eastern England, the climate is generally drier and summer temperatures higher than in the north and west. So, dry and probably hypersaline conditions are likely to occur in the upper parts of Type A saltmarshes and intolerance of these conditions may partly explain the absence of mosses. The other factor involved may be the nature of the vascular plants, which are commonly tall and quite shrubby (e.g. *Limonium* spp.) in Type A marshes but form a short, grassy sward in grazed marshes (Type B) of the west and north. Thus, mosses tolerant of salinity may also require open, unshaded conditions and be unable to persist beneath tall plants.

Question 4.9

Regarding first the suitability of planting *Spartina*, you need to bear in mind that, if successful, the grass is likely to spread out onto open mudflats and reduce the area suitable for birds to feed. So you need to determine by observations and enquiries whether the marsh is of special importance for birds. Success will depend on whether conditions are appropriate for *Spartina*, so you would need to check that: (a) the substrate was muddy and not sandy; (b) coarse silt was being deposited and was present in the water; (c) the area was reasonably sheltered (a subjective judgement based on comparisons with other marshes that supported *Spartina*); (d) there was no more than 6 hours submergence per day at the planting site. These necessary conditions are described in Section 4.5.4.

Question 4.10

(a)   Figure 4.31 shows that surface temperature, rates of water loss and surface salinity are all significantly greater in bare patches (assuming that sponges mimic the soil surface). Measurements of soil surface salinity over six weeks confirmed that it is indeed greater in bare patches. Because plastic straws and shading prevented this rise in salinity, you can conclude that absence of plant cover is the main factor causing raised salinity: it exposes the soil to direct sunlight so that temperatures are higher, and evaporation increased (Figure 4.31). Conditions can be described as 'severe' in bare patches compared with vegetated areas.

(b)   (i) There is actually no evidence to support this statement. From Figure 4.32 and Table 4.17, the patch pioneers are either unaffected by rising salinity or inhibited by it; and they show either increased survival when patches are watered (*Atriplex*) or increased growth (*Salicornia*), so that the hot, dry conditions and high salinity on patches (Figure 4.31) do not increase survival or growth.

(ii)   There is clear evidence to support this statement. From Figure 4.32, growth is strongly inhibited as salinity rises and the highest value tested (30 g kg^{-1}) is below that measured in the soil of bare patches (50 g kg^{-1}). Both survival and growth are much increased when bare patches are watered (Table 4.17) to reduce salinity and increase the supply of water. Thus, knowing that conditions are less saline and less dry in vegetated areas, it can reasonably be *assumed* that zonal dominants grow better here – but there are no data to show that they actually do.

(iii)   There is evidence to support the first part of this statement but none for the second part – although it is a reasonable assumption. Growth and seedling survival of zonal dominants are more strongly inhibited in patches than they are for patch pioneers (from (i) and (ii) above). So one can reasonably say that conditions are more favourable for patch pioneers. There are no data about the growth of patch pioneers in the undisturbed vegetation but given that they grow as well or better at lower salinities and on moister soil, their rarity here is very probably due to competition from the dominant perennials.

Question 4.11

(a) is generally true because the phytoplankton would be outside the euphotic zone (which lies at or above the thermocline but virtually never below it) and unlikely to be able to swim back from the circulating hypolimnion to the epilimnion – so they would starve (Sections 4.6.1 and 4.7.1).

(b) is partly false. The first part of the statement is true (Section 4.6.2) but the second part is not: rivers have a water column compartment, shallow in headwaters but substantial in lower reaches and the main habitat for fish (nekton).

(c) is false. Although chemical *conditions* and oxygen supplies are relatively invariant in both space and time for the open oceans, chemical *resources* vary in space vertically, with higher levels of nutrients below the thermocline. There are also some localized upwelling regions where surface waters are enriched with nutrients (Section 4.7.3).

(d) is true and explains why anoxia is common here (Section 4.7.3).

(e) is generally true: there is very little variation in temperature in polar oceans compared with temperate oceans (Figure 4.36) so polar fish are *likely* to be more stenothermal (tolerating little variation in temperature – Chapter 3). The only possible exceptions are temperate ocean fish restricted to the deep sea, where temperature also varies little.

Question 4.12

Mainly top-down control. The uniform algal cover in grazed areas, despite great variation in algal growth potential, and the appearance of green bands in ungrazed areas indicate that catfish grazing determined algal biomass. If bottom-up control had been operating, you would expect to see variation in the amounts of algae in shaded and unshaded pools, with corresponding variation in the growth or survival of catfish. Bottom-up control operated to the extent that catfish *numbers* were greater in sunny compared with shaded pools.

Question 4.13

1   The species may have been killed or prevented from reproducing by the low pH or by raised aluminium levels. This is a direct, abiotic effect and implies neither top-down nor bottom-up control.

2   Since the number of species as well as biomass of phytoplankton decrease in acidified lakes (Section 4.7.2), the species may have disappeared because of shortage of food (especially if it was a highly selective grazer and its preferred food disappeared). This would imply bottom-up control.

3   A slightly different version of hypothesis 2, again implying bottom-up control, is that the species disappeared because reduced amounts of phytoplankton led to greater competition with other species of zooplankton.

4   The species may have disappeared because acidification caused an increase in predation (top-down control). This could arise because, for example, fish disappeared and larger, predaceous zooplankton which ate the small species increased. Alternatively, loss of rooted macrophytes (a common effect of acidification) could have exposed the species to greater predation without any change in the numbers of predators.

Question 4.14

(a)   Either (iv) and (v) (but not (vi) which dominate in oceans) would be correct here.

(b)   (i) then (vii).

(c)   (x).

Question 4.15

(a)   The green nanoplankton is represented by the green curve because it shows the following characteristics: it grows very poorly at low light intensity (Figure 4.50a) but shows a rapid burst of growth at high light intensity before declining as cyanobacterial biomass increases (Figure 4.50b). Such behaviour would be predicted for a small-type species (Section 4.9.1). The cyanobacterium (black curve) grows well at low light intensity (probably by keeping close to the surface using positive buoyancy from gas vacuoles) and is eventually dominant (after 17 or 18 days) even at high light intensity, i.e. it competes strongly in crowded conditions. This behaviour is typical of a large-type species.

(b)   In early summer, after the spring diatom bloom and before the slow-growing large species (including cyanobacteria) had built up in numbers.

Question 4.16

(a)   The data are consistent with bottom-up control via phosphate but other explanations are still possible; they provide limited support. Figure 4.51a shows that there was a decline in *total* phosphorus (i.e. both inorganic and organic, soluble and colloidal or particulate) from 1972–83/4, but it was rather erratic. Average phytoplankton biomass in summer declined roughly in parallel with P and, therefore, *might* be explained by it but an increase in grazing zooplankton could also have been responsible (though less likely). The spectacular decline in eutrophic marker species (item (i)), including cyanobacteria which are not grazed, is the best indicator that P level (bottom-up control) and not zooplankton (top-down control) was dominant.

(b) The decline in alewife (planktivorous fish) as numbers of piscivorous (fish-eating) walleye increased suggests top-down control but the decline in large cladocerans (Figure 4.51e) might also have contributed and would have been bottom-up control. Overall, bottom-up control appears dominant for phytoplankton and top-down control for most fish and zooplankton. The appearance of a large herbivorous *Daphnia* in 1983 (item (ii)) and its subsequent dominance coincide exactly with minimum alewife numbers and strongly suggest previous top-down control by alewife on the *Daphnia*. The appearance of the predaceous cladoceran and subsequent crash of *Daphnia pulicaria* again suggest top-down control.

Question 4.17

(a) Within a river, current velocity varies with *distance from the bed* (highest at the surface, lowest adjacent to the bed). Mean velocity is influenced by gradient, mean depth, bed roughness and, often, discharge. (Section 4.9.1)

(b) In stream (ii). Mean velocity, $\overline{U}$, in stream (ii) is obtained from $18 = 6 \times 1 \times \overline{U}$, so $\overline{U} = 3$ m s^{-1}. From the same calculation velocity in stream (i) is 2 m s^{-1}.

(c) The most likely change is that depth will increase. There may be some increase in current velocity but width cannot change within a rocky gorge.

Question 4.18

(a) Flow refugia are sites within a river where current velocity and turbulence are much less than in the main channel; finer particles may sediment out here and persist for longer than in the main channel. Examples are backwaters or slacks (Figure 4.53), or behind obstructions such as large boulders or fallen trees. They are significant in the upper reaches of rivers as places where invertebrates may shelter during periods of high flow (spates) and thus avoid being swept downstream; and also as places where rooted plants may avoid being swept away in spates (Figure 4.53). In the deeper, lower reaches of rivers, they are significant as places where phytoplankton and zooplankton may sustain high populations which feed into and replenish the main channel (Section 4.9.2).

(b) Conditions in riffles are turbulent, shallower than other areas, well-oxygenated and with a high current velocity and bare rock or large stones as the substratum. The only plants likely to occur here are periphyton attached to rocks. Strong-swimming nekton (fish) may occur in the water column but other invertebrates will all be benthic, living among stones and either firmly attached to them (e.g. *Simulium*, Figure 4.56a) or with mechanisms such as claws for clinging to them. These animals are likely to require high oxygen levels.

Question 4.19

(a) In habitat (ii). Any plant with floating leaves is unlikely to occur in turbulent water, even if it could root at the edges; so habitat (i) is unlikely. It is also unlikely to occur in water as deep as 1 m if the current is even moderate because the long leaf stalks (petioles) would be very susceptible to being broken (hence not habitat (iii)).

(b) Characteristics (i) and (ii) suggest that this stonefly clings to stones in areas of high current velocity and/or turbulence. (iii) suggests that it feeds there and does not swim about in the water column. (iv) and (v) indicate that it probably requires well-oxygenated water and has no means of increasing oxygen supply (e.g. by fluttering external gills). So (1) the stonefly probably lives among stones on the bottom of fast-flowing, turbulent streams and (2) is a shredder (consistent with (vi)) or possibly a grazer on periphyton.

Question 4.20

(a) Shear stress affects drift, especially when there is little or no woody debris present; because the greater the force acting on organisms near the bottom, the greater the chance that they will be washed away. There appears to be a critical value of around 10 dyn cm^{-2} below which shear stress has no effect. The amount of fixed woody debris also affects drift: the more there is, the lower the drift for shear stress values of 10 dyn cm^{-2} or more. The debris must act as flow refugia (Section 4.10.1).

(b) The most striking difference is that whereas *Gammarus* drift losses are approximately linear with time for all three levels of woody debris, *Ephemerella* losses are not linear: they are initially high (for about 12 min) and then fall to a very low rate. The explanation lies in *Ephemerella*'s low mobility: if 'out in the open' when shear stress increases, it cannot move rapidly to a flow refugia and so drift losses are high. Once in a refugium, *Ephemerella* stays there and drift losses are low, whereas the more mobile *Gammarus* moves constantly in and out with a constant risk of drift.

(c) The main reason for this suggestion is that organisms with limited mobility (such as *Ephemerella*) may not be able to reach refugia if they are all located near the edges of the channel. The position of flow refugia within the channel influences their relative value for different species.

Question 4.21

(a) (i) The difficulty of moving laterally to find new prey once phytoplankton have been grazed down; vertical migration where deep currents move in the opposite direction to surface currents effects such lateral movement. (ii) Energy conservation: cooler temperatures deeper in the water column reduce the metabolic rate during digestion. A third factor mentioned was reduction of predation (as for freshwater zooplankton): surface-feeding predators cannot see prey so readily at night when zooplankton move to the surface.

(b) The selective pressures cannot be related to high pressure or to other physical conditions in the deep sea but are most probably linked to food supply: shortage of food is a feature of the deep sea (Section 4.11.2) and probably also of caves.

Question 4.22

The right-hand arm (movement between the nursery site and adult feeding ground). Migration to a separate 'spawning site' and of young to the nursery site are neither possible nor required because the young are not planktonic and must remain with their parents in order to suckle.

Question 4.23

(a) is false: hydrothermal vents are deep-sea communities based on autotrophic (chemosynthetic) production and are also areas where food is plentiful.

(b) is false: some deep-sea organisms have relatively high rates of metabolism and move rapidly (e.g. the ambush predators' final lunge) for *brief* periods.

(c) is true.

(d) is true.

Acknowledgements

Grateful acknowledgement is made to the following sources for permisssion to reproduce material in this Book:

Cover photo Mike Dodd, Open University; *Figure 1.3b* Hassall, M. and Dangerfield, J. M. (1990) Density-dependent processes in the population dynamics of *Armadillidium vulgare* (Isopoda: Oniscidae), *Journal of Animal Ecology*, **59**(3), Blackwell Science; *Figure 2.8* A. Mahdi, R. Law and A.J. Willis (1989) Large niche overlaps among co-existing plant species in a limestone grassland community, *Journal of Ecology*, **77**(2), June, Blackwell Science; *Figure 2.23* C. Mitter and B. Farrell (1991) Macroevolutionary aspects of insect–plant relationships, in E. Bernays (ed.) *Insect–Plant Interactions*, Vol. III, CRC Press, Boca Raton; *Figure 2.4* W.H. Settle and L.T. Wilson (1990) Invasion by the variegated leafhopper and biotic interactions: parasitism, competition and apparent competition, *Ecology*, **71**, Ecological Society of America; *Figure 3.7* adapted from *Field Studies*, **1**(5) (1963); *Figure 3.10* Schonbeck, M.W. and Norton, T.A. (1980) Factors controlling the lower limits of fucoid algae on the shore, *J. Exp. Mar. Biol.Ecol*, **43**, 131–50, © Elsevier/North-Holland Biomedical Press; *Figure 3.14* Farrell, T.M. (1991) Models and mechanisms of succession: an example from a rocky intertidal community, *Ecological Monographs*, **61**(1), 95–113 © 1991 Ecological Society of America; *Figure 3.18* from Charles J. Krebs (1978) *Ecology*, 2nd edn, copyright © 1978 Charles J. Krebs, reprinted by permission of HarperCollins; *Figures 3.19, 3.28a, b* Sutton, S. L. (1972) *Woodlice*, Ginn & Co; *Figure 3.25* Christian, K., Tracy, C.R. and Porter, W.P. (1983) Seasonal shifts in body temperature and use of microhabitats by Galapagos land iguanas, *Conolophus pallidus*, *Ecology*, **64**(3), 466, copyright © 1983 Ecological Society of America; *Figure 3.28 c, d Complete Atlas of the British Isles*, © 1965 The Reader's Digest Association Ltd; *Figure 3.29* Dick Roberts/Holt Studios; *Figure 3.30* Payette, S. *et al*. (1989) Reconstruction of tree-line vegetation response to long term climate change, adapted with permission from *Nature*, **341**, 5 Oct. 1989, copyright 1989 Macmillan Magazines Ltd; *Figure 3.32 b, c* Galen, C. and Stanton, M.L. (1993) Short-term responses of alpine buttercups to experimental manipulations of growing season length, *Ecology*, **74**(4), 1055, copyright © 1993 Ecological Society of America; *Figure 3.33* Celtic Picture Agency; *Figures 3.37 and 4.15* Etherington, J.R. (1982) *Environment and Plant Ecology*, John Wiley, reprinted by permission; *Figures 3.40, 3.41* Pigott, C.D. and Pigott, S. (1993) Water as a determinant of the distribution of trees at the boundary of the Mediterranean zone, *Journal of Ecology*, **81**(3), Blackwell Science Ltd; *Figures 3.42, 3.43* Buxton, P.A. *Animal Life in Deserts*, Edward Arnold; *Figure 3.44* Cloudsley-Thompson, L. and Chadwick, M.J. *Life, in Deserts*, Hunter and Foulis; *Figures 3.53, 3.54* Jordon, P.W. and Nobel, P.S. (1981) Seedling establishment of *Ferocactus acanthodes* in relation to drought, *Ecology*, **62**, 901–6, copyright © 1981 Ecological Society of America; *Figure 3.55* Bjorkman, O.B. (1971) in Hatch, M.D., Osmond, C.B. and Slatyer, R.O. (eds) *Photosynthesis and Photorespiration*, copyright © 1971 John Wiley, reprinted by permission, all rights reserved; *Figure 3.60* Furness, S.B. and Grime, J.P. (1982) Growth rate and temperature responses in Bryophytes, *Journal of Ecology*, © 1982 British

Ecological Society, Blackwell Science; *Figure 3.62* Rundel, P.W. in Lange, O.L. *et al.* (eds) *Physiological Plant Ecology*, **1**, Spinger-Verlag; *Figure 3.64* Stein, S.J. *et al.* (1992) The effect of fire on stimulating willow regrowth and subsequent attack by grasshoppers and elk, *Oikos*, **65**, 190–96, Munksgaard © OIKOS; *Figure 3.67a* Keeling, C.D. *et al.* (1995) Interannual extremes in the rate of rise of atmospheric carbon dioxide since 1980, *Nature*, **375**, p. 667, 22nd June; *Figure 3.67b* Brasseur, G.P. and Chatfield, R.B. (1991) The fate of biogenic trace gases in the atmosphere, in Sharkey, T.D. *et al.*, *Trace Gas Emission by Plants*, Academic Press; *Figure 3.67c* reproduced by courtesy of The Hadley Centre for Climate Prediction and Research; *Figure 3.67d* Briffa, K. R. *et al.* (1990) A 1,400 year tree-ring record of summer temperatures in Fennoscandia, *Nature*, **346**, p. 437, 2nd Aug.; *Figure 4.3 Unit of Comparative Ecology Report*, 1984; *Figure 4.5* C.D. Pigott and K. Taylor (1964) Distribution of some woodland herbs in relation to supply of N and P in the UK, *British Ecological Society Symposium* (suppl. to *Journal of Ecology*, **52**), Blackwell; *Figure 4.5* K.K. Treseder, D.W. Davidson and J.R. Ehleringer (1995) Absorption of ant-provided carbon dioxide and nitrogen by a tropical epiphyte, *Nature*, **375**, 137–39; *Figure 4.13* I.H. Rorison (1985) Nitrogen source and the tolerance of *Deschampsia flexuosa*, *Holcus lanatus* and *Bromus erectus* to Al during seedling growth, *Journal of Ecology*, **73**, 83–90; *Figures 4.17 and 4.18* R.E.D. Snowdon and B.D. Wheeler (1993) Iron toxicity to fen plant species, *Journal of Ecology*, **81**, 35–46; *Figure 4.19* N.J. Grundon (1972) Mineral nutrition of some Queensland heath plants, *Journal of Ecology*, **60**, 171–81; *Figure 4.22* P. Adam (1990) *Saltmarsh Ecology*, Cambridge University Press; *Figures 4.31 and 4.32* M.D. Bertness *et al.* (1992) Salt tolerances of fugitive marsh plants, *Ecology*, **73** (5), p.1848; *Figure 4.37* R.W. Battarbee *et al.* (1988) *Lake Acidification in the UK, 1800–1986*, ENSIS Publishing; *Figure 4.38* D.J.A. Brown (1987) Freshwater acidification and fisheries decline, in *CEGB Research: Acid Rain*, CEGB; *Figure 4.40* S.E. Arnott and M.J. Vanni (1993) Zooplankton assemblages in fishless bog lakes, *Ecology*, **74**, 2361–80; *Figure 4.47* G.P. Harris (1986) *Phytoplankton Ecology*, Chapman and Hall; *Figure 4.50* J. Benndorf (1992) The control of indirect effects of biomanipulation, in D.W. Sutcliffe and J.G. Jones (eds) *Eutrophication: Research and Application to Water Supply*, Freshwater Biological Association; *Figure 4.52* J.D. Allan (1995) *Stream Ecology*, Chapman and Hall; *Figure 4.53* M. Jeffries and D. Mills (1990) *Freshwater Ecology*, Wiley; *Figure 4.58* T.T. Macan (1963) *Freshwater Ecology*, Longman; *Figures 4.61–4.63* R.S.K. Barnes and R.N. Hughes (1988) *An Introduction to Marine Ecology*, Blackwell.

INDEX

Entries and page numbers in **bold type** refer to key words which are printed in **bold** in the text. Pages indicated in *italics* refer to a figure or caption.